T0092025

AGE OF AUTO ELECTRIC

AGE OF AUTO ELECTRIC

ENVIRONMENT, ENERGY, AND THE QUEST
FOR THE SUSTAINABLE CAR

MATTHEW N. EISLER

The MIT Press
Cambridge, Massachusetts
London, England

The MIT Press would like to thank the anonymous peer reviewers who provided comments on drafts of this book. The generous work of academic experts is essential for establishing the authority and quality of our publications. We acknowledge with gratitude the contributions of these otherwise uncredited readers.

This book was set in Stone Serif and Stone Sans by Westchester Publishing Services.

Library of Congress Cataloging-in-Publication Data

Names: Eisler, Matthew N., 1970- author.
Title: Age of auto electric : environment, energy, and the quest for the sustainable car / Matthew Eisler.
Description: Cambridge, Massachusetts : The MIT Press, [2022] | Series: Transformations : studies in the history of science and technology | Includes bibliographical references and index.
Identifiers: LCCN 2021060543 | ISBN 9780262544573 (paperback) | ISBN 9780262372022 (pdf) | ISBN 9780262372039 (epub)
Subjects: LCSH: Electric automobiles—Technological innovations. | Electric automobiles—Political aspects. | Electric automobiles—Public opinion. | Sustainable engineering—Social aspects.
Classification: LCC TL220 .E26 2022 | DDC 629.22/93—dc23/eng/20220215
LC record available at https://lccn.loc.gov/2021060543

For Hannah

CONTENTS

ACKNOWLEDGMENTS

Books are collective efforts, and this one would not exist without the direct and indirect help of very many people. I thank all those who provided insight and inspiration during a project that proved far more challenging than I could have imagined at the outset. I began research around 2013 as a sessional instructor at the University of Virginia's Department of Engineering and Society and received stimulation and support from colleagues including Catherine Baritaud, Jack Brown, Bernie Carlson, Mike Gorman, Deborah Johnson, Jongmin Lee, Lisa Messeri, Peter Norton, Tolu Odumosu, Bryan Pfaffenberger, Hannah Rogers, and David Slutzky.

I am grateful for the enduring mentorship of Robert Smith of the University of Alberta and Patrick McCray of the University of California at Santa Barbara (UCSB), without whose support I would not have been in a position to begin this project. My earliest thoughts about the relationship between materials science, energy and environmental policy, and electric cars were informed by discussions with Patrick, Gwen D'Arcangelis, Yasuyuki Motoyama, and Barbara Herr Harthorn at the Center for Nanotechnology in Society (CNS) at UCSB, and Jongmin, Carin Berkowitz, Hyungsub Choi, Ben Gross, Cyrus Mody, and Jody Roberts at the Chemical Heritage Foundation (now the Science History Institute). My knowledge of the history of the practices and institutions of materials sciences and engineering was greatly enhanced by Lillian Hoddeson and Catherine Westfall, as well as by John Goodenough, whom I was fortunate enough to interview in 2013, before he achieved global celebrity as a colaureate of the 2019 Nobel Prize in Chemistry. I thank them all for sharing their experiences and wisdom.

I am also indebted to Bob Kolvoord of James Madison University and Kathleen Vogel of Arizona State University, for their advice and advocacy at pivotal junctures in my career. I am grateful to the School of Humanities at the University of Strathclyde for giving the project a new home. I thank Ann Bartlett, Patricia Barton, Kirstie Blair, Catriona Ellis, Mark Ellis, Richard Finlay, Laura Kelly, Mo McDonald, Yvonne McFadden, Arthur McIvor, Jim Mills, Martin Mitchell, David Murphy, Emma Newlands, Maureen Noor, Jesse Olszynko-Gryn, Rogelia Pastor-Castro, Elsa Richardson, Natalia Telepneva, Ksenia Wesolowska, Niall Whelehan, Manuela Williams, John Young, and especially Matt Smith, for their collegiality and comradeship. I could not ask for better friends and colleagues.

I owe many thanks to the East Bay Chapter of the Electric Auto Association of San Francisco, and especially to the association's former president, Tom Keenan. Tom introduced me to electric car culture in California, and his hospitality and warmth was exceeded only by his depth of knowledge of the technology of electric cars. I also thank the staff of the Bentley Historical Library at the University of Michigan at Ann Arbor for their assistance in navigating through the papers of Stanford Ovshinsky and Robert Stempel. And I am deeply grateful to electric vehicle pioneers, including Alec Brooks, Alan Cocconi, Martin Eberhard, and Wally Rippel, for taking the time to share their experiences with me. I owe particular thanks to Alec for his generosity in providing documents and detailed personal knowledge of the early years of the electric car revival in the 1990s and early 2000s. Alec helped me connect the dots and understand a number of interwoven events, especially the relationship between AeroVironment, AC Propulsion, and Tesla Motors, and it was my great good fortune to be able to work with him.

In addition, the book was crucially informed by the inaugural workshop of the Research Institute for the History of Science and Technology (RIHST) at the California Institute of Technology (Caltech) and the Huntington Library in September 2020. The fact that the workshop had to be conducted remotely owing to the COVID pandemic did not detract from the richness of discussion, which dwelt on the theme of electricity in a contemporary global context and influenced my understanding of the relationship between automobility and electricity as energy conversion infrastructure. I thank Bernie Carlson and Erik Conway for inviting me to be part of a distinguished group including Stathis Arapostathis, Julie Cohn, Amy Fisher, Nathan Kapoor, Cyrus Mody, David Nye, Ruth Sandwell, and Abby Spinak.

My deepest thanks to RIHST director Jed Buchwald, who promoted the book for his Transformations series of the MIT Press. I am indebted to MIT Press acquisitions editor Katie Helke for believing in the book and tirelessly pushing it forward to assistant acquisitions editor Laura Keeler for her consummate professionalism, and to Christine Marra for her unfailing judgment. All mistakes in this book are my own.

Finally, I am thankful to the Eisler and Rogers families, to Mom, Dad, John, Sara, Christian, Bailey, and Ruffalo, and to Mrs. Crystal, Dr. Hugo, Hiram, and Dora, for their unflagging support over the years. I am especially grateful to Hannah Rogers for her advice and critical insights in a project that often seemed as if it might never come to fruition. I dedicate this book to her.

1 INTRODUCTION

The auto as a particular form of transport technology is not only the basis for a physical world constructed for its uses, it is also embedded as a cultural artifact in our personal experiences and belief systems. It is seen as an inevitable and desirable feature of life.

—Peter Freund and George Martin, *The Ecology of the Automobile*, 1993

If there is an overarching theme in the history of the electric car, it is the reversal of fortune. In 2005, there were a little over one thousand electric vehicles of various types on US roads. By 2020, there were nearly two million, and the global stock of electric passenger vehicles numbered more than 10 million, with China accounting for nearly half the fleet.[1] This dramatic growth represents one of the marked events of the early twenty-first century and one of the most remarkable industrial-technological and social developments of recent times. In its own ways, the revolution in electric cars is as disruptive and as sweeping as the revolution in information technology that preceded it, with important implications for the material culture of the personal passenger automobile, for infrastructures of industry and energy conversion, and for the environment.

Yet the case of the electric car is perhaps singular in the history of technology. The electric vehicle revolution was preceded by not one but two abortive efforts to develop and deploy technology at scale, the first unfolding at the turn of the twentieth century and the second unfolding almost exactly 100 years later at the turn of the twenty-first century. Much academic attention has been devoted to understanding the so-called golden age of the electric car from the early 1890s to the early 1920s. This book addresses the revival, or the second, age of the electric car, a period conventionally dated from the early 1990s but with origins in post-World War II developments in many areas of science, technology, and public policy that did not always directly relate to the personal passenger automobile.

Perhaps the defining event of the initial phase of this revival was the *false start*, the premature demise of promising electric vehicle programs in the early 2000s, an episode that left an indelible impression on popular culture and laid the foundation for the second phase of the revival in the later 2000s and 2010s. This book seeks to understand the causes of the electric car revival, its false start, the subsequent new age of auto electric that emerged out of it, and what it means for the ways people build and drive automobiles, use energy, and reshape environments in the process.

In understanding the vicissitudes of the electric car it is instructive to review its two origin stories, the narratives that explain the technology's rise and demise in its golden age and its rise, demise, and resurrection in its revival, or second, age. The first origin story holds that in the 1890s and early 1900s, the electric motor mated to the lead-acid rechargeable battery was the most reliable form of automobile propulsion technology. Most of the then-small fleet of passenger automobiles consisted of electrics, which were favored because they were easy to start, relatively durable, and operated with little noise and no emissions. In contrast, cars powered by the gasoline-fueled internal combustion engine (ICE) technology of the day were unreliable and unpleasant to ride in and to be around. They were noisy, rickety, fragile, difficult and often dangerous to start, and belched noxious emissions. Practically the only advantage the ICE format offered over the battery electric format was greater range. Around World War I, however, the balance of technological capability and convenience abruptly shifted. Engineers like Charles Kettering, the inventor of electric ignition and leaded gasoline, helped improve the reliability and manners of ICE vehicles, and arch industrialists like Henry Ford masterminded their mass manufacture. When tens of millions of improved ICE cars poured out of Detroit factories onto US roads in the 1920s, the lead-acid battery-powered electric passenger car lost most of its comparative operational advantages and became extinct.

The second origin story ascribes the electric car revival to the intersection of environmental policy and improved technology. Strict new regulations developed by California air quality regulators in the 1990s required the auto industry to market vehicles that produced zero emissions, a public-policy objective that industry was eventually able to accomplish, according to the origin story, thanks largely to new advanced batteries that narrowed the performance gap with ICE technology.[2] This narrative was ratified by no less august an authority than the Royal Swedish Academy of Sciences (RSAC) in 2019, when it awarded the Nobel Prize in Chemistry to John B. Goodenough, M. Stanley Whittingham, and Akira Yoshino for inventing the lithium-ion rechargeable battery. It was this technology, held the RSAC, that enabled the long-range electric car and paved the way for a fossil fuel–free society.[3]

The idea that the second age of the electric vehicle was launched largely by better battery technology may be intuitively attractive but it glosses the anomaly of the false start. A decade before automakers began adopting the lithium ion rechargeable battery in the late 2000s, they had experimented with the nickel-metal hydride rechargeable, then widely regarded as the greatest advance in battery technology in a century. In the 1990s, then, automakers had a relatively better battery, but they argued that the technology could not fully cure the neurosis of *range anxiety* said to afflict pioneering electric motorists accustomed to the capabilities and affordances of ICE technology. The prospect of running out of stored energy and becoming stranded, claimed the car companies, deterred consumers, consigning the first generation of revival-era electrics to the dustbin of history.[4]

In short, both origin stories of the electric car figure technology as a—if not the—primary causal agent. Studies of science, technology, and society (STS) have long problematized narratives of technological progress and accounts of the contemporary electric car are ripe for critical analysis. Scholars have hitherto been reluctant to engage these accounts, in part because analyzing recent events poses a host of methodological problems, not least of which is the risk of becoming hostage to fortune.[5] Recognizing these problems, this book builds on STS–informed historical studies that challenged the idea that the early electric car, with its lead-acid rechargeable battery, was a failure. Hewing to the view that technology is a reflection or extension of social interests, David Kirsch and Gijs Mom argued that around the turn of the twentieth century, a variety of actors regarded battery electric vehicles as very useful in particular contexts, especially in densely populated urban centers where short range was not necessarily a handicap. Range anxiety, suggested Kirsch and Mom, is relative, not absolute, and manifests only in context.[6]

There is much to commend this perspective. On the other hand, it does not address certain peculiar qualities of the technology of battery electric propulsion that have complicated the enterprise of electric cars. Batteries tend to have shorter life spans than electric motors, a temporal mismatch that rewards battery-making. Historically, batteries and electric motors have generally been manufactured by distinct business interests, so early entrepreneurs of electric cars had to find ways to absorb battery maintenance and replacement costs. Many drew on established management practices in the horse and electric trolley systems, treating electric vehicles as a centrally managed service. Operators used such vehicles in taxi fleets and leased them to wealthy drivers, while electricity-generating stations maintained fleets of electric delivery trucks as a means of storing off-peak electricity.[7] The service-based business model insulated users of electric vehicles from the costs and inconveniences of battery replacement and maintenance and passed them onto the service providers.

Recognizing these facts while hewing to the constructivist position, Kirsch and Mom argued that it was shortcomings in business management rather than technology than ended the early use of electrics. They held that when Ford and other manufacturers started selling millions of cheap and reliable ICE cars, the leased electric car could not compete. Marketing the electric car as a consumer durable, held Kirsch and Mom, might have prevented the technology's premature demise.[8]

From this counterfactual proposition arises the nettlesome question of whether particular technologies are compatible with particular social orders, a perennial bugbear of social constructivism.[9] As the historian Cyrus Mody has suggested, the social facts of technology are never purely social but rather are intimately linked with interpretations of physical phenomena.[10] In this book, I give consideration to the beliefs of certain actors that the material properties of the commercial battery electric car had important organizational and operational implications for the business of automaking. Electrochemical energy storage, especially the rechargeable battery, is an archetypal black box. For most of their history, rechargeables have been regarded by innovators and users alike as a terra incognita of mysterious and ungovernable chemical reactions. The historian Richard H. Schallenberg argued that after electric passenger cars disappeared from American roads in the 1920s, electrochemical energy storage became an orphan discipline. For years afterward, the field remained recondite, a *dark art* more empirical than scientific. Ideas for advanced batteries and power sources tended to be regarded as solutions in search of problems.[11]

The Cold War roused battery technoscience from its torpor in service of military applications. In this context, the capacity of a battery to store energy (known as energy density, typically expressed in terms of watt-hours per kilogram) and release it on demand (power density, or the rate of energy flow, typically expressed in terms of watts per kilogram) was valued over its durability and cost-effectiveness. On the other hand, the idealized commercial electric car presupposed batteries that were robust and affordable as well as energetic and powerful, qualities that required an even deeper fundamental knowledge of electrochemistry. Only with the emergence of air pollution as a public health emergency in the early 1960s did policymakers identify a problem that justified a commercial electric car, and, thence, sustained research in suitable batteries and other power sources that met all the criteria for this application.

ADVANCED MATERIALS, BETTER BATTERIES, AND THE ELECTRIC SUPERCAR

These efforts took decades to yield practical results and fueled expectations that became the basis of better battery discourse, a way of speaking about the future electric car that played a determining role in the electric vehicle revival. Discourses of expectation are an integral part of technological futurism and sociotechnical imaginaries, defined as rhetorics of the new that, as STS scholar Sheila Jasanoff put it, conjure forms of social life and order "attainable through, and supportive of, advances in science and technology."[12] Anticipatory discourses of science and technology have deep roots in the Western intellectual tradition. In contemporary technological futurism, expectations expressed in imaginaries derive from routine practices of prediction-making that date to the professionalization and institutionalization of science and engineering from the mid-nineteenth century.[13] In their initial forms, such expectations constitute what the historian David Nye termed "little narratives" of the future, a kind of tactical social capital used to frame theories, stake priority claims, mobilize resources, and build prestige.[14] When amplified by promoters and patrons of science and technology,

such expectations may inform grander narratives of the technological utopia, a rhetorical genre often associated with and pressed into the service of
socially conservative politics.[15]

Better battery discourse and the idealized electric supercar that it implied
originated in narratives of the future that emerged around materials science
and engineering, an interdisciplinary field devoted to the study of complex
physical systems. Science-based materials research first emerged in industrial
laboratories at the turn of the twentieth century and was greatly expanded by
the federal government during World War II and the Cold War in the wake of
the Sputnik shock.[16] William O. Baker, a top executive of Bell Laboratories and
an influential Cold War-era science policymaker, became a leading promoter
of materials determinism, a worldview that understood history as a quest
for ever-superior substances with civilization-defining properties. Materials
determinism can be traced to the three-age (stone-bronze-iron) periodization
of human prehistory, an idea often credited to the Danish museum cura-
tor Christian J. Thomsen as a foundational element in the development of
science-based archeology in the 1830s.[17] Baker drew on the three-age system
in arguing for expanding federal government support for advanced materials,
invoking humanity's progress through the material ages up to industrial steel
in suggesting that the new compounds then under development for electronics and missiles would have similarly dramatic consequences for contemporary society at large.[18]

The teleology of materials determinism was paralleled by the belief
that national programs of undirected basic science informed the development of technology and national economic growth in stepwise fashion.
The intellectual origins of the contemporary linear model of research and
development is often traced to the scientist and science administrator Van-
nevar Bush, who developed the concept in the context of the science-based
weapons programs that he helped set up and manage during World War II.
The dramatic success of these programs crystallized the institutionalization
of basic research in high-energy physics.[19]

However, it was virtually impossible to predict outcomes of linear
research and development, especially in state-sponsored enterprises of
basic research and early-stage technology development.[20] Institutional patrons
of basic science came under increasing political scrutiny, with policymakers faulting not the linear premise, but the manner of its application. They
launched a series of institutional-organizational reforms designed to facilitate

the translation of science into technology.[21] Part of this effort involved the growth of federal support for a specialized infrastructure devoted to materials research at universities and for branches of physics devoted to the study of solid materials that promised technological applications.[22] These efforts paralleled the rise of the semiconductor sector in the early 1960s in service of the military market for missile control systems.[23]

The solid-state revolution in electronics also had the effect of reviving power-source electrochemistry through solid-state ionics, the science and technology of moving, inserting, and storing charged particles inside solid materials without changing the fundamental structures of those materials. This emerging field yielded new compounds that enabled the development of potent new power sources, but it also introduced the reductive view, prevalent in the semiconductor field, that the material constituted the device.[24] This way of thinking tended to conceive the power source as a discrete thing rather than a component of an appliance, whose duty cycle, or proportion of time of operation, had important consequences for the chemistry of the power source. Such thinking reinforced the existing preference for high energy and power shaped by government and military imperatives and greatly complicated the innovation process. Researchers who needed to maintain government funding and wished to make inroads in the civilian market found that it was relatively easier to boost the energy and power of devices than it was to make them cheap and durable. Moreover, demonstrations of energy and power made for far more dramatic and convincing displays to potential patrons than did demonstrations of durability and cost-effectiveness. The social selection of energy and power as the most important measures of performance informed the political economy of the power-source field and little narratives of the future that importantly influenced the electric vehicle revival.[25]

Materials thinking and better battery discourse began to intersect with the electric car thanks to the emergence of the neoliberal interventionist state in response to the environmental, energy, and socioeconomic crises of the 1960s and 1970s. The rise of this state represented in part an attempt to *save the phenomenon* of linear innovation and, more broadly, the classical liberal self-regulating marketplace, one that acknowledged that the market did not always address issues of crucial public interest, and advocated politically correct modes of intervening to correct market failure.[26] These interventions aimed to augment linear innovation by mixing support for the research and development of technologies deemed capable of enabling social policy with regulations and incentives designed to encourage and even force industry to adopt them.

Sociologists sometimes refer to the decentralized complex of institutions responsible for managing these policies as the *national developmental state*.[27] This complex can be seen as an expansion of what the historian Brian Balogh referred to as the *proministrative state*, a product of the fusion of an emerging class of professional expert, often trained in science disciplines, with federal administrative capacity in the form of new institutions of science and technology with roots in World War II, notably the Atomic Energy Commission (AEC) and the National Science Foundation (NSF).[28] The national developmental state was constituted of these and other institutions as the federal government expanded over time and added new priorities. This complex was staffed by a class of professional civil servant whose missions and resources were determined by the legislative and executive arms of the federal government, and also often importantly influenced by the judiciary (and, as we shall see, the state of California), but who operated with substantial autonomy in tenures that often spanned multiple presidential administrations.

The responsibilities of the national developmental state included the regulation of energy and the environment, and the administration of research and development related to energy and the environment. To avoid the appearance of overt winner-picking, the proministrative elite aligned energy and environmental regulations and technology-forcing regulations in a purposely uncoordinated fashion, producing a sometimes-contradictory mix of measures that can be defined as *quasi-planning*.[29] From the 1970s, policymakers and planners began deregulating capital and deconstructing organized labor, while tightening environmental regulations and deepening and expanding existing collaborative arrangements between industry, the academy, and the federal government.[30] They also reformed the patent process to hasten the commodification of academic research performed with state funds, placed federal research infrastructure at the disposal of industry, and refined the public-private partnership as the preferred vehicle for applying science to industrial technology.[31]

Policymakers first raised the prospect of the electric car as one of several possible solutions to energy and environmental problems in the early 1960s. In the early 1970s, new energy and environmental policies provided some federal resources and programmatic guidance for research and development on the components of electric propulsion and also stimulated industrial work in these areas. These efforts continued at a somewhat diminished level through the 1980s, before intensifying with the advent of the Zero Emission Vehicle (ZEV) mandate, a landmark technology-forcing initiative created by the state of California in 1990. Car companies that had grudgingly participated in federal and state efforts to cut the emissions of gasoline-fueled ICE automobiles since the late 1960s were happy (along with some oil interests) to study advanced propulsion technology, but they were deeply hostile to the mandate because it implied the commercialization of the all-battery electric car, a scenario that posed a host of unknowns that would have important implications for commerce. As automakers built up their fleets of ICE cars, they derived an increasingly large proportion of their profits from supplying spare parts and servicing vehicles.[32] Electric cars were widely assumed to have higher up-front costs and lower operating costs than conventional vehicles, so in principle an electric fleet threatened the parts and service revenue model. However, the temporal mismatch of battery and motor raised the possibility that the battery would have to be replaced at some point in the vehicle life cycle, suggesting that the operating costs of electric cars might be higher than assumed. In theory, the commercial electric car presented an opportunity for automakers to supply parts and services relating to replacement batteries, a potentially lucrative enterprise.

At any rate, the temporal mismatch had significant implications for business practices that automakers had developed around ICE cars over nearly a century. Car companies responded by trying to neutralize the mandate through conventional legal and political means. To buy time, they enlisted better battery discourse. Automakers argued that existing batteries could not match the energy and cost of gasoline-fueled ICE propulsion, and therefore industry's mandate responsibilities should be limited to the research and development of advanced propulsion systems. To be sure, some emerging power source technologies were very powerful and energetic, but few had been designed specifically for electric vehicles and required considerable effort to be adapted for this application. When automakers claimed that the development of ZEVs could not be rushed, they made a case that regulators regarded as reasonable.[33]

Through the 1990s and into the early 2000s, car companies treated their all-battery electric car programs essentially as a large-scale experiment. They produced small demonstration fleets, including some equipped with the nickel-metal hydride rechargeable, before suppressing these programs on grounds that the technology underperformed and was too expensive. The auto industry's preferred ZEV was the electric car powered by the fuel cell, an electrochemical energy conversion device that in theory offered many advantages over the conventional galvanic storage battery in the electric traction application, especially in terms of range. For this reason, California based its standards of ZEV performance on the capabilities of the notional fuel-cell electric vehicle.

Another important technological development in this period was the hybrid electric car. Alone among the major automakers in the 1990s, Honda and Toyota decided to commercialize hybrids, primarily with a view to reconciling the industrial-technological demands of US emissions and fuel efficiency regulations. In principle, hybrid electrics also resolved the temporal mismatch of battery and motor.

These enterprises involved a range of actors with disparate interests and objectives. Materials and power source specialists, electric vehicle enthusiasts and end users, and automobile and electronics engineers (both independent and corporate) generated new knowledge on an array of technologies relating to advanced propulsion systems, sometimes in collaborations that could be variably tacit, explicit, amicable, and antagonistic. Over time, these groups helped foster an incipient advanced automobile manufacturing complex that possessed characteristics both of the traditional heavy industries associated with automaking and the newer so-called *high technology* industries associated with the electronics sector, including the capacity to fabricate new and complex compounds and components to very high tolerances.[34] Over time, tactical narratives of the future told by materials and power source researchers were retold by policymakers and the ICE/fossil fuel industrial establishment as grand narratives of automobile techno-utopias. Backed by regulators, lawmakers, and the oil industry, car companies counterposed cancellation of their battery electric car programs with promissory visions of a fuel cell–enabled hydrogen economy as the ultimate form of sustainable energy conversion for personal automobile transportation.

Those who regarded the battery electric vehicle as a moral imperative, including legions of enthusiasts, however, were alienated by this vision. They became part of a movement committed to using the lithium ion battery to solve the perceived shortcomings of electric propulsion technology, one that gained strong public policy support in the wake of the near-collapse of US automaking during the Great Recession of 2007–2009. Most enthusiasts of electric cars had hitherto favored relatively small vehicle platforms, in part owing to the limitations of extant power source technology, but the advent of lithium rechargeables enabled the construction of larger and more capable electric platforms approaching standards of comfort, performance, and utility of premium ICE automobiles. Pundits, promoters, and marketers of these electric supercars in turn invoked a lithium economy as the latest materials-based automobile techno-utopia.

AUTOMOBILITY, ENERGY CONVERSION, AND ENVIROTECHNOLOGY

These stories of user initiative, public policy, and industrial enterprise are rooted in a broader story of the national developmental state's efforts to coordinate environmental, energy, and economic policies over time. As the analysts Richard Chase Dunn and Ann Johnson noted, the history of efforts to control ICE automobile emissions is characterized not by the linear application of the principles of environmental science in regulatory controls and pollution-mitigation technologies, but by the nonlinear coproduction of knowledge of how to track and transform the molecules of a host of pollutants, whose generation and interaction in environmental and sociotechnical context was only gradually understood by scientists and engineers.[35]

The point of departure in engaging and making sense of the politics of automobile pollution are the politics of energy conversion. From the early twentieth century onward, and especially after World War II, the prime imperative of US energy policy, uncodified but tacitly understood by political, military, and economic elites, was to secure as much primary energy as possible, initially in the form of petroleum but subsequently in all other forms as well, and to subsidize its development so that energy was both profitable for business and cheap for users. In effect, energy was decoupled from the supply-and-demand signals of the classical liberal model marketplace and became part of a quasi–command economy.

By the late 1960s, policy elites were becoming aware that this system, with all its contradictions, was unsustainable. They had no interest in radically changing it, but they did try to reform it in the manner of the progressive-era conservation movement of the early twentieth century, whose worldview traced social problems to the improper management of natural resources.[36] Problems of energy resource exploitation then-defined mainly in terms of hasty and wasteful recovery and that by mid-century were believed to have been largely resolved by means of *rational management* and had faded almost entirely from policy view in the prosperous post–World War II years were brought back into focus and reframed as elements of an environmental crisis from the late 1950s on, following a succession of high-profile disasters including oil spills (exemplified by the 1969 spill in Santa Barbara) and radioactive contamination from nuclear reactors and weapons tests, as well as endemic industrial pollution from fossil-fueled ICE technology. The association between the energy and environmental crises deepened in the 1970s with the advent of the Middle East oil shocks and efforts by three successive presidential administrations to develop technocratic solutions, which had the practical effect of eliding energy efficiency with clean energy. By the end of the 1970s, the national developmental state's position was that energy had to both be conserved and efficiently (and hence cleanly) converted, but not at the cost of corporate profits, consumer convenience, or the structural commitment to energy plenitude. In succeeding years, energy policy substantially subsumed environmental policy, increasingly so from the Clinton administration.

The resulting composite energy-environmental policy equated to nothing less than a quest for limitless, clean, and efficient energy. In the decentralized institutional order of the US polity, the national developmental state apportioned responsibility for securing plentiful, clean, and efficient energy to discrete sociotechnical regimes whose distinct interests in the fields of science and technology associated with plentiful, clean, and efficient energy often clashed, triggering a cascading series of unintended consequences. Among these was the commercial electric car, which became a key prop in the limitless, clean, and efficient energy imaginary.

The story of the contemporary electric car hence compels a reconsideration of the automobile system not only as an infrastructure of transportation, but also as an infrastructure of energy conversion in environmental context. A useful concept in understanding these relationships is *envirotechnology*, the idea that the construction and operation of infrastructure produces a hybrid entity in geophysical space that is neither purely social nor purely natural.[37] The historian Richard White referred to such entities as *organic machines*, an expression he coined to describe human activities and artifacts along the Columbia River.[38]

The idea of the organic machine readily applies to automobility and electricity as energy conversion infrastructures and helps elucidate the envirotechnical contexts of the electric car revival. The commercial electric car implies some level of integration between the automobile and electricity systems, the two largest parts of the legacy energy conversion complex in the US. The term *automobility* is sometimes employed by scholars to express the pervasive social and cultural effects of this form of infrastructure. I use the expression to encompass both the sociocultural aspects of the automobile fleet as well as this fleet's function as a type of energy conversion infrastructure in relation to *electricity* as another type of energy conversion infrastructure.[39]

Automobility and electricity emerged in parallel in the late nineteenth century, but aside from the case of electrified auto manufacturing plant, these infrastructures remained essentially discrete, with largely incompatible technologies and business models. Whereas automobility was constructed as a system of privatized public transportation around a massively scaled durable consumer good converting petroleum-derived carbonaceous fuels to motive force by means of ICE on state-subsidized roadways, electricity was constructed as a service utilizing all forms of primary energy converted by a range of energy conversion technologies in large, centralized complexes.[40]

The problems for which the commercial electric car was framed as the solution issued in part from the asymmetry between automobility and electricity wherein the automobile fleet came to possess an aggregate generation (and pollution-producing) capacity that was vastly larger than stationary generation plant over the course of the twentieth century.[41] By 1966, it was estimated that motor vehicles accounted for nearly 60 percent of identified air pollutants in the US.[42]

This asymmetry crucially informed the construction of institutions responsible for regulating energy supply, energy efficiency, and environmental quality. The envirotechnical context in this process was primarily California, where a permanent temperature inversion exacerbated the effects of emissions both anthropogenic and natural. In the 1950s, the discovery that the primary cause of persistently poor air quality in the Los Angeles basin was automobile effluent, which reacted photochemically to produce smog, led to the creation in the 1960s of institutions of environmental regulation at both the state and federal level devoted mainly to controlling mobile sources of pollution. In the 1970s, federal energy efficiency measures were similarly directed mainly at the automobile fleet, largely because the energy crisis manifested mainly as a severe, albeit temporary, shortage in the supply of petroleum, automobility's chief source of primary energy.

Over time, analysts, regulators, scientists, and engineers came to believe that emissions controls on and fuel efficiency improvements to ICE technology worked together to reduce pollution.[43] In 1965, the first federal auto emissions regulations were instituted, and from the early 1970s, the California Air Resources Board (CARB) and the federal Environmental Protection Agency (EPA) cooperated in compelling automakers to cut smog-producing emissions, defined as fine particulate matter of carbon monoxide, hydrocarbons, and nitrogen oxides, by means of a succession of pollution-mitigation technologies. In 1971, the EPA also launched a campaign to phase out leaded gasoline. From 1975, the US Department of Transportation's National Highway Traffic Safety Administration (NHTSA) regulated fleet fuel efficiency through the Corporate Average Fuel Economy (CAFE) system, leading automakers to introduce a larger proportion of smaller, more efficient vehicles.[44]

Yet the environmental dividend from the expected synergy of emissions abatement and improved energy conversion efficiency proved elusive. The analyst Sudhir Chella Rajan held that California's system of air quality governance was predicated on two flawed assumptions: first, that the automobile fleet could be policed using an actuarial model that spread risk and quantified emissions, mandated remedies, and monitored individual drivers for compliance; and second, that economic growth would not outpace technological progress.[45] Emissions control technology could and did improve the energy conversion efficiency of the individual ICE, although it took time for automakers to master the integration of systems in producing this result.[46] At fleet scale, however, improved emission controls and fuel efficiency yielded ambiguous environmental outcomes. Even as CARB and the EPA forced the car industry to produce cleaner cars, and the NHTSA forced it to produce more efficient cars, and the aggregate tonnage of smog-forming automobile emissions in the US declined gradually from 1970, severe smog continued to plague cities well into the 1990s and early 2000s, especially in California.[47]

The persistence of smog in this period has been ascribed in part to Detroit's reversion to the old trusted business formula. The transportation analysts Daniel Sperling and Deborah Gordon held that from the late

1980s, US automakers began to offset their gains in fleet fuel efficiency by building heavier, more powerful, and more profitable vehicles that were relatively more efficient and cleaner than earlier generations of similar vehicles but were more polluting than smaller contemporary vehicles.[48] Antipollution technology also had limitations. The first catalytic converter was a two-way device, so called because it oxidized carbon monoxide into carbon dioxide and unburned and partially burned hydrocarbons into carbon dioxide and water, but it did not convert nitrogen oxides. In 1981, automakers introduced the improved three-way catalytic converter, which reduced nitrogen oxides to nitrogen but still produced carbon dioxide from the oxidation of carbon monoxide and hydrocarbons. Moreover, the functionality of catalytic converters degraded over time thanks to constant chemical side reactions that gradually fouled the catalytic surface area with waste by-products.[49]

But the main reason why cleaner and more efficient cars did not significantly improve air quality had to do with the envirotechnical qualities of automobility and its enabling industrial complexes in the context of US capitalism. Every year, the regulated vehicle fleet grew larger, used more energy overall, and produced more pollution overall, increasingly in the form of carbon dioxide, despite emitting less particulate matter from the late 1960s to the late 1980s.[50] Moreover, the cumulative effects of pollution were difficult to analyze and predict and presented policymakers with almost insurmountable management challenges.[51]

The failure of fuel efficiency and emission controls to substantially mitigate smog in Californian cities by the late 1980s provided the policy rationale for the ZEV. All the advanced propulsion vehicles that appeared in the wake of the mandate had important implications for manufacturing that rippled throughout the industrial value chain, creating demands for unfamiliar materials, components, and techniques of fabrication and assembly. The notional commercial all-battery electric car also implied a radical renovation of the energy conversion complex because it shifted some of the function of primary energy storage and conversion from the ICE automobile system, which held a large quantity of fuel in hundreds of millions of vehicle fuel tanks, to the various primary energy systems that supplied the electricity system. Accordingly, a fleet of battery electric vehicles of appreciable size raised problems of systems integration with grid electricity that policymakers did not consider much in the early 1990s.

Nor did policymakers anticipate the broader effects of smog and fuel efficiency controls on automobility in envirotechnical context. Ironically, smog began to be mitigated not by all-battery electric propulsion, but by improved ICE technology from around the turn of the millennium. And while the automobile fleet produced less smog by the turn of the millennium, it produced increasing quantities of greenhouse gases, overtaking industrial plant in the mid-1990s and electricity generation in the mid-2010s as the leading source of these pollutants.[52] Greenhouse gases were much harder to control than the particulate matter associated with smog, and the federal government, under strong pressure from the auto and oil industries, long refused to identify them as directly harmful to human health and the environment. Mounting evidence of the correlation between greenhouse gas emissions, climate change, and a panoply of environmental harms made this position increasingly untenable by the late 2000s. As the national developmental state responded to climate change, the environmental rationale for deploying electric cars shifted from preventing smog to preventing greenhouse gas emissions.

Energy policy correspondingly began to incentivize renewable energy sources in this period, especially solar and wind, which created further unintended consequences for the organic machines of energy conversion.[53] Solar, wind, and other intermittent forms of primary energy are difficult to exploit, and they complicate management of the electricity grid because they become available when demand is not always high and their periodicity induces variabilities in net load that have to be quickly filled with other forms of generation.[54] Some policymakers and entrepreneurs perceived that the expanding electric car fleet could serve as an important resource in resolving these problems. They expected that with further technological development, electric cars would be able not only to directly store and use renewable electricity, but also feed it back to homes and businesses in decentralized power systems and to the grid itself.

These imagined systems, known as *vehicle-to-grid* and *vehicle-to-everything*, implied the deep integration of automobility and electricity as business enterprises and energy conversion infrastructures.[55] A host of sociotechnical barriers stood in the way of this project. Nevertheless, efforts to develop the electric car as a utility power plant on wheels implied important shifts in the American lifestyle around expectations for reliable grid electricity and personal vehicle ownership and use.[56]

AGE OF AUTO ELECTRIC

The electric vehicle revival was hence determined not by better batteries, but by the interplay between changing envirotechnical and socioeconomic conditions, energy and environmental policies, systems of energy conversion and industrial production, and material practices of innovation. These multifaceted relationships are the subject of this book. Embedded within this discussion are elements of business history, of batteries and electronics as well as automobiles in global context, the history of institutions of energy and environmental regulation and science, technology, and industrial development, and the history of makers and users of electric vehicles.

Another key theme of this book is the interdisciplinary nature of knowledge production in complex sociotechnical enterprises. Interdisciplinary collaboration has long been widely believed to be conducive of innovation, but its practice has been inhibited by a number of sociocultural and institutional factors, especially accounting rules that divide labor into activities designated as *science* and *engineering* and compel linear, stepwise problem-solving.[57] These factors were certainly at play in many of the institutions involved in the creation of electric vehicle technology and shaped the material practices of invention and innovation. But interdisciplinarity did exist on the margins of large corporations and federal institutions, as well as in many smaller enterprises where accounting and managerial oversight was weaker and where actors from different backgrounds could cooperate in solving problems in unconventional, nonlinear ways that lay outside disciplinary norms.

This book explores these relationships in several exemplary episodes and enterprises spanning seven decades following World War II. Chapters 1 and 2 use primary sources relating to power source research drawn from the National Archives and Records Administration and the NASA Headquarters Library, published primary texts on power-source technoscience and environmental policy, and secondary texts on the social relations of Cold War-era science and technology to trace the emergence of air quality as a public policy concern in the late 1950s and early 1960s, the identification of the electric car as one of several possible solutions, and the resulting debate as to whether its technologies were at hand and had only to be applied or whether they had to be invented. Some builders of electrics looked to adapt the methods, materials, and components of industrial automaking, while others tried to adapt technology produced for defense aerospace and electronics and still others sought to develop wholly new systems. The contemporary electric car grew out of the engagement of and tension between approaches often conventionally defined as *low* and *high* technology, categories with a mystifying moral connotation. I instead refer to relations between communities of practitioners associated with established and emerging technoscience.

Chapter 3 explores interpretations of electric vehicle systems in the context of the environmental and energy crises, the counterculture, and the appropriate technology movement of the 1960s and 1970s. Persistent air pollution provoked enthusiasts and activists to mount experiments with existing electric vehicle technologies while disruptions in the global supply of petroleum stimulated researchers supported by the US and foreign national developmental states as well as affiliated corporations and universities to make significant progress in power source technoscience and advanced componentry relevant to electric propulsion. While these sets of practices tended to be exclusive to these communities, they sometimes overlapped to produce new knowledge and hardware. The appropriate technology and counterculture movements also had an important indirect influence on energy and environmental policies as they related to electric automobility. As the historian W. Patrick McCray observed, the political establishment's reaction to the counterculture argument of the limits of economic growth had the effect of embedding techno-utopian thinking in public policymaking.[58]

Dreams of a limitless future informed many enterprises of science and technology in the last third of the twentieth century, and they also informed the technological politics of air quality and energy conversion. Chapter 4 explores how the ZEV mandate accelerated the infiltration of *technofuturist* ideas and counterculture enthusiast-experts into the affairs of mainstream automakers. An important part of the strategy of the car companies in buying time for lobbying to take effect against the mandate consisted of exploiting the ambiguity of the ZEV classification, a category constructed by California air quality bureaucrats as a consequence of the state's limited legal authority on envirotechnical questions. California had the power to compel automakers to produce automobiles that yielded zero emissions but it did not have the power to specify the kinds of energy conversion technologies equipping these vehicles because energy questions were the purview of the federal government, a loophole that allowed car companies to reinterpret the technological identity of the ZEV.[59] Automakers problematized the equation of the ZEV with the all-battery electric vehicle and emphasized the shortcomings of existing power sources while promoting emerging advanced power source technologies and alternative propulsion platforms. Where air quality bureaucrats defined automobile performance solely in terms of emissions, automakers privileged other metrics, primarily range but also comfort, convenience, cargo capacity, and acceleration. Car companies sought to demonstrate the inability of the all-battery electric car to meet these standards through public rituals of presentation that framed questions and ratified knowledge claims.[60]

These rituals were defined primarily by General Motors (GM) and Toyota, archrivals with large ZEV commitments whose competition in commerce and green automobile dramaturgy was complicated by tacit cooperation in undermining the mandate. The efforts of these giant automakers converged in the Ovonic Battery Company (OBC), a division of Energy Conversion Devices (ECD), a materials research company founded in the mid-1960s by the inventor Stanford R. Ovshinsky. OBC dominated the intellectual property of the nickel-metal hydride rechargeable battery, and the company's sweeping claim to this technology, in a regulatory context where air quality bureaucrats were increasingly willing to consider and promote (if not mandate) specific emerging technologies they believed enabled the zero-emission outcome, was interpreted by the automaking establishment as a threat, one that GM and Toyota sought to mitigate and manage for their own purposes. OBC had extensive ties to the global consumer electronics and automaking communities and was an important technological gatekeeper. Accordingly, the company is an ideal proxy of broader attitudes on electric cars.

I probe these relationships using the personal papers of Ovshinsky and his friend and business partner Robert C. Stempel, a former chair and chief executive of GM, which are held at the Bentley Historical Library of the University of Michigan at Ann Arbor. This extensive collection of documents sheds insight into corporate thinking about electric propulsion technology at a number of automakers besides GM and Toyota, including Ford, Chrysler, and Honda, as well as Matsushita/Panasonic, for many years the world's most important supplier of battery cells for electric vehicles.

Chapter 5 explores the emergence of the hybrid electric car, a sociotechnical means of reconciling the demands of emissions and fuel efficiency regulations, as well as the economic problems posed by the all-battery electric format, with a focus on Toyota's Prius. In theory, hybrid battery electric technology could narrow or even neutralize the temporal mismatch between battery and motor, solving the economic conflict of interest between automaking and batterymaking. Where all-battery electrics had potential hidden battery replacement costs, hybrid electrics used smaller and less costly batteries designed to last the lifetime of the vehicle. In short, the hybrid electric vehicle aged like an ICE vehicle and could therefore be accommodated within the product-based aspect of the auto industry's business model.

I address GM's mandate strategy throughout the second third of the book, especially in chapters 6 and 9. I argue that the automaker partnered with OBC with a view to monopolizing legal rights to the large nickel-metal hydride rechargeable for the electric vehicle and producing a small number of sophisticated but costly all-battery electric cars that would demonstrate to regulators the intractability of the temporal mismatch.

Automakers further undermined the all-battery electric car by promoting the hydrogen fuel cell electric car as the ultimate ZEV, a subject I address in chapter 7. In the 1990s, the car companies argued that the technology of fuel cell propulsion then under development was in theory capable of giving the electric car a much greater range than the all-battery electric format, and moreover enabled the electric car to use the existing liquid fuel infrastructure. Air-quality regulators accepted this argument and devised new benchmarks for ZEV performance around the theoretical capabilities of the fuel cell, allowing automakers to engage in protracted research and development and delay full implementation of the mandate.[61]

Chapter 8 explores the effects of the competition between GM and Toyota in green car dramaturgy on environmental policy. Toyota's commercialization of the Prius helped shift the frame of reference in environmental discourse from one that linked control of emissions of smog-forming pollutants to the mitigation of local smog, an objective that in principle could be managed by local-regional authorities like CARB, to one that linked fuel efficiency to the control of emissions of greenhouse gases and the mitigation of climate change, objectives that implied far less easily policed geophysical spaces. The story of the commercial hybrid electric car helps elucidate the completely dissimilar technopolitical regimes associated with local air quality and climate change, a crucial issue of contemporary environmental politics.[62]

Chapters 10 through 13 address the reaction of enthusiast-experts to the auto industry's suppression of the all-battery electric format, the decline of the US automaking establishment, and the rise of new enterprises devoted to electric cars equipped with lithium ion batteries in the 2000s. I chart these events using a mixture of open-access primary documents, media accounts, and interviews with principal actors in industry and civil service. When GM cancelled its all-battery electric EV1, it sold its share of its joint venture with OBC to an oil company, an arrangement that restricted the use of large, advanced nickel-metal hydride rechargeables in all-battery electric cars. Enthusiast-experts sought to revive the all-battery electric format around the lithium ion rechargeable, an important enabling technology of mobile telephony and computing in the late 1990s and early 2000s. Facilitated by Silicon Valley expertise and venture capital, this initiative gave rise to the idea of the electric car as a computer on wheels.[63] Concurrently, California's efforts to deregulate and marketize its electricity system created a crisis in grid electricity that promoters of the electric car claimed could be solved using electrics reconfigured for bidirectional power flow.

The reframing of electric vehicles as computers and utility power plants on wheels helped inform changing public policy perceptions of electric propulsion technology. As US automakers lost billions of dollars in the Great Recession, the national developmental state shifted its emphasis from hydrogen fuel cell power to lithium power, both in the all-battery electric format and a new format known as the *plug-in hybrid electric*, a technology that employed a much larger battery than conventional hybrid electrics that could be recharged from the electricity network. The federal government and the state of California supported electric vehicle start-ups including Tesla Motors with stimulus policies that over a period of two decades helped position Tesla as a leading supplier of ZEVs. As a result, the company became a prime instrument of environmental policy and one of the world's most valuable enterprises by market capitalization.

An important aspect of the electric vehicle revival is the experience of users, a subject explored throughout the book, and at length in chapter 14. Hobbyists and enthusiasts played a notable role in knowledge-making in the early years of the electric vehicle revival, and the auto industry's protracted testing and evaluation of electric cars enrolled scores of ordinary motorists in this process as well. A crucial element of experimentation is what the sociologist Trevor Pinch referred to as the *similarity judgment*, or the comparison of classes of artifacts that are superficially similar.[64] Scientists and engineers can hardly avoid making similarity judgments in coming to grips with nature, but such thinking can also produce misleading accounts of the world. Designers and engineers worked assiduously to give electric cars the performance and manners of the best ICE cars, but in real-world operation, electrics often exhibited unexpected phenomena that ordinary users were the first to experience. Contemporary electric cars proved substantially dissimilar from their ICE counterparts, not only in terms of how they converted energy to mechanical motion, but also in terms of their manufacturability and economics.

In this, as in so many other contexts of science and technology, boundaries separating scientists, engineers, and laypersons, and laboratory, factory, and public spaces, frequently blurred.[65] Over time, some enthusiasts (and even ordinary users) of electric cars became enthusiast-experts on electric propulsion technology. In this chapter, I use oral histories to explore how such actors tested similarity judgments and understood the car-driver interaction as a knowledge-making practice.[66]

For all its technological novelty, the electric vehicle revival was informed by race, class, and gender dynamics not dissimilar to those unfolding at the outset of the automobile era. In the late nineteenth and early twentieth centuries, automobiles were generally the preserve of white men of means, and the unregulated and often irresponsible use of these powerful vehicles caused social instability that was then regarded as more than a nuisance.[67] Critics of the electric supercar note the same "arrogance of wealth" that Woodrow Wilson (as president of Princeton University) indicted in his 1906 denunciation of the ICE automobile.[68] On the other hand, one might expect the explicit gendering that then characterized the marketing of the electric car to be a relic of a bygone era. The separate-spheres campaign mounted by now-extinct marques like Argo, Baker Electric, and Pope Manufacturing, which framed the electric as the "woman's car," died with the initial demise of the technology, a victim, suggested the historian Deborah Clarke, of its own limited appeal.[69] Contemporary electric vehicle advertising tends to mobilize symbols of environmental and technological virtuosity, but it would be a mistake to assume that its subtexts are necessarily neutral in terms of race, class, and gender, and without consequence for social relations.

This book challenges the equation of social, socio-technological, and environmental progress, distinct categories that all too often are conflated not only by policymakers, but sometimes also by historians and social scientists. Writing in 2007, the political scientist Matthew Paterson held that the project to green the automobile fleet may never completely succeed.[70] Whatever its ultimate trajectory, the electric vehicle revival tracks important changes unfolding in industrial society since the 1970s. It emerged in the context of the rise of the information technology revolution, the decline of the classic, vertically integrated industrial corporation and its in-house research laboratory, the trend toward outsourcing and offshoring, and the increasing marketization and financialization of civil society. Rooted in conditions of deepening socio-economic as well as environmental crisis, the contemporary electric vehicle can be seen as the product of a prolonged reconsideration of the automobile as an "inevitable and desirable feature of life."[71]

In some ways, the contemporary electric car may be interpreted as a California car. The performance parameters of many models were shaped by the state's regulatory regime, and the technology of some of the most important, especially those produced by Tesla, was also crucially informed by the state's culture of invention and innovation. But if the contemporary electric could be said to have gestated in the organic machine of Californian automobility, the technology owed its continued development as much to chronic air pollution in places outside the Golden State, to the efforts of other regulatory regimes to control it, as well as to the contributions of a score of high-technology enterprises, both American and foreign, in places across the US and in many countries around the world, increasingly in East Asia. Electric automobility has become a global experiment in regulated technological change, one that has forever altered lived experience of the personal passenger vehicle. This book shows how the affinities and conflicts between established and emerging industries, environmentalism and capitalism, and the public and private interest served to construct a new age of auto electric.

2 RECONSIDERING THE AUTOMOBILE

The electric car does not mean a new way of life, but rather it is a new technology to help solve the new problems of our age.
—Senator Warren G. Magnuson, March 14, 1967

In 1960, virtually the only passenger electric vehicle available for sale in the United States was built by a company that manufactured vacuum cleaners. The Henney Kilowatt was the brainchild of C. Russell Feldmann, an entrepreneur whose interests bridged the worlds of automobiles, power sources, and electrical appliances. In the 1920s, Feldmann anticipated the market for car radios and founded the Automobile Radio Corporation, collaborating with the industrialist Walter P. Chrysler in this enterprise. Feldmann became president of the parent company, known as the National Union Electric Corporation, whose divisions would include Eureka, then a household name in vacuum cleaners. Feldmann built the Kilowatt around existing automobile and electric technology, applying the same physical principles used to suck up dust from the suburban domicile to drive passenger cars in suburban spaces. With an eye to Henry Ford, Feldmann sought to integrate vertically. He acquired Henney, a builder of custom ambulances, limousines, and hearses, and selected the Renault Dauphine subcompact as a conversion platform. Eureka furnished the motor, but Feldmann needed help for the rest of the drivetrain. He brought in the Exide company, which supplied lead-acid rechargeable batteries, and in 1962 consulted the electrical engineer Victor Wouk, a graduate of the California Institute of Technology

(Caltech), who had founded a research company devoted to electric power controls. Wouk told Feldmann that he could build him a speed controller but advised that the system would not be very efficient because the lead-acid rechargeable gave so little energy (some 500 times less than gasoline). Moreover, the equipment would be expensive because Feldmann did not plan to convert a large volume of vehicles.[1]

The Kilowatt demonstrated the limits of the commerce and technology of electric vehicles in the years after World War II. There are several ways to configure an electric vehicle. Innovators who favored the all-battery electric format could use existing batteries and adapt an automobile chassis to the capabilities of these power sources, either by converting a conventional commercial automobile, as Feldmann did, or by building a new dedicated and lightweight chassis and body. Both approaches represented the path of least engineering resistance. Another means of maximizing the potential of existing power sources that required more complicated engineering was to develop a hybrid battery electric system and embed it either in a repurposed or purpose-built chassis, an approach that Wouk came to favor. There was also the option of building an all-battery electric car, either a conversion or purpose-built, around a new and more capable power source, an approach that required advances in materials science and electrochemistry. In the late 1950s and early 1960s, most observers thought that building a better battery for electric cars was a costly and long-term proposition.

In the early 1960s, the only type of roadworthy passenger electric vehicle available in the US was the conversion. Conversions varied in ride quality and drivability but were invariably inferior to the original ICE vehicles in range and speed, in good measure because battery electric propulsion systems were underpowered for the weight and mass of the body and chassis. Most American cars of the late 1950s and early 1960s were large and heavy, in part a function of the availability of cheap gasoline, which enabled large and powerful ICEs. It was for this reason that Feldmann had chosen to convert a smaller European car. Most American motorists could not then consider these trade-offs for themselves. Feldmann built only around 100 Kilowatts and sold only a handful, which were among the few dozen electric vehicles operating on US roads in the early 1960s. For policymakers and most consumers, the electric car was an object of curiosity at that time.

Only a few years later, however, the electric car would feature in an emerging discourse of environmental policy triggered by the emerging public health crisis of air pollution, especially in California. This discourse turned largely on a debate involving scientists, regulators, and industrialists on the role of road vehicles in this crisis, as well as possible solutions. Automobile pollution is a complex nonlinear problem with many variables, including the type of gasoline used in ICEs, the temperature of the engines, how and where automobiles are driven, and the temperature, altitude, and composition of the ambient air.[2] California's climate and topography exacerbated emissions of all types. The state is famed for its crystal clear sunshine, and yet for much of the year, large parts of it are not sunny at all, even when the air is not fouled by human activity. In spring, summer, and fall, hot air from the southwestern deserts moves over the cold waters of the Pacific Ocean, where, hemmed in by the San Gabriel, San Bernardino, and San Jacinto mountains, it corrals cooler air and creates a temperature inversion that traps moisture and particulate matter.

The result is that California's coastal regions are often shrouded in fog and marine layer. In such conditions, smoke and other emissions seem to blend with fog in a dense haze that in the early 1900s was dubbed "smog." The inversion phenomenon was well known to the region's indigenous peoples and European colonialists. Chumash Indians referred to what is now known as the Los Angeles Basin as the "Valley of the Smokes." The Spanish explorer Juan Rodriguez Cabrillo observed a smoke-trapping inversion event in 1542 from a schooner in what would become San Pedro Bay. Contemporary coast-dwelling Californians refer to the "May gray," "June gloom," "no-sky July," and "Fogust."[3]

The problem of automobile pollution was initially approached solely as a question of the chemistry of air quality, not the chemistry of the ICE as a technology of energy conversion in an envirotechnical context.[4] This was partly because the science of air quality did not originate in efforts to comprehend industrial pollution. Credit for pioneering the science of smog is usually given to Arie Haagen-Smit, a chemist at Caltech whose primary interest was plant biochemistry. Haagen-Smit's story is well known.[5] In the mid-1940s, the chemist was investigating the chemical basis of flavor in pineapple, and when he condensed fruit vapors in ambient air in an effort to obtain a sample for analysis, the sample also yielded drops of a foul-smelling brownish fluid. Haagen-Smit linked this substance to emissions from automobiles and industrial infrastructure, a claim strongly contested by the oil industry. The view of the Los Angeles County Air Pollution

Control District (APCD), formed in 1947 as the first organization of its type in the US, was that smog was caused primarily by sulfur dioxide emitted by industrial plants, especially refineries. This theory was plausible because sulfur dioxide had been implicated in previous smog events, including one in St. Louis in 1939. Lewis C. McCabe, the first director of the APCD, launched a campaign to regulate sulfur dioxide and other industrial efflu-ent, as well as municipal garbage incineration.[6]

Nevertheless, smog persisted, and Haagen-Smit suspended a promising career in biochemistry to devote himself to solving the mystery. In 1950, he took a leave of absence from Caltech to lead the research program at the APCD and several years later determined that smog was the product of a photochemical reaction between sunlight and nitrogen oxide and unburned hydrocarbons, produced mainly by road vehicles.[7] The air qual-ity regulatory apparatus that subsequently emerged understood its mission primarily in terms of controlling tailpipe emissions, rather than under-standing the industrial ecology of automobility. In the succeeding decades, efforts to engineer a clean ICE would proceed largely empirically through a series of incremental technological fixes that progressively neutralized many (but not all) of the by-products of internal combustion.[8]

While it could be said that the science of automobile emissions origi-nated in California, it would be a mistake to assume that the state's air pollu-tion problems stemmed from a uniquely fragile environment. Automobile emissions became a public health emergency in California mainly because the center of US industrial gravity began to shift to the state following World War II and the start of the Cold War. The massive expansion of Cali-fornia's military industrial complex, and the stimulating effect this had on petroleum, petrochemicals, aviation, aerospace, and electronics, brought prosperity in the postwar years, in turn stimulating Detroit's automobile industry, for whom the Golden State became the most important US mar-ket. In these conditions, the organic machine of automobile energy con-version burgeoned, complicating efforts to understand how it produced pollution and how to control it. It was in this context that policymakers began to consider the electric car as a potential technological fix.

WEIGHING THE OPTIONS

The immediate consequence of the revelation that the automobile fleet was the primary source of smog was the rapid expansion of air quality legislation and institutions at the municipal, state, and federal levels. Initial efforts focused on regulating the use of the positive crankcase ventilation (PCV) valve. Invented in World War II to insulate engines of armored vehicles to enable such vehicles to ford water obstacles, this simple device could also be used to capture unburned gases generated by the engines of passenger automobiles that were normally vented to the atmosphere and return them to the combustion chamber for more thorough combustion. In 1960, California created the Motor Vehicle Pollution Control Board, empowered to issue certificates of approval to control mobile sources of pollution based on standards set by the state's Department of Public Health. In 1961, California passed legislation that required automakers to adopt the PCV valve for the 1963 model year.[9] The federal government added a PCV valve mandate in 1966, and subsequent federal initiatives reinforced California's mobile emissions control program. In 1963, Congress passed the Clean Air Act (CAA) and added amendments in 1965 to set emissions standards for certain types of vehicles for the model year 1968. In 1967, Congress made a key amendment to the CAA that allowed US states to apply for a waiver of federal emissions standards for new motor vehicles, enabling them to set stricter standards if they could demonstrate "compelling and extraordinary conditions."[10] California policymakers were the first to argue that their state met such conditions. In 1967, Ronald Reagan, the state's otherwise socially conservative governor, signed legislation that unified California's air quality regulatory functions in the Air Resources Board (CARB) and appointed Haagen-Smit as its first chair.[11]

Dunn and Johnson held that the PCV valve was low-hanging fruit as far as the technology of mobile source emissions control was concerned. The PCV valve campaign reduced the volume of unburned hydrocarbons generated by the individual motor vehicle but did not neutralize the nitrogen oxides, carbon monoxide, and carbon dioxide it emitted.[12]

By the late 1960s, policymakers were willing to discuss more radical approaches. In March 1967, Democratic senators Warren G. Magnuson and Edmund S. Muskie introduced a pair of bills to further amend the CAA by supporting the research and development of electric vehicles and other ostensibly nonpolluting alternatives. The bills proposed only $5 million for studying technological options, but they triggered a wider discussion of the role of public policy in shaping these options.[13] In five days of Congressional hearings, battery makers, automakers, policymakers, and representatives from electric utilities engaged in sometimes heated debate on the question of whether the technology of the electric car already existed or whether it had to be invented, a problem that turned on differing interpretations of this automobile system.

Authoritative voices including that of Lee C. White, chair of the Federal Power Commission (FPC), held that electric vehicle technology was available, but suitable only in contexts where long range or high speed was not necessary. A chief criterion of vehicles used in enclosed spaces such as factories and warehouses was zero-emission operation, and small freight and forklift trucks became the primary applications of electric drive technology after it disappeared from public roads in the interwar years. A promising environment for electric propulsion was the urban core, a space that electric utilities and makers of rechargeable batteries saw as an obvious market for city cars, or "urbmobiles." One British executive invited to testify noted that a fleet of nearly 50,000 battery electric delivery trucks, the world's largest stock of roadgoing electrics, plied the narrow streets of densely populated British cities. Production of roadgoing electrics was never interrupted in the UK in the twentieth century, unlike in the US, held Horace Heyman.[14]

But American urban spaces were different. In the 1960s, metropolitan cores were being hollowed out by *white flight* and suburbanization, and to automakers like General Motors (GM), available electric propulsion technology was wholly unsuited to the new system of automobility being built around the sprawling, freeway-linked suburbs. To exist in this part of the automobile system, argued Harry F. Barr, GM's vice president for engineering, electric cars needed to be all-purpose. Such vehicles, he maintained, required new, advanced power sources that were cheap, safe, and durable, criteria that were satisfied by no technology at the time.[15]

From these conflicting interests came conflicting views of the role of government in stimulating technology. Transportation Secretary Alan S. Boyd voiced the conventional wisdom that the role of the federal government was solely to set standards, promote research, and let industry determine the pace of innovation. In a sometimes tense exchange with Indiana senator Vance Hartke, a proponent of immediate, robust action, Boyd noted that some government-funded science and technology programs relevant to the electric car were already underway.[16] The published text of the joint hearings appended a report by the FPC indicating that industry and the federal government each spent around $9 million annually on the research and development of components relating to electric vehicles, with most of the federal contribution coming from the US Department of Defense. But there was no systematic effort to develop a purpose-built electric passenger automobile.[17]

Some battery executives believed that existing technology could be effectively applied if electric car programs were properly organized and funded. One held that innovators of civilian electrics were forced to scrounge a hodge-podge of components and assemble them "in the manner of a teenager-built hot rod."[18] Another opined that all that was necessary to make the electric car an instrument of emissions control policy was a federally–subsidized program substituting ICE vehicles with electric vehicles in transit and delivery fleets and in the federal government's own vast vehicle fleet.[19]

It was to this relatively modest vision that Magnuson alluded when he remarked that the electric car did not mean a new way of life for Americans.[20] At that very moment, to be sure, industry and the federal government were radically changing the American way of life through suburbanization. Racialized redlining, the provision of low-interest home loans, and the underwriting of the interstate freeway system all served to massively stimulate ICE automobility.[21] But Boyd clearly articulated the federal line on emissions control policy. The proper role of the government, held the transportation secretary, was to support industry in cleaning up the ICE automobile, although he did hold out the possibility of more robust federal action if results were not forthcoming.[22] In the short term, however, there would be no electric car for civilians that was built from the ground up.

POWER SOURCE MATERIALITY

Policymakers limited federal efforts in the science and technology of electric propulsion to electronic controls and advanced batteries and other power sources, primarily for nonautomobile military applications. This had important implications for the historical development of power-source technoscience and the contemporary electric car. Power-source systems built to military specifications offered higher performance than existing systems, but none were designed specifically for automobile applications and presented performance trade-offs that posed serious challenges to efforts to apply them in commercial civilian vehicles.

Military imperatives in turn informed better battery discourse, a way of imagining the idealized battery that traced to materials science and engineering, a boundary-straddling discipline originally built around industrial research in metallurgy, and later in ceramics and plastics. Materials science and engineering received fresh impetus from the federal government in the wake of the Korean War.[23] In response to this costly and stalemated conflict, the Eisenhower administration devised the New Look doctrine of qualitative technological superiority, calling for the development of a wide array of nuclear weapons, new jet aircraft and rockets to deliver them, spacecraft, a nuclear-powered navy, and soldier-portable sensors. The Sputnik shock accelerated this program. New Look in turn created requirements for new power sources that were energetic, powerful, and lightweight.[24] Fuel cells, photovoltaic arrays, and exotic batteries were on the agenda, and policymakers hoped these technologies could be also adapted for civilian use.[25]

New power sources in turn created requirements for new materials. In the early 1960s, William O. Baker, then vice president of research for Bell Laboratories, held that such materials were the "means through which man realizes his dreams of well-being on earth, or failing that, liberation into space."[26] Solid-state electronics and power-source technoscience were both given impetus by the revolution in military materials but had substantially different developmental arcs. Where semiconductors found wide use in civilian applications from the late 1960s, advanced power sources using solid-state components were not widely applied until decades later. This was due partly to the significantly different physical characteristics of electronics and power sources. It was much more difficult to release and store electrons in sustained cycles of oxidation and reduction than it was to utilize those electrons in electronic switches.

In the main, though, the knowledge gap between power sources and their applications was socially constructed. Properly understood as an allied but estranged field of electronics, power-source technoscience (comprised of electrochemistry, and subsequently solid-state ionics) was a solution in search of a problem for much of its history. For these and other reasons, the research and development of batteries occurred at a great social and intellectual distance from the research and development of consumer devices.[27]

The estrangement of batteries from their applications, and of solid-state electronics from solid-state power sources, had important implications for how researchers understood the technology of rechargeable batteries. Knowledge-making in this context was characterized and complicated by scaling and testing in various applications. As electrodes and cells are integrated into batteries and mated to applications, it becomes increasingly difficult to predict how reactions and side reactions will unfold over time. At each new stage of integration, fresh problems arise, with consequences for durability and cost that are further complicated by the duty cycle.[28] Different applications will have substantially different effects on the same battery chemistry.

In short, power-source technoscience implied the cooperation of heterogeneous experts embedded in a host of institutional homes. But there were then few incentives for would-be systems builders to organize all the relevant actors into enterprises of civilian commercial advanced power sources, owing partly to lack of demand and partly to the formidable knowledge deficits related to attaining durability, safety, and low cost, the key desiderata of such technologies. Accordingly, actors tended to treat power sources primarily as materials rather than as components of complex sociotechnical systems.[29]

The auto industry's first post-World War II experiments with electric propulsion technology unfolded in this reductive context. Electronics and materials researchers working at a distance from automobile designers took energy and power to be the defining properties of batteries, in part because these qualities were the traditional objectives of battery technologists. Unlike improvements in durability, safety, and cost, improvements in energy and power could be relatively quickly and dramatically demonstrated. Such advances in turn constituted valuable social capital for isolated researchers beholden to demanding and fickle state, academic, and industry patrons. Responsibility for integrating components usually lay with other research communities with varying degrees of contact with automakers, and who often had their own ideas about applications. Few of the resulting technologies met all the requirements of commercial electric drive.[30]

MILITARY INDUSTRIAL PROPULSION

In the late 1950s and early 1960s, technologically conservative automakers were philosophically aloof from the world of materials science and engineering. On the other hand, the jet age and the space race fired the imaginations of car stylists and marketers, enthusiasm that found further expression in the willingness of the auto industry to diversify into New Look military industry. In the early years of the Cold War, each of the Big Three automakers founded defense divisions linked to aerospace, acquiring new technologies and encountering opportunities to investigate advanced power sources. Both Chrysler and GM applied turbine engines in concept cars used to promote corporate innovation and spice up ICE product lineups.[31]

Detroit's defense connections also informed research in battery electric propulsion that was of more practical significance. GM initiated the largest of these efforts around 1963, partly in response to rising concerns over air quality. Executives were also curious about the potential of some of the company's defense activities, especially in solid-state electronics, and how that might relate to its other enterprises relevant to electric propulsion. In his Senate testimony, Barr held that GM had considerable resources and experience in the latter field, noting that the company manufactured both electric motors and lead-acid battery rechargeables, including for golf carts. What that experience told GM, according to Barr, was that while it was feasible to produce an electric vehicle around existing technology that would be something between a golf cart and a full-sized car, such a vehicle would be suitable only for special purposes involving limited distances to local amenities.[32]

The question of electric drive that GM professed interest in was whether the latest technology, including equipment developed for military applications, could be adapted in a general-purpose vehicle platform. The first project, managed out of the GM Research Laboratories in Warren, Michigan, involved a conversion of the Chevrolet Corvair, the corporation's lightest production vehicle.[33] Dubbed the "Electrovair," the car used a Delco Products Division induction motor, a machine that is simpler, more robust, and more amenable to miniaturization than the direct current electric motor. The traditional direct current motor operates on the principle of an interaction between a rotating device called an "armature" housed within a cylindrical magnet called a "stator." In the resulting magnetic field, the armature generates alternating voltage, and to prevent the device from uselessly alternating 180 degrees back and forth, a notched or ringed cylinder called a "commutator" is attached to its shaft in contact with a set of spring-loaded brushes. When the commutator rotates past the brushes, the electrical polarity of the armature is switched, converting alternating current to direct current and maintaining rotation in one direction.

Over time, this mechanical system of power control is prone to wear and tear. The induction motor does away with these physical linkages and the permanent magnet. It has only two primary elements: a cylindrical stator fitted for three-phase alternating current, and a rotor shaft subsumed within it. When current passes into the stator, it induces a rotating magnetic field that turns the rotor at a speed controlled by varying the input power frequency. Invented independently in the late 1880s by the Italian physicist Galileo Ferraris and the celebrated Serbian electrical engineer Nikola Tesla, the induction concept had long dominated most industrial applications of electric motors.

Using induction motors to propel cars offered a number of advantages, but it also raised new problems. In theory, the speed of such motors is highly amenable to control, enabling the use of a single-speed transmission that saves weight and volume. But exploiting the induction phenomenon for use in an automobile requires sophisticated control systems as well as an inverter to convert the direct current produced by the battery to the alternating current used by the motor. The military revolution in solid-state electronics had yielded some solutions. The control systems for the Electrovairs were products of GM's Defense Research Laboratories, which had facilities in Goleta, California, a hotbed of military innovation not far from Vandenburg Air Force Base. Electrovair I used a discrete modulator and inverter, while Electrovair II employed a more advanced control system using an integrated modulating inverter.[34]

The Electrovairs' power sources also had a military connection. GM equipped the cars with packs of silver-zinc cells, a battery chemistry then used mainly in torpedoes, submarines, missiles, portable military radios, aircraft, and spacecraft. Developed during World War II by the French-American weapons inventor Michel N. Yardney and his collaborator Henri André, the silver-zinc battery was reliable and safe and had the highest energy density and peak power of any rechargeable of the day. On the other hand, the silver-zinc battery had a short life span and was expensive.[35] In the military, these trade-offs were acceptable; in a commercial automobile, they were not. The Electrovair battery packs were said to cost more than $10,000, more than the sticker price of a conventional Corvair. Silver-zinc chemistry was lighter than lead-acid, so GM engineers crammed as many cells as they could in the trunk and engine bay of the Electrovair in order to maximize its range. Even so, the modified cars were 600 to 800 pounds heavier than the conventional model and had a range of only 40 to 80 miles, performance that did not compare well to the 250 to 300 miles that a conventional Corvair got from a full tank of gasoline.[36]

With these problems in mind, GM engineers also investigated fuel cell propulsion. In the late 1950s and early 1960s, many experts considered fuel cells to be among the most exciting and promising new developments in the power source field. Fuel cells are a family of devices that produce electricity by electro-oxidizing hydrogenous fuels and are typically classified by electrolyte, each of which offers distinct operating system dynamics and advantages and disadvantages in terms of material costs and performance. The basic concept dated to the mid-nineteenth century, when European scientists experimented with reversing electrolysis. Instead of using electricity to dissociate water into hydrogen and oxygen, they attempted to combine oxygen and hydrogen to produce electricity, a reaction that yielded water as the waste product. Working independently in the late 1830s, the Welsh lawyer and amateur scientist William Robert Grove and the Swiss physicist Christian Friedrich Schönbein used costly platinum foil to catalyze the reaction.[37]

It took another century before the first practical fuel cell appeared. It was developed by an English mechanical engineer named Francis Bacon in the late 1940s and early 1950s and employed a liquid alkaline electrolyte at high pressure, allowing the use of cheap nickel as a catalyst. The Bacon cell produced very high current density, and in the early 1960s, the aerospace manufacturer Pratt & Whitney began to develop a variant for use in the *Apollo* spacecraft. Bacon hoped to adapt his technology for terrestrial applications, but its potassium hydroxide electrolyte was poisoned by carbon and thus required relatively expensive pure hydrogen, limiting the technology's commercial attractiveness. For this reason, researchers sought to develop fuel cells equipped with acidic electrolytes, which were resistant to carbon poisoning. In principle, such devices could operate on cheap carbonaceous fuels, but they required an expensive platinum catalyst. General Electric (GE) developed a fuel cell for the *Gemini* spacecraft using an acidic electrolyte based on a solid polymer, enabling a light and compact power source that was also suitable for road vehicles. The company hoped to market the device in civilian applications, but early versions suffered from a host of teething troubles, including membrane dehydration and high electrical resistance. Fuel cells using electrolytes of molten carbonate and solid oxides at high temperatures were in principle capable of directly using the dirtiest, most carbon-rich fuels, but they were not well suited for automobile applications.[38]

Researchers believed that progress in hydrogen fuel cells for spacecraft suggested that solutions for a cheap and reliable carbonaceous fuel cell for terrestrial applications were within reach.[39] In theory, a battery or stack of fuel cells gave an electric vehicle a far greater range than galvanic batteries and afforded greater convenience as well because such a vehicle could be refueled with a liquid or compressed gas, much like a conventional car. To be sure, early fuel cells of all types were fragile and finicky. Hydrogen fuel cells for spacecraft required complex plumbing to handle reactants and dispose of wastewater, were expensive, and were only as durable as the lifetime of the vehicle (on the order of a few weeks at most).

As with batteries, the criteria for fuel cells for commercial applications were much more demanding than for military and paramilitary applications. Such devices had to be cheap and robust and also needed to be capable of using cheap fuels. In the early 1960s, the US Army partnered with the Department of Defense's Advanced Research Projects Agency (ARPA) and industrial contractors, including GE and Esso, in developing acidic electrolyte fuel cells intended for vehicular applications.[40] Researchers soon discovered that if such devices were directly fed chemically complex fuels like diesel, kerosene, and gasoline, they required very high platinum loadings to catalyze the reaction, and even then they did not operate very well. Engineers then sought to develop an indirect fuel approach, using reformers to convert carbonaceous fuel into a cleaner, hydrogen-rich fuel stream.[41] But reformers added a great deal of complexity to fuel cell systems and were plagued by operational problems, even with alcohol fuels like methanol that were logistically less valuable but considered the easiest of the liquid carbonaceous fuels to electro-oxidize.[42]

With the technoscience of the carbonaceous fuel cell in its infancy in the early 1960s, GM opted to use a hydrogen system in its experimental fuel cell electric vehicle. The company selected the GMC Handivan as its platform because this vehicle had the volume necessary to accommodate the bulky fuel cell power plant and its cryogenic tankage, which completely filled the cargo bay. The most appealing feature of the Electrovan was its range. On paper, the vehicle could go 120 miles between refuelings. But its Union Carbide–built power plant could not be started with a simple turn of the key. It required a three-hour preliminary checkout to test for leaks, purge impurities, and ensure uniform distribution of oxygen and hydrogen to all cells so as not to reverse their polarity. Weighing more than 7,000 pounds, the vehicle was nearly twice as heavy as the standard GM van.[43]

The project suggested that, for all its theoretical advantages, fuel cell electric drive was even less practical than battery electric drive. The Electrovair/Electrovan experiments furnished ample reason not to commercialize electric cars. While GM successfully demonstrated the technological feasibility of electric propulsion, stated Barr in his Senate testimony, it did not demonstrate economic feasibility, warranting further research on advanced batteries.[44]

For its part, Yardney Electric thought it saw a way to deploy commercial electric cars without requiring a technological breakthrough. The company proposed to lease its batteries in a scheme underwritten by the federal government. Federal silver bullion would literally be put to work in silver-zinc cells for electric vehicles in government fleet service: when the batteries were spent, the silver would be recovered and recycled into new batteries. Yardney and the federal government already had such an arrangement with the silver-zinc batteries used in US Navy torpedoes and submarines.[45] Nothing came of the proposal, but it was a way of obviating the problems issuing from the temporal mismatch of battery and motor, one that recalled the fleet lease model used by the first wave of electric vehicle entrepreneurs at the turn of the twentieth century.

Experiments with all-battery electric vehicles continued through the 1960s, all based on existing rechargeable technology. The most substantial of these was a project sponsored by the UK's Electricity Council known as the Enfield 8000, a purpose-built city car similar in size to the Austin Mini and equipped with lead-acid batteries. Between 1966 and 1976, some 120 units were built.[46] In the late 1960s, Victor Wouk developed a converted electric car for Gulton Industries, an enterprise that he had gone to work for as head of electronic research after selling it his own company. Gulton produced nickel cadmium batteries for the US Air Force to power starters for jet engines and hoped to apply the power source to electric cars. The Big Three automakers all had their own electric vehicle research projects and refused to cooperate, so Gulton and Wouk negotiated a deal with American Motors to convert one of its station wagons.[47]

The experience convinced Wouk that existing battery technology was inadequate and was not likely to be improved in the short term, and therefore the all-battery electric format made no sense. He was increasingly drawn to the hybrid battery electric concept, a technology that did not interest Gulton because, according to Wouk, it did not interest the federal government and also had no industry support at that time. In 1970, Wouk and his colleague Charles Rosen left Gulton and founded Petro-Electric Motors to develop hybrids.[48]

MATERIALS MOMENT AT FORD

Meanwhile, important developments in power-source technoscience came from one of the auto giants that declined to work with Wouk. In 1966, Joseph T. Kummer and Neill Weber, researchers at Ford's Scientific Laboratory in Dearborn, Michigan, invented what they called a "sodium-sulfur battery," a project initially motivated primarily by interest in basic questions of solid state science that broke new ground in battery technology. Where the lead-acid rechargeable battery used a liquid electrolyte and solid electrodes, the sodium-sulfur system reversed the arrangement. It employed molten sodium and sulfur for electrodes and a solid electrolyte known as "beta-alumina," a prosaic ceramic then commonly used as insulation in industrial furnaces. When the Ford researchers applied beta-alumina in the sodium-sulfur configuration, they discovered that the substance efficiently conducted ions. Kummer and Weber had invented a battery that, in principle, promised to be far more energetic than any lead-acid rechargeable.[49]

However, the device proved devilishly difficult to manage. Its electrodes had to be maintained at the relatively high temperature of around 350°C and kept isolated from each other, posing a host of engineering problems. The sodium-sulfur battery was prone to thermal expansion and corrosion, and if its molten electrodes ever breached containment, the result was fire, or even an explosion.[50] Packaging this volatile chemistry in a practical, rechargeable battery required a host of other innovations in materials and control systems that Ford then had no intention of pursuing. Nevertheless, the sodium-sulfur battery aroused great scientific interest thanks to its novel application of beta-alumina. Up to then, electrochemists believed that reactions occurred primarily on electrode surfaces in relation to liquid electrolytes. The discovery that charge-carrying ions could be reversibly inserted inside bulk solids led to a major shift in thinking about how powerful, energetic new rechargeables might be built.[51]

This shift would signal a convergence of electrochemistry with solid-state ionics and a new era in power-source technology. One of the most important figures in these developments was John B. Goodenough, a theoretical physicist and materials researcher employed at Lincoln Laboratory. This facility had origins in MIT's storied wartime Radiation Laboratory and was established by the university in 1951 on a commission from the Department of Defense to develop the Semi-Automatic Ground Environment, the country's first networked air defense system. Goodenough was supervised by Jay Forrester, the chief of the project's digital computer division, and worked in the unit responsible for computer memory, where the young researcher became an expert in the properties of metal oxides. When the project wound up in the late 1950s and passed the technology of what was known as the Whirlwind II computer to IBM for manufacturing, many researchers left to take jobs in industry.[52]

But Goodenough was able to stay. Leveraging his original research contributions, the scientist managed to inherit Lincoln Laboratory's now-redundant ceramics facility, where he was able to study metal oxides without the burden of research targets. Over the next twenty years, Goodenough would perfect a mode of interdisciplinary work that he had developed during Whirlwind II: with research axes staked out by "engineering targets," he, the physics theorist, designed experiments for chemists to execute and pursue with no milestone constraints.

In the meantime, Goodenough received an assignment that would change the course of his career. In the late 1960s, the Atomic Energy Commission asked him to investigate the sodium-sulfur battery because beta-alumina contained spinels, a class of gemlike mineral on which he was an authority. Goodenough's work would bring him into conversation with a handful of other researchers who would help pioneer solid-state ionics, informing the development of a new type of power source that would eventually play an important enabling role in consumer electronics and electric automobiles.[53]

3   DEFINING APPROPRIATE TECHNOLOGY

People want to go directly to a product, they want to see a car. I was saying it is not the car, it is the battery, and you are not going to make a good battery until you have the scientific underpinning.

—Wally Rippel, electric vehicle pioneer, 2019

Painstaking progress in the technoscience of electric vehicle propulsion unfolded in the context of deepening national crises. From the early 1960s through the early 1970s, the war in southeast Asia and compounding social, environmental, and economic problems throughout the US sparked dissent, cultural revolution, and wide-ranging reappraisal of basic elements of American life, including automobility. While many activists developed implicit and explicit critiques of capitalism imperialism as the underlying causes of these problems, policymakers tended to view the social problems of automobiles through the lens of consumer protection, generating what the political scientist Stan Luger referred to as a "politics of compromise."[1] Bureaucrats gained increasing influence over how automakers built automobiles, and the California Air Resources Board (CARB) and the Nixon administration's Environmental Protection Agency (EPA)—both products of socially conservative governmentality—became the public face of a new emissions control regime. Regulators collaborated with industry to extend emissions controls throughout the notional internal combustion engine (ICE), following up the positive crankcase ventilation valve with technological fixes of the tailpipe and carburetor. For young activists like Ralph

Nader, such measures did not go nearly far enough. Nader criticized reformers like Edmund Muskie for framing automobile pollution as a product of
individual consumer choice rather than a systemic effect of industry.[2]

Indeed, the energy and environmental crises were increasingly interpreted
as crises of American automobility itself. The perception that there was something fundamentally wrong with the ICE car opened other avenues of reform
besides emissions controls, sharpening the now-or-later debate on the question of alternative technologies, including electric cars, and strengthening
the public policy rationale for investigating them. Bureaucrats in the national
development state still believed that these technologies required much more
research and development before they could be considered for practical
application. Nevertheless, the intensification of environmental and energy
problems in the 1970s led the federal government to increase support for
the technoscience of advanced electric propulsion systems and bolstered the
argument that the electric car could be a useful instrument of public policy.
Such efforts took place in the context of the reorganization and consolidation of the federal energy complex over the course of the 1970s, beginning
with the dissolution of the US Atomic Energy Commission (AEC) and its
reconstitution into the Energy Research and Development Administration
(ERDA) and its successor, the Department of Energy (DOE). The DOE was
an exemplary institution of the national developmental state, combining
the functions of research and development with some aspects of energy
regulation. The DOE controlled the national laboratories of the former
AEC, and it was also the seat of the Federal Energy Regulatory Commission
(FERC), the former Federal Power Commission (FPC), an independent body
responsible for regulating interstate energy carrier systems (electricity and
pipelined oil and gas). The DOE would play a key role in the revival of the
electric vehicle.

A different approach to the question of sustainable transportation was
offered by the appropriate technology and ecological design movements.
Rooted in traditions of communitarianism and boosted by the 1960s counterculture, these movements emphasized local experience and knowledge
as solutions to the social alienation and pollution caused by large, centralized industrial systems. These ideas were popularized by activists such as
Stewart Brand and the British economist E. F. Schumacher, whose credo
of "small is beautiful" influenced a new generation of environmentally-
minded thinkers.[3]

Appropriate technology and ecological design functioned both as reformist critiques of industrial capitalism and as practical guides in using simple tools to facilitate decentralized living. Schumacher was vague about how technology could be instrumentalized to this end, but devotees of the "small is beautiful" philosophy tended to focus on energy-efficient buildings and building materials and prosaic renewable energy conversion devices like the bicycle, the windmill, and the passive solar heater. Rejecting the conservationism of the traditional environmental movement and conventional political identification, appropriate technologists promoted ecologically-sensitive design as a lifestyle worldview, exemplified by Brand in his *Whole Earth Catalog*. The historian Andrew Kirk held that the "outlaw designers" and "tool freaks" befriended by Brand reclaimed a frontier tradition of amateur innovation that counteracted the authoritarian tendencies of technocracy. For some appropriate technologists, this approach constituted a third way, or "hybrid politics," as Kirk put it.[4]

Beneath their disparate styles, however, appropriate technologists and elite planners shared the assumption that technology could solve social problems. When appropriate technologists considered automobility and how it might be reformed, they thought of changed social behavior like carpooling, as well as technical fixes such as smaller and more efficient ICE cars and the means of producing them. Such fixes were in the vein of those then being enforced by the auto emissions control regime. Notable efforts were made to bridge the worlds of grassroots appropriate technology and elite technocracy by figures such as the physicist Amory Lovins, who would become an influential proponent of "natural capitalism," and the lawyer Jerry Brown, who instituted the Office of Appropriate Technology as California governor in 1976.[5]

The environmental and energy crises also stimulated efforts to develop alternative technologies relevant to electric automobile propulsion. Outsider activism in appropriate automobile technology paralleled the work of actors situated on the margins of institutions such as John Goodenough, Joseph Kummer, Neill Weber, and Victor Wouk on advanced materials, power sources, and power controls. The activities of electric auto enthusiasts, entrepreneurs, and semi-autonomous corporate scientists and engineers intertwined with the efforts of the air quality regulatory apparatus to clean up the ICE automobile and the efforts of the national developmental state to promote advanced science and technology in support of this objective.

Together, these communities enriched the body of experience of and knowledge about the electric car, foreshadowing a larger role for outsiders in the politics of green automobility in the 1990s.

RACING FOR CHANGE

Persistent smog in US cities drove the continued development of institutions of air quality control through the late 1960s and into the early 1970s. California established the first tailpipe emissions standards for hydrocarbons and carbon monoxide in 1966, and for nitrogen oxides in 1971. With the establishment of the EPA in 1970, the Nixon administration created an important tool of regulation and enforcement that would act as an ally— and at times an antagonist—of CARB. To regulate emissions standards more stringent than federal standards for new motor vehicles, California had to obtain a waiver from the EPA, an independent executive agency with a much more complex mission than CARB, of a Clean Air Act (CAA) clause that prohibited states from possessing these powers. National environmental priorities did not always align with California's environmental priorities, and this asymmetry in vision and power eventually provoked tension and conflict between state and federal environmental regulatory apparatuses.

In the beginning, however, CARB and the EPA cooperated relatively effectively. They compelled automakers to adopt a pair of mitigation devices that were more sophisticated than the positive crankcase ventilation (PCV) valve and more difficult to integrate into the ICE: engine gas recirculation, a system that reduced the temperatures at which nitrogen oxides were formed, introduced in 1972; and the catalytic converter, introduced in 1974. Implementing these technologies required considerable planning, capital expenditure, and engineering at industrial scale and took years to yield results.[6]

In the interim, appropriate automobile technology activism found its expression in do-it-yourself electric car conversions. In San Jose in 1967, a retired research engineer and inventor named Walter V. Laski founded the Electric Auto Association (EAA), which a decade later had amassed a membership of around 300 enthusiasts, mostly in California.[7] Electric car activism also emerged farther south around the California Institute of Technology (Caltech). Arie Haagen-Smit's tireless work in air quality science and regulation had given some progressive luster to a largely conservative bastion noted for nuclear physics and connections to the oil and aerospace industries.

Caltech would earn further renown as the stage for a celebrated event sometimes interpreted as the beginning of the era of the modern electric car. In the mid-1960s, choking air pollution in Pasadena inspired a physics student named Wally Rippel to consult with Haagen-Smit and begin an investigation of electric propulsion. Rippel concluded that the main limiting factor was the battery. Aside from the lead-acid rechargeable, there weren't many options. Some of the biggest news in the power-source field at that time was the fuel cell technology that powered the *Gemini* and *Apollo* spacecraft, but when Rippel spoke with an employee of the National Aeronautics and Space Administration's (NASA) Caltech-managed Jet Propulsion Laboratory, he learned that such devices could not easily be adapted for electric cars. Aerospace fuel cells cost millions of dollars and required teams of specialists to set up and maintain.[8]

The underlying problem, Rippel believed, was that the battery field was more empirical than scientific. In order to build his own electric car, he had little choice other than to convert an ICE vehicle using proven components. Using savings from his day job fixing logic circuits at the aerospace contractor Litton Industries, Rippel purchased a 1958 Volkswagen van and components including a forklift truck motor and $600 worth of lead-cobalt batteries, and built a conversion known variously as the "Voltswagen" or the "Caltech Electric." In the process, the student became a technoevangelist, known locally as the university's resident promoter of the electric automobile. But Rippel wanted a larger platform to raise awareness and stimulate scientific battery research. He conceived a plan to stage a race of student-built electric vehicles, and in January 1968, he issued a challenge to the Massachusetts Institute of Technology (MIT), Caltech's chief rival for the title of the most prestigious school of science and technology in the US.[9]

The resulting Great Electric Car Race, begun August 26, 1968, illustrated both the potential and the limitations of existing electric propulsion technology. Where Rippel and his team took a minimalist approach, the MIT team, led by Leon Loeb, opted for high performance. Expending considerably more resources than the Caltech crew, they crammed $20,000 worth of nickel cadmium cells supplied by Gulton Industries into the body of a Chevrolet Corvair donated by General Motors (GM). The resulting vehicle, known as the "Tech I," was technologically superior to the Caltech Electric, at least on paper, but the MIT team soon found that performance came at a price. The automobile duty cycle turned out to have significant unintended consequences for nickel cadmium, a rechargeable chemistry mainly used in portable appliances, power tools, and specialized military applications. When the 120-volt motor drew on the 100-volt battery, it overheated. The battery pack also overheated, having been packaged in such a way that the air-cooling system could not evenly distribute air. The Tech I repeatedly broke down and even caught fire at one point.[10]

The Caltech Electric also had its problems. Its batteries overheated, and the team learned that lead-cobalt chemistry did not respond well to fast charging. The vehicle broke down, burning out its motor just east of Seligman, Arizona, but Rippel's team ordered a replacement unit via airmail from New York. As crew member Dick Rubinstein recalled, they had "no trouble" fitting it into the van. Seven and a half days later, the Tech I made it to Pasadena before the Caltech Electric made it to Cambridge, but under tow. Under the rules of the race, only all-electric mileage counted, and scoring thirty minutes more electric operating time than the Tech I, the Caltech Electric was declared the victor.[11]

The race was a genial affair that deepened collaboration within the small community of electric vehicle enthusiasts. The student teams communicated with each other throughout the race and received support from electric vehicle luminaries including Robert Aronson, owner of Electric Fuel Propulsion, one of the leading converters of electric vehicles in the US, and the Caltech alumnus Wouk. The race attracted a good deal of media attention, but Rippel's broader hopes of promoting battery technoscience went unrealized. The press focused on the event "as if it were a football game," and Caltech did not add power source studies to its curriculum.[12]

On the other hand, the race demonstrated to the largest audience yet that the basic components of battery electric automobility were at hand. Electricity was cheap, with each car having consumed only $25 worth of it. So was the lead-acid rechargeable, a consequence of its co-development as an auxiliary power source for ICE propulsion. General electrification had proliferated electric motors, and the vast expansion of the postsecondary academic complex after World War II produced ever-larger numbers of capable and ambitious science-trained college graduates. Moreover, the Caltech and MIT students had converted their cars in the span of only a few months, suggesting that the industrial challenges of producing electric cars were manageable.

If Rippel failed to stimulate a revolution in academic battery science at Caltech, he succeeded in stoking interest in alternative propulsion technology around the university and beyond. He was part of an emerging movement of enthusiasts that viewed the ICE vehicle and its corporate interests as representative of all that was regressive in the world. To them, the battery electric vehicle was the symbol of a brave new world of environmentally and socially sustainable automobility.[13] Among those inspired by Rippel was a young Caltech-trained engineer named Alec Brooks. Brooks's fascination with advanced, lightweight, human-powered vehicles intersected with Rippel's interests in advanced battery electric cars and led the two engineers to make important contributions to the reform of mainstream automaking in the late 1980s, 1990s, and 2000s.

THE PEOPLE'S ELECTRIC

A somewhat different vision of appropriate automobile technology was pursued by Bob Beaumont, a successful Chrysler-Plymouth dealer based in Kingston, New York. Beaumont's business was prosperous but by the late 1960s the salesman had become disillusioned by the pollution and inefficiency of American-style big car automobility. Beaumont believed that the basic technologies of the battery electric car as they existed in the late 1960s were just waiting to be assembled into a commercial business. In 1967, he sold his dealership and briefly teamed up with Aronson. However, Aronson specialized in converting the large and heavy gasoline-fueled ICE cars that Beaumont had come to despise. Beaumont would find what he was looking for in Augusta, Georgia. In this golfing hotbed was headquartered a company called Club Car, which manufactured golf carts equipped with a lead-acid rechargeable battery and a direct current brushed motor. In this humble platform, Beaumont saw the basis of a two-seat runabout that would be far more efficient for local commuting than a full-sized automobile.[14]

This was technology that GM's Harry Barr had warned Congress was completely unsuited for the US automobile system. Beaumont purchased several carts to run his own testing program, and then he asked Club Car's chief engineer, Bill Lindenmuth, to improve the design. At Beaumont's direction, and with the help of the Terrell Machine Corporation, Lindenmuth added a more powerful motor, higher, thinner tires, a new transaxel, and a fiberglass top.[15]

Dubbed the Vanguard, the vehicle had an aluminum frame and no side windows, and it was barely roadworthy. Beaumont produced a handful of

Vanguards in 1972 using his own money and donations from friends. Relentless promotion and some initial favorable press attracted a group of Florida investors who wanted Beaumont to produce these cars in a disused industrial park in the city of Sebring that was part of a former army air base that had won some fame as the site of a motorsport endurance race. The new company, called Sebring-Vanguard, produced an improved version of the Vanguard that was equipped with Exide lead-acid batteries and a redesigned wedge-shaped body made of polymer plastic. Beaumont needed a specialist to develop the electrical system so he hired Ronald Gremban, a Caltech graduate and member of the Caltech Electric team. The upgraded vehicle was called Citicar. Between 1974 and 1977, Sebring-Vanguard manufactured around 2,200 of them, the largest production run of any post–World War II battery electric passenger car up to that time. For a while, Sebring-Vanguard was the sixth-largest automaker in the US, after Checker Motors, maker of the iconic Checker Taxi.[16]

As a form of no-frills, zero-emission transportation, Citicar met most of Beaumont's performance objectives. A road test in Los Angeles by the *Motor Trend* journalist and editor Mike Knepper in November 1976 illuminated the car's strengths and weaknesses. In a telling measure of the socialization wrought by the ICE, Knepper was confused by the lack of noise upon starting; thinking that the machine was not operating, the journalist inadvertently stepped on the accelerator and almost caused an accident. With a top speed of forty miles an hour, Citicar could not negotiate freeways, and it had a short range. In theory, the 520-pound battery pack gave the vehicle a range of around fifty miles, but Knepper learned that the actual range could be much less depending on driving conditions. In the denser urban spaces of Los Angeles, on the other hand, Knepper reported that Citicar had admirable qualities. It handled well, even in hilly terrain, and its small footprint facilitated ease of parking. And it was fun to drive. Knepper reported that he caught himself "grinning like a fool" more than once as he zipped through parking lots and darted into tiny parking spaces.[17]

Citicar was essentially a sunbelt car that did best on secondary roads. In a northern winter, the vehicle was a less appealing proposition. Cold weather sapped battery capacity and range and the car lacked a heater and a robust defogger, making it unbearably uncomfortable for all but the most zealous enthusiasts.[18] Overall, however, Citicar met the criteria of appropriate

technology. The vehicle was small and cheap, required relatively few material resources, and replacement costs for its basic lead-acid battery pack were
low. The car was simple enough that owners could (and often did) make
their own repairs. And of course it produced no emissions at the point of
use. In principle, the scaled Citicar represented a plausibly alternative form
of American automobility.

POWERING UP PUBLIC POLICY

Beaumont had been motivated by a desire to improve air quality, but it was
the energy crisis of the 1970s that stimulated demand and thrust the electric car further into the realm of public policy. With the oil embargo of 1973
and the quadrupling of the price of petroleum, would-be entrepreneurs had
reason to hope that a nascent electric car industry might be in the offing.[19]
Indeed, the federal government seemed to be compelling automakers to
rethink propulsion technology on a massive scale. In 1975, Congress passed
the Corporate Average Fuel Economy (CAFE) regulation, forcing the car companies to improve the fleet fuel efficiency of passenger sedans from around
15 miles per gallon to 27.5 miles per gallon by 1985. The measure omitted
light trucks, a loophole that would help alter the character of American automobility in the 1990s.[20]

   Then in 1976, for the first time, Congress acted to directly support
research in electric vehicles. Over President Gerald Ford's veto, it passed the
Electric and Hybrid Vehicle Research, Development, and Demonstration Act
(Public Law 94–413). The language of the law acknowledged the utility of
battery electric propulsion technology of the sort that Feldmann, Aronson,
Rippel, and Beaumont had already developed. Congress found that because
most urban driving consisted of short trips, the "expeditious introduction"
of electric cars would both substantially reduce dependence on petroleum
and be "environmentally desirable." From Sebring-Vanguard's perspective,
the legislation's most important provision was that it supported the manufacture of thousands of electric vehicles to be purchased or leased by the
federal government for use in raising public awareness.[21]

   But the company was denied access to the program. While Congress
accepted that battery electric technology existed and could be applied, ERDA
was empowered only to promote research, especially on advanced new

propulsion and control systems, not to support existing commercial enterprises.[22] The program generated a handful of experiments sponsored by the DOE involving one-off concept cars that government officials argued did not demonstrate the desirability of electrics so much as highlight the need for new enabling technologies, including batteries, motors, and controllers.[23]

In 1977, Sebring-Vanguard went bankrupt, an indirect casualty of the gap in public policy between science and technology stimuli, with their bias toward capital-intensive solutions, and industrial stimuli.[24] The episode also spoke to the federal government's difficulty in reconciling its imperatives in energy, environment, and transportation in an increasingly complex world. In the first two decades after World War II, federal domestic energy policy essentially consisted of securing as much energy as possible, leading to the exploitation of natural gas and nuclear power at a time of plentiful cheap oil, coal, and hydroelectric power. In the age of the conjoined environmental and energy crises, the missions of clean energy and efficient energy conversion were added to an injunction to restore energy plenitude and secure energy independence.[25]

This array of imperatives generated a host of contradictory planning initiatives. Lawmakers called for increasing the production of all energy forms, as well as conserving energy and improving energy efficiency. Federal energy research and development investigated both renewables and nonrenewables.[26] The Ford and Carter administrations combined voluntary and mandatory conservation measures with existing price controls that for years had fixed the cost of energy below its replacement value and overstimulated demand. Conservation incentives and subsidies were available for some dwelling and transportation systems, but not all of them, allowing consumers to spend their energy savings in other sectors.[27]

In effect, federal reforms favored incumbent economic interests. They helped crystallize a governmentality of alternative energy and propulsion as public policy solutions (at least in principle) to the energy and environmental crises. The reforms also helped foster new research communities devoted to developing advanced power sources for the electric car, marking the tentative beginning of science-based battery development. The power-source renaissance would be a slow, circuitous, and relatively costly process that was at odds with the principles of appropriate technology, but it would have important implications for how actors thought of the technologies of the electric car.

MATERIALS MILESTONE: EXXON BUILDS A BATTERY

In the late 1960s, Ford had inadvertently helped push the state of the art in power sources with its sodium-sulfur battery. A few years later, the field advanced again, thanks to the patronage of another equally unlikely establishment enterprise. For some years, Exxon had supported research in power sources. Its antecedent, the Standard Oil Company of New Jersey, studied fuel cells in the late 1950s and early 1960s through its Esso Research and Engineering division in hopes of supplying specialized fuels in case the Army decided to adopt the technology for electric drive.[28] In the wake of the oil shocks of the early 1970s, the corporation (renamed Exxon in 1973) had to consider the possibility that automakers might be compelled to build electric vehicles. Indeed, GM was then planning the Electrovette, a conversion of the Chevrolet Chevette subcompact equipped with nickel-zinc batteries that the automaker intended as an emergency solution in the event of a catastrophic rise in oil prices.[29]

Exxon saw research in energy storage materials as a wise hedge that aligned well with its petrochemical interests. It was under the sponsorship of Exxon (formerly Esso) Research and Engineering that the chemist M. Stanley Whittingham made a major contribution to a new rechargeable chemistry of unprecedented power and energy. Whittingham's work began with a consideration of problems that Goodenough's research had raised in solid-state ionics. Goodenough was not a technologist, strictly speaking, and he did not design complete power-source systems. He was instead motivated by fundamental questions of solid-state science and materials design guided by "engineering targets," theoretical problems that arose from devices, for which he would design experiments for chemists to execute.[30]

As the federal government expanded energy research in the years of expensive petroleum, Goodenough considered zirconia-based solids in light of his earlier work on the sodium-sulfur battery. He believed that such solids could be applied as the electrolytes of a solid oxide fuel cell, a high-temperature power source that in principle was capable of directly using even the dirtiest carbonaceous fuels. First developed in the late 1930s, the solid oxide fuel cell also obviated the limitations of water-based electrolytes. Water decomposes into oxygen and hydrogen at 1.23 volts, which constrains batteries with aqueous electrolytes to operating at relatively low power. The trade-off was that the solid oxide fuel cell operated at around

1,000 degrees Celsius, making the device slow to start, prone to thermal expansion and corrosion, and generally unsuited to vehicular applications.

Whittingham was aware of the problems of solid oxide systems, so he pondered the virtues of low-temperature, nonaqueous electrolytes. In 1976, he unveiled a lithium titanium-disulfide cell, which employed a liquid organic electrolyte at room temperature. Whittingham demonstrated that lithium ions could be reversibly inserted into the spaces between the sheetlike layers of the titanium-disulfide cathode. The reversible storage of ions in a layered structure, known as "intercalation," is the fundamental operating principle of a rechargeable lithium battery.

At the time, Whittingham suggested that he had developed a practical battery, but this was not exactly the case.[31] The chemist had focused his efforts on developing the titanium-disulfide cathode, pairing it with a metallic lithium anode for his proof-of-concept tests. It was a dangerous combination. When such a cell was repeatedly recharged, lithium ions plated unevenly on the anode, forming treelike growths called "dendrites," which could bridge the electrodes and cause a short circuit. In such circumstances, the cell's organic electrolyte could ignite, transforming the device into an incendiary.[32]

MATERIALS MOMENT II: OPPORTUNITY COSTS AND CONSEQUENCES

Nevertheless, Whittingham's research marked a major advance in power-source technoscience. The lithium titanium disulfide cell in turn posed interesting research questions for Goodenough, who was at a pivotal point in his career in the mid-1970s. One of the consequences of the reorganization of the federal energy complex was the consolidation of all energy programs into a single agency, a process that in 1976 led to the transfer of Goodenough's fuel cell materials program to ERDA. With Goodenough's work at Lincoln Laboratory "dead in its tracks," the physicist accepted an offer from Oxford University to chair its Inorganic Chemistry Laboratory, and it was here that he considered problems of the lithium titanium disulfide system.[33] Goodenough did not intend to design a complete battery for a specific application, so he thought about the problem in abstract terms, privileging energy and power over safety, cost, or durability. He reasoned that a layered sulfide cathode mated to a metallic lithium anode could not yield much more than 2.5 volts. A safer anode, Goodenough believed, would yield even less voltage, to the point that the device would not be able to compete with existing rechargeable batteries that used non-flammable aqueous electrolytes.

Aware of the limitations of sulfides, Goodenough looked to metal oxides as a potentially more energetic and powerful material for the cathode. He worked with the Japanese physicist Koichi Mizushima to determine how much lithium could be reversibly extracted from a variety of transition metal oxides. Generally, the more ions that can be extracted from a cathode, the greater the voltage it will deliver in a complete cell. In protracted experiments, Goodenough, Mizushima, and their colleagues showed that they could extract about 60 percent of the lithium from a lithium cobalt oxide cathode when it was paired with a metallic lithium anode, which was sufficient to generate 4 volts. They extracted 80 percent from a lithium nickel oxide compound, but that material was unstable and difficult to prepare.[34]

Despite his professed interest in basic research, Goodenough wanted to sell the new cathode. However, battery manufacturers were not interested because there was no suitable safe anode. At any rate, lithium cobalt oxide was too expensive to be produced in the quantities needed for commercial electric vehicles. Goodenough had no money, and with few options, he patented his cathode through the UK Atomic Energy Research Establishment's Harwell Laboratory. The arrangement required him to relinquish all his patent rights, a Faustian bargain that would return to haunt him. From 1985, Sony worked to integrate the lithium cobalt oxide cathode with a safe graphitic anode in a project to replace the nickel cadmium battery in consumer electronics, an enterprise that owed a good deal to the contributions of Akira Yoshino of Asahi Kasei, a multinational chemical concern. Sony succeeded in pioneering the commercialization of a lithium cobalt oxide rechargeable battery in 1991, initially intended for the Kyocera cellular telephone.[35] This technology would become the most popular rechargeable for mobile electronic devices.[36]

Goodenough had begun his research on the lithium cobalt oxide cathode during the years of the oil crisis, a time when the battery electric car began to seem to many like a plausible solution. But the crisis was resolved by conventional politics, not technology. An agreement with the US compelled Saudi Arabia to forswear the use of petroleum as a political tool, sending petroleum prices to preembargo levels.[37] Goodenough's research was applied not in electric propulsion but in consumer electronics, generating billions of dollars in royalties, of which he received nothing. Still, the physicist emerged from the episode as one of the world's foremost experts on spinels and lithium insertion compounds. In the early 1980s, he attracted the attention of Michael Thackeray, a young South African chemist motivated by the quest for a better battery for electric cars with whom he would collaborate in developing new rechargeable lithium cell formulas.

ENERGY PLENITUDE AND ELEMENTAL PANACEAS

The political and economic climate for this collaboration was not auspicious. The Reagan administration was hostile to the principle of renewable energy and cancelled the electric and hybrid vehicle research program. The executive policies took time to filter down to the civil service, however, and the federal government continued to support research relevant to the components of electric vehicles, but in an uncoordinated fashion. The DOE inherited projects exploring high-temperature lithium iron sulfide batteries, hydrogen systems, and several types of fuel cell, not all of which were suitable for electric vehicles.

Electric vehicles were suddenly politically out of favor. Nevertheless, the DOE continued to fund research with industry partners and nonprofit organizations, including the Electric Power Research Institute (EPRI), an independent organization set up by electric utilities in the wake of the Great Northeastern Blackout of 1965 to conduct research and development in support of the electricity industry. Utility companies were leading supporters of electric vehicle development in the US and saw an electric fleet as a large market for off-peak electricity. Their favored platform for this role was the van because such a vehicle could accommodate large, heavy, and bulky rechargeable batteries based on existing chemistries for the purpose of energy storage and did not need to have especially good road performance. Vans also served as test beds for new motors and controllers, especially alternating current systems that were increasingly feasible thanks to the rapid commodification of microprocessors.[38] As in the past, the US national developmental state made no concerted effort to develop a dedicated electric vehicle battery in the 1980s.

On the other hand, academic interest in propulsion systems utilizing hydrogen and lithium as mediums for storing energy persisted. To some scientists, engineers, and policymakers, these elements, especially hydrogen, had world-historical implications. As the most abundant substance in the observable universe, hydrogen had fascinated researchers since the late nineteenth century as a kind of cosmic master key. They imagined hydrogen as a way of making the best use of Earth's primary energy resources because all of these resources could, in theory, be converted to this element, which would serve as a super-efficient universal fuel. This imaginary began to be articulated in increasingly sophisticated ways from the 1960s by scientists who popularized the idea of the hydrogen economy. There is no source of pure hydrogen on Earth, so proponents of this scheme envisioned producing much of the hydrogen by splitting it out of water. For them, a central question was what sources of primary energy were to be used to accomplish this.

Here, there were several schools of thought. One argued for solar hydrogen, an idea first promoted by the physicist and fuel cell engineer Eduard Justi. Another imagined hydrogen produced by nuclear power and counted the physicist Cesare Marchetti and the electrochemist Derek P. Gregory as its adherents. Gregory outlined a particularly detailed vision of a hydrogen economy in collaboration with the Institute of Gas Technology (IGT), the nonprofit research and development association of the US gas utility industry. In the late 1960s, gas utilities began to worry about diminishing US reserves of natural gas and imagined a scenario in which hydrogen, produced initially by a number of means and ultimately mainly by nuclear power, replaced natural gas in the gas pipeline network. They believed that pipelined hydrogen would be more efficient than electric power transmission over long distances because pipelines, in principle, lost less energy than power lines. Such a system, argued Gregory and his collaborators, would enable gas utilities not only to continue their role of delivering gas for cooking and water and space heating but, by virtue of a decentralized network of fuel cells installed in buildings of all sorts that would convert hydrogen to electricity, also directly compete with electric utilities.[39]

Hydrogen futurists envisioned using hydrogen to supplement electricity as an energy carrier and replace fossil fuels as a medium of energy storage. They also considered hydrogen as a vehicular fuel, but they tended to emphasize non-vehicular systems. They were motivated by a certain environmental sensibility, although the hydrogen economies they proposed were large, complex, and capital-intensive sociotechnical regimes generally antithetical to the precepts of appropriate technology. Perhaps for this reason, their ideas gained an audience in the energy research and development establishment. Through the 1970s into the early 1980s, hydrogen research was supported by EPRI, the gas utilities, some oil interests, and the DOE, with the federal agency spending between $20 million to $30 million annually in this field.[40]

Such resources were not nearly sufficient to stage demonstrations of the large-scale hydrogen production and distribution infrastructure that preoccupied the hydrogen futurists. Research sponsored by the national developmental state instead concentrated on problems of storing and using hydrogen as fuel. Compressed gaseous and liquid hydrogen are notoriously difficult to store because the diatomic hydrogen molecule is the smallest of all molecules and easily diffuses through materials. An important focus of US hydrogen research involved developing advanced materials that could efficiently store this substance. Researchers also explored energy conversion devices that used hydrogen as fuel, especially in vehicles. A favorite experiment in this period involved converting ICE automobiles to run on compressed hydrogen, a relatively simple exercise justified as a means of understanding the performance qualities of hydrogen fuel until an inexpensive means of producing hydrogen could be developed. The DOE also supported some work on hydrogen fuel cell electric drive.[41]

The other elemental energy storage solution was lithium. In the 1980s, as we have seen, research on lithium ion rechargeable chemistry was in its infancy, but important work was beginning outside the consumer electronics context in relation to other types of advanced rechargeables based on molten salts along an axis linking Oxford University with the South African materials research community. The oil crisis ended in the early 1980s for most of the developed world, but it continued in oil-poor South Africa, exacerbated by increasing international isolation and a trade embargo provoked by the country's apartheid policies. In response, South African policymakers identified the development of advanced batteries as part of a self-sufficient industrial strategy that emphasized transportation applications. In 1977, the country's Council for Scientific and Industrial Research (CSIR) began work on the sodium-metal chloride battery, a chemistry similar to the sodium-sulfur battery but less corrosive and easier and safer to manufacture.[42]

Thackeray began his career in this context. In the mid-1970s, he was a doctoral student at the CSIR's main laboratory in Pretoria, working with the chemist Johan Coetzer. The sodium-metal chloride battery originated in a project to build a safer sodium-sulfur battery by immobilizing its molten cathode in a matrix of zeolite, a microporous mineral commonly used as an industrial catalyst and for which the project took its code acronym (Zeolite

Battery Research in Africa, or ZEBRA). When that configuration was judged to be too heavy, Coetzer instead mated a molten sodium anode with an iron chloride cathode, a component that he developed as an alternative to the iron sulfide cathode of the high-temperature lithium battery created by the DOE's Argonne National Laboratory. Like the sodium-sulfur battery, the sodium-metal chloride battery featured a solid electrolyte made of beta-alumina, as well a liquid sodium anode. Unlike its antecedent, however, the sodium-metal chloride battery could be assembled in the discharged state, enabling workers to avoid the hazards of handling hot materials. Table salt and metal powders (initially iron) were mixed into the cathode, and the molten salt anode formed after the cell was charged. The sodium-metal chloride battery utilized cheap materials and was almost twice as energetic as the lead-acid rechargeable.[43]

As Coetzer's work progressed, Thackeray contemplated metal oxides as a less corrosive alternative to the iron sulfide and iron chloride cathodes of hot lithium cells. He noted the potential of certain spinels to absorb and release lithium ions, and with this realization, he contacted Goodenough and arranged to work with him at Oxford as a postdoctoral fellow. Supported by the CSIR, its affiliated South African Inventions Development Corporation, and mining giant Anglo American, Thackeray demonstrated the insertion of lithium into the two spinels magnetite and hausmannite between the fall of 1981 and the end of 1982. This work informed Thackeray's subsequent demonstration of lithium insertion into a lithium-manganese oxide cathode. In 1985, Goodenough and Thackeray patented their research on the spinel frameworks for use as battery components.[44]

With relatively limited resources, the duo had considerably advanced the technology of the lithium ion rechargeable battery. However, the CSIR was preoccupied with the sodium-metal chloride system. By 1986, basic research on these materials was completed, and the council transferred most of its staff to Anglo American. Shortly thereafter, the mining firm partnered with Daimler-Benz, which would begin testing sodium-nickel chloride batteries in electric vehicles in the late 1980s and early 1990s.[45]

HYDROGEN, HYDRIDES, AND NEW POWER SOURCES

Daimler-Benz was interested in other advanced propulsion systems besides the molten salt battery. The automaker also had an abiding fascination with hydrogen dating to the early 1970s, when it began experimenting with converting the engines of conventional automobiles to run on hydrogen fuel stored in metal hydrides. Between 1984 and 1988, Daimler built several sedans equipped with dual hydrogen/gasoline fuel systems as well as several all-hydrogen vans, displaying the vehicles in a high-profile demonstration in Berlin.[46]

Hydrogen storage also motivated research on metal hydrides at Energy Conversion Devices (ECD), a small engineering research company based in a Detroit suburb. It had been founded in 1964 by Stanford Ovshinsky, an inventor who began his career as a working-class machinist in the automobile parts industry in Akron and Detroit in the 1940s and 1950s. Ovshinsky made a number of innovations in the field of machine tools before his interest turned to advanced materials, a science-based field in which he became a self-taught expert. His signature invention was amorphous or disordered materials, notably chalcogenides and amorphous silicon, an early form of nanotechnology that imparted useful electronic and catalytic properties to prosaic, cheap substances. The historians Lillian Hoddeson and Peter Garrett described ECD as an Edisonian-style "invention factory."[47] Ovshinsky performed research and development for much larger enterprises and derived royalties from materials that often improved existing technologies.[48] ECD received grants from the Advanced Research Projects Agency (ARPA; renamed the Defense Advanced Research Projects Agency, or DARPA, in 1972) and later from corporations that contracted with ECD in the hope of commercializing amorphous materials for application in electronics and power sources.[49]

In the early 1980s, ECD won a contract from the oil giant Atlantic Richfield Company (ARCO) to research hydrogen systems. Ovshinsky organized groups studying hydrogen generation, storage, and utilization, including a unit devoted to fuel cells, but the research concentrated on metal hydrides as a medium for storing hydrogen.[50] While Daimler-Benz used metal hydrides to store hydrogen for use in ICEs, Ovshinsky's group realized that these materials could also be used in the negative electrode of an electrochemical couple with a nickel hydroxide cathode. This produced a nickel-metal hydride cell, a device that in effect electrolyzed and dissociated water into oxygen and hydrogen upon charging and electro-oxidized hydrogen upon discharging.[51]

General Electric (GE) and Philips patented similar technology in the 1970s, but the version developed by ECD and patented in 1986 would become widely recognized as the first practical such power source. What was different about the ECD cell was that the metal hydride alloy of its negative electrode was constituted of disordered materials with additional surface area that could accommodate additional hydrogen. Disordered materials used a wide variety of elements that, in concert, allowed the maximum number of hydrogen atoms to be stored per metal atom. This elemental mash-up was designed to balance oxidation and reduction, the discharge and recharge reactions respectively, in such a way that gave cells the longest possible life span.[52]

With further refining, disordered materials would represent the most important advance in rechargeable battery technology in a half-century, especially in terms of durability. In 1982, Ovshinsky set up the Ovonic Battery Company (OBC) as a wholly owned unit of ECD for the purpose of developing these compounds. This small research enterprise would go on to play a pivotal role in the electric automobile revival of the 1990s.

Daimler was also interested in hydrogen fuel cell power owing to the potential of proton exchange membrane fuel cell technology in automobile applications. Unlike other types of fuel cells, membrane fuel cells were relatively light, compact, and versatile, combining electrodes and electrolyte in a single membrane-electrode assembly. Operating at low (below 100 degrees Celsius) temperatures, membrane fuel cells avoided the corrosion and thermal expansion experienced by their high-temperature cousins. Moreover, fuel cells using acidic polymers were somewhat tolerant of carbon and thus could operate on ambient air instead of pure bottled oxygen, suggesting to researchers that such devices might also be able to operate on hydrocarbon fuels.

Like the nickel-metal hydride battery, the membrane fuel cell was a technological orphan adopted by third parties who applied new materials or repurposed existing ones in these systems in ways that dramatically boosted their performance. In the mid-1950s, GE researchers discovered that prosaic polymer membrane, then used mainly as a water softener, was an excellent conductor of ions. They believed the compound could serve as an effective electrolyte in fuel cells operating at 50 to 100 degrees Celsius.[53] In the early 1960s, GE developed this technology for use in NASA's *Gemini* spacecraft and hoped to employ it in civilian applications as well. Despite early promise, the device did not work particularly well. It produced low current density and the membrane was prone to dehydration and cracking, especially at higher temperatures, allowing hydrogen and oxygen to directly react and potentially combust or explode.[54]

GE eventually solved the problem of membrane dehydration and crack-ing, but a more important advance came with the development of a new membrane technology. This was Nafion, invented by DuPont in the early 1960s for a number of industrial electrochemical processes including chlor-alkali production. When used in a fuel cell, this acidic polymer delivered greater power than the GE membrane and proved more durable as well. In the mid-1960s, the two companies began collaborating in applying the new material.[55] However, these developments came too late to benefit *Gemini*. With the space race winding down in the late 1960s and early 1970s, the membrane fuel cell essentially disappeared from use.

In the mid-1980s, however, the technology was revived by a little-known engineering research company similar to ECD. This was Ballard Power Systems, founded by a retired geophysicist named Geoffrey Ballard. Like Beaumont, Rippel, Goodenough, Ovshinsky, and so many other inno-vators in fields linked to electric propulsion, Ballard was something of an outsider with an eclectic résumé. An engineer and scientist with dual US-Canadian citizenship, Ballard worked for the oil industry, the US Army, and then the federal government as an energy advisor in Washington, DC. With the onset of the energy crisis, Ballard became interested in develop-ing advanced power sources for electric vehicles. In 1974, he quit his job and, with an electrochemist from the University of Texas at El Paso named Keith Prater, set up a small lab in Arizona devoted to researching recharge-able lithium chemistries. In 1977, Ballard relocated to Vancouver, British Columbia, and reorganized the enterprise first as Ultra Energy and then as Ballard Research. The company worked under contract to Shell and Amoco but failed to develop a practical lithium battery. Ballard Research was near-ing bankruptcy when it was commissioned by the Canadian Department of National Defense in 1983 to build prototype membrane fuel cells and investigate their potential for military and civilian use.[56]

Ballard would assemble the components for what would become one of the most potent power sources yet applied to electric drive. In 1987, his company acquired a Dow Chemical polymer membrane and tested it in a fuel cell. Thinner than Nafion and with a higher sulfonic acid concentration (and hence a greater ion exchange capacity), the material enabled over four times as much current density as Nafion.[57] The results astonished J. Byron McCormick, a former deputy division leader at the Electronics Division of the Los Alamos National Laboratory and then an engineer with GM's Delco

Systems Operations, a supplier of defense, aerospace and advanced automobile systems. According to the historian Tom Koppel, when McCormick learned of the Ballard test, he exclaimed that the company had "made the electric vehicle possible." McCormick would go on to manage the development of power electronics, batteries, and fuel cells at GM.[58]

Ballard soon expanded its work on membrane fuel cells, helping subsidize its failing business in disposable lithium batteries. Initially, Ballard explored a number of applications but was increasingly drawn to electric vehicles after Daimler-Benz made overtures in 1989. Under the leadership of Firoz Rasul, who replaced Geoffrey Ballard as chief executive officer in 1989, the company promoted its fuel cell as a revolutionary prime mover for the electric automobile, adopting the appropriately ambitious motto "Power to Change the World."[59]

PRELUDE TO A SALTATION

Twenty years of research in materials and devices of energy conversion and storage motivated by the sequential and compounding public policy emergencies of the Cold War and space race, chronic air pollution, and the energy crisis, as well as the demands of the revolution in consumer electronics and personal computing, substantially produced the result that Wally Rippel had set out to achieve in staging the Great Electric Vehicle Race. By the late 1980s, the field of power source technoscience had substantially progressed and could boast several new technologies: the sodium-metal chloride, nickel-metal hydride, and lithium ion rechargeable batteries and the improved proton exchange membrane fuel cell.

The innovators of this new generation of advanced power sources occupied space somewhere between the appropriate technology movement and the industrial establishment. On the one hand, their work was relatively capital-intensive and dependent on patronage from the national developmental state and industry, especially the consumer electronics sector. But they were still outsiders invested in new and untried technologies that required vast amounts of capital to be further developed and commercialized. These researchers selected combinations of materials primarily for their propensity to yield the highest possible energy and power and gave secondary consideration to durability, cost, and safety. And while the makers of the revolution in advanced power sources often imagined electric propulsion as a possible application, their situation in isolated laboratory contexts meant that relatively little of their work was coproduced with experts in electric vehicle systems. While Detroit with its concept electrics, enthusiasts with their homebrewed electrics, and entrepreneurs with their semicommercial electric fleets had accumulated substantial operational experience with classic rechargeable battery technologies, hardly anything was known of how the new power sources might function in the electric vehicle duty cycle. Moreover, each had implications for manufacturing and marketing that could be known only when they were produced at scale.

In the late 1980s, advanced power sources for electric vehicles were still largely solutions looking for problems. That was about to change. If it was still unclear to environmental policymakers whether better power sources made for better electric vehicles, automakers could no longer credibly argue that there were no alternatives to the classic rechargeable battery formulas as their excuse for inaction on advanced propulsion cars. By the end of the decade, air quality regulators were preparing drastic new measures to solve persistent smog, which would compel the auto industry to consider all approaches, involving established as well as emerging technology, and to engage in a debate on what exactly was appropriate for envirotechnical and commercial contexts. A new phase in the history of the automobile was about to begin.

4 FORCING THE FUTURE

Impact is a genuine full-performance machine with capabilities that rival those of today's internal combustion–powered cars. It is an experimental vehicle and is not now ready for production. On the other hand, it was designed to be producible.
—Roger Smith, General Motors chief executive officer, January 3, 1990

By 1990, the governance of automobile pollution was at a crossroads. Statistically, the air quality control apparatus could claim major successes. In the span of a quarter-century, under regulatory duress, automakers were able to integrate a series of pollution control devices into commercial internal combustion engine (ICE) technology, progressively improving fleet energy conversion efficiency and significantly reducing the three main chemical constituents of smog. In 1990, the US highway vehicle fleet produced about 45 percent less volatile organic matter, 32 percent less carbon monoxide, and 24 percent less percent nitrogen oxide than in 1970.[1] Moreover, all this had been accomplished by a highway fleet that nearly doubled in size from around 111 million to 193 million vehicles in this period.[2] On its face, the emission control campaign was a singular achievement of regulation, planning, and industrial innovation.

But to Californians, smog seemed as bad as it had ever been, and the repercussions of this protracted crisis extended beyond public health into the air quality control regime itself. By the early 1980s, the political landscape was tilting to the right at both the state and federal levels, seemingly threatening a host of programs in alternative energy. The incoming

Reagan administration ostentatiously removed solar panels that the Carter administration had installed on the White House, and similar developments unfolded in California following George Deukmejian's defeat of Jerry Brown in 1982. Among Deukmejian's first acts as governor was to shutter the Office of Appropriate Technology, a pet project of Brown's. These symbolic gestures were born of a broader reaction against the idea of limits to growth, encouraged by the political resolution of the oil embargo and the return of cheap petroleum, which in turn led federal policymakers to shift the emphasis in energy research and development from large-scale demonstrations to long-term precompetitive projects.[3] These measures did little to resolve pollution problems. The ironic reality was that pathbreaking California, the state that had pioneered the world's most stringent automobile emission policies, was unable to meet federal clean air standards, and risked losing federal funds for highway infrastructure as a result.[4]

In response, the Deukmejian administration tried to make emissions controls more effective, yet friendlier to business. In the old system, regulators at all levels applied a single uniform emission standard to new cars. In the new Low Emission Vehicle (LEV) program, regulators defined automobiles in terms of the quantity of effluent they produced and created three certification categories that required automakers to produce and make progressively cleaner types of automobiles available for sale.[5] Although the new rules complicated enforcement, they represented continuity with the previous regime because they assumed improvements in ICE technology. However, there was one crucial exception. Among the provisions of the LEV was a requirement that automakers produce small numbers of Zero Emission Vehicles (ZEVs). Jananne Sharpless, chair of the air resources board from 1985 to 1993, held that the ZEV mandate (popularly known simply as the "mandate") was planned as a mere "footnote" in a large and complex regulation.[6]

Yet the mandate implied far more onerous obligations than any previous technology-forcing instrument. The California Air Resources Board (CARB) could not stipulate the energy conversion technologies that would achieve air quality outcomes because energy was a federal prerogative. In 1990, the only practical ZEV was the all-battery electric vehicle, a technology that mainstream automakers long insisted could not compete in the marketplace.

But air quality regulators believed that industry was signaling that it *was* capable of building such technology. The most notorious such signal was the Impact electric car, introduced by General Motors (GM) chief executive officer Roger Smith at the Los Angeles Auto Show in January 1990. Impact is widely interpreted as the proximate cause of the mandate but there is little evidence Smith conceived the project as a specific response to California air quality politics. Impact was rooted partly in corporate intramural politics. It was a concept car, a public relations device used to demonstrate engineering principles and corporate vitality, something that was increasingly important to US automakers as they lost market share to foreign competitors through the 1980s.[7] Smith hoped Impact would serve as an example of technology transfer from Hughes Aircraft in order to justify his much-criticized purchase of the aerospace company in 1985. Another important consideration for Smith was that cutting edge automobile technology should also be seen to be environmentally friendly, a view informed by sharpening popular environmental sentiment. The executive took pains to emphasize that the car was merely a concept but Robert C. Stempel, Smith's successor, committed to commercialize it, generating publicity that engaged federal science and technology resources in ways that further entrenched the mandate. In turn, GM's peers felt compelled to expand their own hitherto limited research into alternative propulsion vehicles.

California's air quality crisis hence intersected with GM's internal politics and the broader environmental sensibilities of the day in setting into motion a cascading series of unintended consequences that would forever reshape automaking. These forces exposed car companies to the influence of an eclectic host of outsider practitioners specializing in advanced materials, power sources, lightweight structures, and electronics, including enthusiast-experts and formally trained experts from other industries. GM itself initiated much of this technopolitical ferment, first by acquiring Hughes and then by contracting AeroVironment, a research and design company specializing in lightweight experimental aircraft, and directing it to collaborate with the aerospace giant to help create Impact. Public policy also importantly contributed to the cross-pollination. One of the most significant initiatives of the national developmental state in the wake of the mandate was the extension of crucial support to the Ovonic Battery Company (OBC), which helped insinuate this small enterprise in the affairs of GM and the wider automaking community. OBC's nickel-metal hydride rechargeable battery represented the sort of technological advance that car companies long claimed was the prerequisite for a commercial electric car. The company's strong claims to this chemistry and ambitions to dominate the nascent electric vehicle market caused particular concern to an automaking establishment already anxious about its ability to set the technological agenda in the mandate era.

Car companies initially responded to the mandate with a mixture of grudging compliance and lobbying, taking their cues largely from GM. Automakers sought to reclaim the initiative from CARB and roll back the mandate by exploiting the ambiguities of LEV nomenclature in a way that problematized the association between the ZEV and the all-battery electric vehicle. The objective was to convince regulators that the all-battery electric vehicle was commercially infeasible. GM came to believe that an alliance with OBC would both serve this goal and prevent competitors from using the nickel-metal hydride rechargeable. While automakers were of one mind when it came to the undesirability of the all-battery electric car, GM could not ignore the possibility that competitive pressures might cause one of its peers to break ranks.

INSIDERS AND OUTSIDERS

Impact had origins in an earlier project that illustrates the growing influence of the appropriate technology and environmental movements in the affairs of the automaking establishment by the mid-1980s. In late 1986, the Danish-Australian adventurer Hans Tholstrup informed Smith's office that he planned to stage a race of solar-powered cars across Australia in November 1987. Smith's office passed the information to Hughes, and a Hughes executive named Edmund Ellion in turn asked AeroVironment founder Paul MacCready if he would be interested in developing an entry for GM. Mac-Cready was enthusiastic. Hughes vice president Howard Wilson had overall control of the project, and the project study was managed by Alec Brooks, a friend of electric vehicle pioneer Wally Rippel and one of AeroVironment's chief engineers. The team considered the technological requirements of a race defined by the direct use of solar energy. They envisioned an electric vehicle that would make the best use of available power sources through aerodynamics and light weight, echoing the ethos that had informed Bob Beaumont's Citicar. In every other way, however, the proposed vehicle was unique. It was a one-seat racer employing the very latest technologies. Following a joint presentation by the team in late March 1987 to GM vice chair Don Atwood and Stempel, then serving as head of GM Truck and Bus and GM International Operations, the proposal was quickly approved.[8]

The project was called Sunraycer and it embraced a diverse array of engineering talent representing the automobile, electronics, and aerospace communities. Graduates of the California Institute of Technology and other California post-secondary institutions played a notable role through employment at AeroVironment and Hughes. Stempel himself held a degree in mechanical engineering from Massachusetts's Worcester Polytechnic Institute and an MBA from Michigan State University and was involved in many notable firsts at GM including its catalytic converter program in the mid-1970s.[9] Wilson earned a degree in electrical engineering from the University of California at Berkeley, managed radar systems at Hughes, and was an accomplished pilot. In some ways, MacCready bridged the appropriate technology movement and the aerospace establishment. Possessing a doctoral degree in aeronautics from Caltech, MacCready built the first human- and solar-powered aircraft in the late 1970s and early 1980s and developed a solar-powered spy drone for the Central Intelligence Agency (CIA) in the later Reagan years. Brooks more closely fit the profile of the appropriate technologist. He was interested in advanced human-powered vehicles of all sorts and built and raced advanced bicycles as an undergraduate in civil engineering at the University of California at Berkeley and as a doctoral student at Caltech. In 1983, Brooks and the sports medicine physician Allan Abbott designed, built, and successfully tested the Flying Fish, the first human-powered hydrofoil to maintain flight.[10]

Design and construction of Sunraycer was the responsibility of AeroVironment's Aerosciences Division, headed by Caltech graduate Peter Lissaman, with Brooks serving as overall project manager. Brooks enlisted collaborators from similarly unconventional backgrounds, bringing in colleagues from the human-powered vehicle community, including Abbott and Chester Kyle, to help develop the car's ergonomics and wheels and brakes respectively. For the power electronics, Brooks recruited Rippel and an electronics wizard named Alan Cocconi, both of whom served as consultants. All three engineers were graduates of Caltech and knew each other as habitués of its machine shop.[11] Cocconi's main engineering interests related to aerodynamics, but he believed that a career in the aerospace industry would mean working for the military, an institution whose values he opposed, his parents having lived through World War II in Italy.[12] Brooks met Cocconi as a student in the late 1970s and was impressed with his plans for testing propellers in wind tunnels, research that Brooks thought was "quite astonishing for somebody to be doing on their own."[13] In 1980, Cocconi graduated with a bachelor of science degree in engineering and applied science and joined a Caltech start-up called TESLAco, where he worked on a government-sponsored project to design a solar power inverter. Cocconi did some of this work in the Caltech machine shop, where he encountered Rippel, who was then employed by the Jet Propulsion Laboratory and had been assigned by the government to monitor the progress at TESLAco.[14] Rippel also used the machine shop to tinker on bipolar batteries. In 1981, Brooks earned a doctoral degree in civil engineering and went to work for AeroVironment.[15] A few years later, Brooks helped Cocconi get consulting work at the company on a project to build a flying model pterodactyl for an Imax film called *On the Wing*, where Cocconi provided advice on the aircraft's stability systems, servodrives, and radio links.[16]

Together, Brooks, Cocconi, and Rippel would figure prominently in the electric vehicle renaissance, their careers intertwining in a series of projects over a span of more than twenty years. In Sunraycer, Brooks was in charge of driver controls and was involved in many other aspects of the construction, systems integration, and testing of the vehicle. He also served as one of its drivers. Cocconi made major contributions to Sunraycer's exotic propulsion unit, a blend of silver-zinc batteries and Hughes solar cells mated to a prototype direct-current brushless motor designed and built by GM Research Laboratories. Working with Hughes and GM, Cocconi helped improve and integrate the electronic components of the system, notably the peak power tracker, a device designed to orient solar cells to the Sun to enable the most efficient energy conversion, and the inverter, a device that converted direct current produced by the battery to the alternating current used by the brushless motor.[17] Rippel contributed advice on Sunraycer's electric drive systems.[18]

In four-and-a-half months of concentrated effort, AeroVironment, Hughes, and GM produced two low-slung, teardrop-shaped vehicles, with the second one incorporating improvements retrofitted to the initial prototype.[19] Sunraycer handily won the inaugural World Solar Challenge and admirably served Smith's goals. Basking in the favorable publicity, Smith made clear that the car was intended only to demonstrate GM's "expertise" in new technologies.[20]

But Brooks wanted to create a practical battery electric car. He had the support of Wilson, who provided funds to keep the Sunraycer team together, as well as Stempel, who convinced Smith to lend support.[21] In the AeroVironment camp, there was some dispute about what kind of vehicle to build. MacCready wanted a "bread and butter" delivery vehicle, while Cocconi and Rippel preferred a sports car, a view that ultimately prevailed.[22] In their July 1988 report to Hughes and GM, Brooks and Wilson argued that all the requisite technologies were available to build a two-place vehicle with a range of 120 miles, adding that Japanese and West German manufacturers were actively investigating electric propulsion and planning on developing an electric car market in the near future.[23]

Brooks managed the project, dubbed Impact, a two-seat sports coupe designed to have the performance of a commercial car in this class. Rippel and Cocconi collaborated to develop the electronics. The vehicle was equipped with an induction motor, designed by Rippel with help from Cocconi, and built to specification by an aerospace components company called Lucas Western. As constructed, the car had two induction motors, one for each front wheel. Cocconi devised the power electronics using solid-state components. He built another inverter, and he and Rippel devised a charger integrated into the motor controller.[24]

Sunraycer's solar-battery system was deemed unsuitable for practical automobile applications, so the Impact team sought a rechargeable battery that was energetic and powerful as well as durable and cost-effective. For Electrovette, GM had utilized the nickel-zinc rechargeable, a chemistry that was powerful and relatively energetic but not especially robust. For Impact, Brooks wanted an advanced lead-acid rechargeable and looked to Delco Remy, the GM division that manufactured starters, alternators, and batteries for ICE vehicles. The parts supplier had to adapt its experience of battery technology to a completely new application. In an all-electric vehicle, batteries are subjected to much more punishment than in their auxiliary role of starting and lighting an ICE vehicle. In a lead-acid rechargeable, this causes an electrolytic reaction in the aqueous electrolyte that depletes water through the evolution of oxygen and hydrogen. The water has to be replaced, and the so-called flooded lead-acid battery accomplished this with plumbing and venting that added weight and complexity.

Delco Remy researcher Bob Bish took an alternative approach to the water problem. He isolated the electrolyte in a glass fiber matrix, enabling the recombination of oxygen in the water and dispensing with the water circulation system. The resulting sealed lead-acid battery was denser, more compact, and more energetic than its flooded cousin. With 843 pounds of such batteries, the Impact was a veritable sports car, capable of accelerating to sixty miles an hour in about eight seconds.[25]

Smith was delighted with the results. He lauded Impact as yet another technological triumph for GM. Four months after the Los Angeles Auto Show, in a speech delivered at the National Press Club in Washington, DC, days before Earth Day, Smith again touted Impact. Some media claimed that the executive promised to produce the automobile commercially, but the evidence suggests that Smith made no such explicit promise. What is known is that Smith followed his praise with a remark that the Impact's weak link was its lead-acid battery, and that developing a suitable replacement would be a major collaborative undertaking involving both industry and government.[26]

THE ACCIDENTAL REVOLUTION

In essence, Smith made the same argument that automakers had made for years in rejecting the electric car. The call for better batteries had long resonated in light of the problematic electric concept cars that automakers had built after World War II but the technopolitical context had changed drastically. The lead-acid Impact was a very capable purpose-built automobile, and it appeared at precisely the time that California air quality regulators were drafting the new emissions control regime. The Impact's makers were not reticent about promoting their achievement. Tom Cackette, the CARB deputy director in charge of mobile source emissions regulation between 1982 and 2012, recalled that AeroVironment invited his team to view the prototype in 1989. Around the time of the Los Angeles Auto Show, GM gave CARB members including Don Drachand, then-chief of the Motor Vehicle Emissions Control division, an opportunity to test-drive the car.[27]

With Smith's retirement in July, the Impact moved deeper into the institutional structure of GM and air quality technopolitics. In August, Stempel succeeded Smith as CEO and made plans to commercialize the Impact. A few weeks later, on September 28, after two days of public hearings, CARB adopted a resolution approving the LEV and Clean Fuels program. The mandate was a relatively late addition to this regulatory package and required automakers with annual California sales of at least 35,000 light duty units (passenger cars and light trucks) to deliver ZEVs amounting to 2 percent of their sales fleets for each model year from 1998 through 2000, 5 percent from 2001 through 2002, and 10 percent from 2003 and subsequent model years. Air quality regulators believed that they provided automakers with plenty of time to deliver results. In the old system, regulators worked with equipment suppliers and the Environmental Protection Agency (EPA) with a view to what automakers could accomplish, in terms of technological fixes, in three to five years. The mandate gave the car companies eight years to produce thousands of ZEVs a year in a market where they sold nearly two million ICE vehicles annually.[28]

The mandate did not immediately arouse opposition from the car companies. Sharpless observed that automakers were then preoccupied by the efforts of California state senator Bill Leonard to press CARB to compel the use of methanol in California. Among the proposed provisions of the LEV was a trigger that required oil companies to produce alternative fuels if automakers passed a certain sales threshold. Leonard succeeded in passing a study bill, and CARB devoted a good deal of attention to the issue before dropping it in the face of industry lobbying. Sharpless believed that the debate over alternative fuels distracted attention from the ZEV mandate and enabled it to survive.[29]

One early opponent of the mandate was a member of the air resources board itself and the sole dissenter on the question of battery economics. During the September hearings, Andrew Wortman raised the specter of the temporal mismatch of battery and motor, citing a study by the US Department of Energy (DOE) indicating that operators of electric cars equipped with standard lead-acid batteries would have to replace their packs every 15 months, at a cost of between $3,000 to $4,000 for each pack.[30] Stempel opposed the mandate because he wanted the Impact to succeed in the marketplace on its merits, not as a result of state intervention.[31] But the problem of battery replacement costs was not on his list of priorities. The immediate task was to convert AeroVironment's concept car into a production prototype, a process overseen by Kenneth Baker, the GM executive who had managed the Electrovette over a decade earlier. AeroVironment played an important role in the early phases of this project. Brooks participated in GM's production study of the Impact, and the aviation company supplied the automaker with custom-built battery cycling equipment, allowing it to build a laboratory to test power sources under real-world conditions. GM would also take an equity position in AeroVironment.[32]

As this process unfolded, the design choices that had informed the original Impact were substantially modified. Baker's team determined that the production prototype required features like air conditioning and a sound system, setting off a chain reaction of trade-offs between weight and capability. Such amenities required more energy and power and more lead-acid cells from Delco Remy. That made the car heavier, leading the team to replace the original fiberglass frame with one made of aluminum.[33] Delco Remy built copies of Rippel's induction motor, and Hughes collaborated with Cocconi to make his idiosyncratic inverter manufacturable. Cocconi found the work tedious, and in the fall of 1991, he and Rippel left the project and with the Hughes engineer Paul Carosa cofounded AC Propulsion (ACP), a research and development company that would play a pivotal role in electric automobile technopolitics in the years to come.[34] By December 1990, Baker had developed a plan to start production by late March 1993 at a rate of 20,000 vehicles per year for four years, at a cost of around $1.5 billion.[35]

But the Impact divided GM's leadership elite, and Stempel's position as CEO was not strong. Deteriorating economic conditions sharpened opposition to the electric car project, reducing Stempel's freedom of action and drastically altering Baker's plans. GM lost $4.5 billion in the post–Gulf War recession, the largest loss in corporate history up to that point, and Stempel responded by slashing tens of thousands of jobs, closing twenty-one plants, and cutting the Impact's budget. With time running out, Baker fast-tracked the production prototype and delivered it on May 1, 1992. By then, Stempel was facing a coup led by former Proctor and Gamble chief executive John Smale, a member of GM's board of directors.[36] In April, the board of directors rallied behind Smale, who replaced Stempel as head of the board's executive committee. Six months later, GM's powerful management committee forced Stempel to drastically downgrade the Impact to a few dozen hand-built proof-of-concept vehicles. In October, Stempel was pushed out entirely. Smale assumed the chair and the accountant John "Jack" Smith became CEO.[37]

POWER SOURCE POLITICS AND THE RISE OF OVONIC BATTERY

Sacking Stempel did not make the Impact disappear. The program was conjoined with the politically popular mandate, making it difficult to cancel outright. In late 1993, Jack Smith moved the program to GM's research and development division where it became known as PrEView, a production run of fifty cars that were to be loaned for a few weeks to users selected by lottery, ostensibly for testing in real-world conditions. Media observers noted that GM officials emphasized the Impact's limited range and, for the first time, battery replacement costs, seemingly with a view to lowering expectations.[38]

At the same time, the Impact gained favorable new publicity through Stanford Ovshinsky. The inventor paid close attention to air quality technopolitics and was looking for opportunities to apply OBC's nickel-metal hydride battery to electric vehicles. In the summer of 1990, representatives of OBC and GM discussed the possibility of collaboration. William B. Wylam, Delco Remy's chief engineer for technology development, had ambitions to dominate the supply of electric vehicle components to GM and showed particular interest in these discussions.[39]

Little of substance came of these talks.[40] Nevertheless, OBC was attracting attention in government and industry circles. Researchers at Argonne National Laboratory began testing cylindrical Ovonic C cells for their suitability in electric vehicles. OBC was already working on prismatic cells, a configuration that some experts regarded as more suitable for electric traction.[41] Rectangular prismatic cells have a larger surface area than cylindrical wound cells, making them more energetic and powerful and enabling a slimmer and more compact battery form factor. Prismatic cell modules were also more difficult to manufacture than cylindrical cells and had to be built in such a way as to allow cells to swell during recharging without warping the boxlike battery can.

In September 1991, OBC found the first customer for its prototype electric vehicle battery in Honda and worked out a deal that included provisions for either a joint venture or licensed manufacture.[42] In May 1992, two weeks after the debut of the Impact production prototype, OBC received an $18.5 million grant to develop prismatic cells from the United States Advanced Battery Consortium (USABC).[43] Set up in January 1991 and comprising the DOE, the Big Three US automakers, the Electric Power Research Institute (EPRI), and several independent battery manufacturers, the USABC was a product of the prevailing view in policy circles of the efficacy of collaborative precompetitive public-private research. The exemplar was Sematech, a consortium of US semiconductor manufacturers organized by the Defense Advanced Research Projects Agency (DARPA) in 1987 to help the industry improve its production techniques and ability to compete with Japanese enterprise.[44] The Clinton administration saw Sematech as a model for developing other strategic technologies, including advanced automobiles.[45]

But where US semiconductor manufacturers more or less agreed that collective research could produce useful practical results for their industry, US automakers only agreed that they did not want to develop advanced batteries and electric cars. Consequently, the car companies tried to use the USABC as an instrument to collectively control innovation rather than collectively stimulate it. Ford and GM each had its own longstanding advanced battery program (sodium-sulfur and nickel-zinc, respectively) that neither company was in any hurry to commercialize. Moreover, OBC's involvement in the USABC posed a host of problems to US automakers. As the battery company's star rose, the car companies feared that air quality regulators would be encouraged to further entrench the mandate and even make the nickel-metal hydride rechargeable the industry standard in the process. For these reasons, US automakers sought to treat their disparate research initiatives as the shared property of the USABC, a difficult task for an entity that was loosely centrally managed. They were particularly concerned by Ovshinsky's penchant for breaching USABC protocol. The researcher-entrepreneur saw an opportunity to test OBC batteries in the TEVan, a converted Chrysler minivan that had been jointly developed with the DOE, Southern California Edison, and EPRI. Chrysler managed OBC's USABC file, perhaps because the automaker lacked its own advanced battery program, and in August 1993, fifteen months after receiving its grant, OBC installed its first prototype pack of prismatic cells in the minivan.[46]

Ovshinsky promoted the vehicle as the first electric to be powered by a nickel-metal hydride battery. This audacious maneuver annoyed John Williams, the GM executive who chaired the USABC's management committee, and it also brought Ovshinsky to the attention of Stempel, who was greatly impressed. The former CEO still had influence at GM and helped furnish the Impact team with an OBC prototype battery pack. In a secret test in ideal conditions at the GM Desert Proving Ground in Mesa, Arizona, in January 1994, the car ran over 200 miles on one charge, about twice what Delco Remy's lead-acid rechargeable was capable of affording.[47]

REINTERPRETING THE ZERO EMISSION VEHICLE

The test represented a milestone in the development of electric vehicle technology and signaled the beginning of a partnership between Ovshinsky and Stempel to promote OBC technology to the broader automaking industry. The stakes were becoming higher. Auto executives had not paid much attention to the mandate in the weeks following its introduction, but they became increasingly concerned about it after northeastern states moved to adopt similar regulations. This had the potential to greatly expand the quotas and also prevent automakers from designing mandated electrics solely for California, a generally dry and mild environment well suited for battery electric automobility. Car companies now faced the prospect of having to develop a four-season national electric car, a considerably more complicated and costly enterprise.[48]

GM executives began to plan to contain both OBC and the Impact, with a notable role played by Harry J. Pearce. Pearce had an unusual professional profile for an auto industry executive, combining expertise in engineering, corporate law, law enforcement, and national security. In 1964, he took a bachelor's degree in engineering sciences from the US Air Force Academy and a juris doctor degree from Northwestern University in 1967. Pearce served in the Air Force as a military lawyer and was certified as a military judge before transitioning to civilian life and serving as a police commissioner, municipal judge, and US magistrate. He maintained close ties to the military throughout his life.[49]

Pearce built his career defending industry in product liability cases and represented a number of companies, including GM, through the 1970s and into the early 1980s. In 1985, Pearce joined GM as associate general counsel responsible for product litigation, and in 1987, he was promoted to general counsel. In this period, Pearce purged GM's legal cadre, eliminating job titles and firing dozens of lawyers.[50] In the leadership shakeup of 1992, Pearce was promoted to executive vice president with broad powers, joining Smale and Jack Smith in a de facto triumvirate. In addition to responsibility for GM's government and industry affairs, Pearce managed the corporation's nonautomobile enterprises, including Hughes and Electronic Data Systems (EDS), assets acquired by Roger Smith in the mid-1980s. Pearce had a mandate to transfer technologies from Hughes and EDS to GM's automobile business if possible and, some observers thought, to liquidate GM's nonautomobile businesses if necessary.[51]

In effect, Pearce oversaw GM's electric car program. In February 1994, the automaker moved to bring OBC on board and redefine the Impact as a demonstration program with the ultimate goal of developing alternatives to battery electric propulsion. The details of the plan, as recorded by the journalist Michael Shnayerson in *The Car That Could*, were presented by Kenneth Baker (appointed vice president of GM's research and development laboratories in early 1993) in a meeting with the GM President's Council, the automaker's principal decision-making body. Baker held that while the test results of the Ovonic battery were promising, the technology required much more development before it was ready for market. Consequently, suggested Baker, Impact would serve largely as a test bed, initially for hybrid battery electric and later for fuel cell electric propulsion, the basis of the ultimate ZEV.[52]

In March, the GM President's Council voted tentative approval of Baker's proposal and the Impact was upgraded to precommercial status. GM reiterated the promise of 20,000 vehicles, including some equipped with OBC batteries when they became available. In the meantime, GM and its peers planned to meet mandate quotas with what became known as "compliance cars," a pejorative expression for money-losing all-battery electrics that mostly were converted from existing conventional models.[53]

Pearce, Stempel, and Ovshinsky then negotiated a partnership to produce advanced electric vehicle batteries called the "Joint Manufacturing Entity," later dubbed GM-Ovonic. The automaker owned 60 percent of the enterprise and would supply start-up capital while OBC held the remaining stake and would supply materials, components, and the initial assembly facilities. The promise of 20,000 electric vehicles represented an unprecedented scaling in production of electrics and the batteries to equip them and GM suggested that GM-Ovonic would also supply the other USABC members as well as other automakers, implying an even larger market. Yet the partnership emphasized research and development over manufacturing. GM had operational control over the joint venture's plant facilities and OBC had to meet cost targets set by the USABC.[54] As events would transpire, GM would insist that OBC first cut costs before moving to production, a tall order made more difficult by the fact that the battery company had to adapt a formula initially designed for electronics to electric vehicles.[55] In effect, the arrangement enabled GM to control OBC's intellectual property as it related to the manufacture of commercial batteries for electric cars. In turn, the automaker would use GM-Ovonic, the Impact, and its production variant, the EV1, as instruments in a campaign to limit CARB's ability to impose its will on the auto industry.

5 HYBRID POLITICS

Some say hybrid vehicles are a bridge to the future. We think it could be a long bridge, and a very sturdy one.
—Takeshi Uchiyamada, Toyota chairman, September 30, 2013

The mandate gave further impetus to the popular technopolitics of the electric car. While established automakers counseled caution and tempered expectations, electric car activism and enterprise mushroomed. In 1991, a Santa Rosa–based conversion specialist called Solar Electric Engineering supplied an electrified Ford Escort to a Dartmouth professor and environmentalist named Noel Perrin, whose attempt to drive from California to Vermont became a celebrated moment in enthusiast culture.[1] Renamed US Electricar, this company targeted the corporate fleet market and rapidly expanded in the early 1990s, becoming the largest producer of electric vehicles in the US.[2] On the East Coast, the Boston-area start-up Solectria also sold converted electrics to corporate fleets, but it had more ambitious plans. Founded in 1989 by MIT graduates James Worden and Anita Rajan Worden, Solectria was also developing a ground-up electric car called the Sunrise with the support of the Defense Advanced Research Projects Agency (DARPA). The vehicle used an advanced induction motor, and Solectria sought to equip it with Ovonic Battery Company (OBC) batteries with a view to competing with established automakers.[3]

At this juncture, a third way between the poles of ICE and battery electric propulsion emerged in the form of the hybrid electric. The hybrid concept was not new. The technology was coproduced alongside battery electric and ICE cars at the turn of the twentieth century as a solution to the limitations of these propulsion formats at that time.[4] The automobile historian Gijs Mom held that hybrid systems allowed the electric vehicle to become the functional equivalent of the long-range ICE touring vehicle, at least in principle. The trade-offs were that hybrid electrics were more difficult to manufacture and more expensive than either ICE vehicles or all-battery electrics.[5] For these reasons, hybrid propulsion became widely used only in rail transport in the form of the diesel-electric locomotive, which replaced steam power after World War II.

The hybrid electric concept remained an orphan until 1969, when the Environmental Protection Agency (EPA) began a three-year study to evaluate the technology for its potential in meeting the 1976 emissions standards as part of the Federal Clean Car Incentive Program. One of the few complete prototypes to be submitted for testing was built by Victor Wouk and Charles Rosen at their own expense; it consisted of a 1972 Buick Skylark equipped with a lead-acid powerpack mated to a Mazda Wankel rotary engine. The EPA tested the vehicle but chose not to support moving it to development, citing concerns about intervening in the marketplace. Wouk and Rosen's Petro-Electric Motors went bankrupt, a fate that recalled the saga of Sebring-Vanguard.[6] The Jet Propulsion Laboratory used the data in a 1975 review that seemed to confirm long-standing assumptions about hybrid technology. The study found that hybrid electrics did not require advanced high-energy batteries and offered higher fuel economy than ICE cars but were likely to be more expensive to maintain and repair than either ICE or all-battery electric cars. Hybrids did promise to be powerful enough to operate on freeways but were projected to offer only marginal advantages in fuel efficiency in that context. From a supply chain perspective, hybrid electric propulsion technology portended the worst of all possible worlds because it required materials that were strategic to both the ICE and the electric car.[7] Nevertheless, Public Law 94–413 (the Electric and Hybrid Vehicle Research, Development, and Demonstration Act) stimulated modest research on hybrids over the next fifteen years, with the US Department of Energy (DOE) sponsoring a number of projects. These interventions produced innovations but no breakthroughs, and hybrid electrics had a lower profile than the experimental all-electric cars of those years.[8]

To the incoming Clinton administration, however, the notional hybrid electric had attractive discursive and political qualities. The technology seemed to offer something for everyone. In principle, hybrids could use practically every known gaseous and liquid chemical fuel more efficiently and with fewer emissions than straight ICE vehicles. And because hybrids used smaller, less expensive batteries than all-electrics, they also mitigated the temporal mismatch.

Hybrids were one of the key technologies promoted by the Partnership for a New Generation of Vehicles (PNGV), launched by the Clinton White House in September 1993 as part of a broader plan announced the previous February to mobilize technology to solve social problems. At a well-publicized event in San Jose, President Bill Clinton and Vice President Al Gore framed the program using metaphors of infrastructure as a means of highlighting the public policy responsibilities of government in the post–Cold War era. Where the "information superhighway" would bring the information technology revolution to the people, the PNGV would make an analogous revolution on America's real superhighways.[9] The partnership was a public-private research consortium in the mold of Sematech and the United States Advanced Battery Consortium (USABC). It was formed by the United States Council for Automotive Research (USCAR), an organization set up in 1992 by the car companies in consultation with the federal government to coordinate all their efforts in collaborative research and development in advanced automobiles.[10] The PNGV was designed to help US automakers modernize their manufacturing base and improve the efficiency, environmental footprint, and performance of ICE technology. In practice, the partnership would focus most of its energy on the technologies of a future supercar, an affordable vehicle that was to have triple the fuel efficiency of the average 1994 family passenger sedan (equating to nearly eighty miles a gallon) without detracting from performance or comfort. The White House favored hybrid electrics, especially those using fuel cells, partly because it thought most of the basic technologies were available and required only further development and application.[11]

However, and the PNGV was not a technology-forcing enterprise in the vein of California's Low Emission Vehicle (LEV) and Zero Emission Vehicle (ZEV) mandates. While Clinton and Gore framed the effort as the automobile equivalent of the Apollo moon program, the PNGV essentially coordinated existing research and development programs at a number of federal agencies. Its legal basis was a declaration of intent and an expectation that automakers would field prototype supercars sometime in the new millennium.[12] Paradoxically, the partnership's performance goals for the hybrid electric supercar were so ambitious that the vehicle required virtually the same sort of advanced power sources that automakers had informed air quality regulators were necessary before development of the all-battery electric car could even be considered. Turbines and advanced diesels were on the agenda along with fuel cells, but so were advanced galvanic batteries, substantially overlapping the mission of the USABC.[13]

The PNGV drew criticism from practically every quarter. For conservatives, it did too much, and for liberals, it did too little.[14] American electric vehicle enthusiasts scorned the hybrid electric car as an unprincipled compromise with ICE propulsion, comparing it unfavorably with the so-called pure all-battery electric car. Nevertheless, the possibility that the federal government and industry might cooperate in developing a commercial hybrid electric drew the attention of Japanese automakers closely monitoring developments in the US. Toyota came to believe that such a vehicle could be an effective technopolitical means of overcoming many of the regulatory obstacles that the company faced in the world's largest market. The corporation's pioneering efforts to commercialize this technology would intersect and conflict with OBC's own ambitions to dominate the market for electric vehicle batteries.

CLIMATE CHANGE AND INDUSTRIAL POLICY

Japanese automakers were no more disposed to commercializing an all-battery electric vehicle than their US and European counterparts. However, they faced a host of challenges in the US market that caused them to consider approaches in advanced propulsion systems that were more radical than anything their competitors imagined for commercial production. As Japanese industry boomed through the 1980s, US lawmakers resorted to protectionism and currency manipulation. Beginning in 1981, a quota was levied on imports of Japanese automobiles, cutting their share of the market from 21 to 18 percent.[15] It became even harder for Japanese industry to compete after the industrialized nations signed the Plaza Accord of 1985 and devalued the US dollar, raising the price of Japanese imports in the US. Japanese manufacturers retrenched, offshoring operations and boosting efficiency, but Japanese automakers faced additional hurdles in the regulations governing fuel efficiency and regulations governing emissions in the US, something that they believed put them at a competitive disadvantage.

In response to Corporate Average Fuel Economy (CAFE), Japanese car companies built the world's most efficient fleets, and Toyota was among the leaders.[16] However, the end of the oil shocks of the 1970s and early 1980s meant that fuel efficiency was less of a pressing issue in US public policy by the late 1980s and early 1990s. With the return of cheap gasoline, Detroit exploited CAFE's light truck loophole and enjoyed a new golden age based on highly profitable large vehicles. When the ZEV mandate was factored into the new energy equation, Japanese industry came to believe its decade-long investment in fuel-efficient ICE propulsion was at risk.

This informed a distinct vision of environmental politics at Toyota. In 1989, US automakers and oil companies founded the Global Climate Coalition to debunk the science of climate change and lobby against the increasingly urgent calls to cut greenhouse gas emissions. German and Japanese companies took a different approach. They tacitly accepted the reality of climate change and discussed a range of technological means of ameliorating it.[17] However, the situation of German automakers was substantially different from their Japanese counterparts. German car companies were not under US trade sanctions, and they were unconcerned with meeting US fleet fuel efficiency rules. Indeed, Daimler-Benz and BMW had among the worst records for fuel efficiency of any automakers doing business in the US, and these companies were willing to pay the trifling fines incurred for selling relatively small numbers of gasoline-thirsty but lucrative luxury vehicles. Moreover, the relatively small market share of German car companies in California exempted them from the ZEV mandate.[18] Substantially immune from US technology-forcing controls, German companies showed an early willingness to experiment with risky and exotic technologies like hydrogen and fuel cell systems and would play an important role in promoting them as alternatives to all-battery electric propulsion.

In contrast, Toyota had both a significant industrial-technological investment in efficient ICE propulsion and its own mandate commitments. The company sought to reconcile these interests by emphasizing climate change over air quality and positing fuel efficiency as the solution. The argument was that as increasingly efficient vehicles burned less fuel, less carbon dioxide was produced, and less carbon dioxide in turn averted climate change. The logic was contentious from the perspective of environmental science, but it exploited an important gap in US pollution control regulations. Federal and state laws regulated the substances that constituted smog. But carbon dioxide was then unregulated, and while the global auto industry grudgingly accepted smog controls in the US market, it successfully lobbied against carbon controls throughout the 1990s. Toyota seized on the inconsistency, arguing that not regulating fuel efficiency was the same as not regulating carbon dioxide, the substance most responsible for climate change.[19]

This thinking grew out of the G21 committee, struck in September 1993 by Toyota's executive vice president of research and development Yoshiro Kimbara and so-called because it was charged with determining the parameters of Toyota's commercial advanced propulsion vehicle of the twenty-first century. By privileging fuel economy, G21 ruled out all-battery electric drive. To be efficient, the car would have to be small, but to be marketable, it had to have the same capabilities as an analogous ICE car, identified as the Corolla. The G21 team thought that boosting efficiency by 50 percent was a realistic goal that could be accomplished using existing direct injection ICE technology and an advanced transmission.[20] The group's subsequent deliberations were influenced also by the Clinton administration's decision to launch the PNGV, and very likely also by the emergence of GM-Ovonic in early 1994.[21]

In late 1994, Toyota executive vice president of development Akihiro Wada stipulated that the G21 car now had to double the efficiency of the best ICE technology. Now, only hybrid electric propulsion could meet all the G21 criteria. The car was named the Prius, derived from the Latin for "prior," in order to connote that commercialization would occur before the twenty-first century. In May 1995, the same month that the governors of Michigan, Ohio, Illinois, and Wisconsin urged California governor Pete Wilson to abolish the mandate lest it raise car prices, eliminate jobs, and disappoint consumers with premature technology, Toyota officially decided that the Prius would be a hybrid electric. The project's identity as a hybrid electric passenger car would remain a closely guarded secret until its line-off, the start of mass production of a new model, a few years later.[22] The company chose December 1998 as the production launch date.[23]

Toyota's new objective of doubling fuel efficiency was more modest than the PNGV target of tripling fuel efficiency, but the automaker was not necessarily taking the path of least engineering resistance. It planned the Prius as a series-parallel hybrid, the most complicated and sophisticated hybrid configuration. Parallel hybrids have an ICE and at least one electric motor, each of which can independently put the vehicle into motion. Toyota's system was designed to exploit the most efficient operating states of each power source, summoning one or the other, or both, depending on circumstances. For starting, the system engaged electric drive, and for high-speed cruising, it transitioned to ICE drive. When the user called for maximum power, the system engaged both power sources at once. Toyota's parallel hybrid could also operate as a series hybrid by using the ICE to drive a generator to recharge the battery or drive the traction motor itself as conditions warranted. The Prius also featured regenerative braking, a technology that used electromotive force to slow the vehicle and produce rotational force (torque) that converted the motor into a generator that recharged the battery.[24] From the perspective of industrial engineering, the Prius would be the most complex commercial passenger automobile yet built.

KEIRETSU, COMMODITY CELLS, AND THE COMMERCIAL ELECTRIC CAR

American environmental and energy regulations and the requirement for alternative propulsion automobiles that they implied imposed important changes in Toyota's material practices, as well as relations in its *keiretsu*, or corporate group. Where US automakers in the early 1990s still possessed internal auto parts empires, the *keiretsu* were loose confederations wherein core firms substantially outsourced from affiliates and suppliers who in principle could work for other core firms. Over time, this system fostered strong "relational skills" between the players and a certain level of inter-industry cooperation and standardization in product development among nominal competitors. Toyota had relations with and an investment in Daihatsu, Japan's leading producer of electric vehicles, mainly small delivery and utility types for the domestic market.[25]

The stakes were far larger for the electric vehicles that Toyota planned to develop for the US market. The automaker did not want its suppliers and affiliates to develop competency in the core electric vehicle technologies, so it sought to design and build most of them itself, both for Prius and for its compliance car, a converted RAV4 sport utility vehicle (SUV).[26] The exception was the battery, which was the responsibility of consumer electronics giant Matsushita Electric Industrial, owner of the Panasonic brand. The arrangement was not outsourcing as it was conventionally understood in the West, or even a *keiretsu* relationship strictly speaking, but an industrial alliance that would depend on intimate relational skill-building. In May 1992, Matsushita announced that it had developed the world's first nickel-metal hydride rechargeable built specifically for electric vehicles and tailored it to the requirements of the RAV4 EV. The vehicle's battery pack used twenty-four EV-95 modules of stacked flat cells, regarded as more suitable for electric traction than cylindrical commodity cells owing to their greater surface area and storage capacity.[27]

However, the power plant was far too big for the Prius. Toyota and Matsushita might have built a pack around a smaller number of EV-95 modules, but they opted for a different approach. Toyota knew that for every battery electric vehicle it produced, the battery supplier was among the first to get paid. As long as vehicle production was low, battery costs would be high. Conversely, the battery supplier had to worry about the automaker's intentions. Purpose-built batteries like the EV-95 could not make money if an automaker produced only a few compliance cars. A commercial hybrid, however, was a different proposition. In the spring of 1995, Yuichiro Fujii, the general manager of Toyota's Electric Vehicle Division, asked Matsushita if it could develop a hybrid battery by the end of the year. Matsushita said that it could, using commodity cells.[28]

The idea was not novel. Malcolm Currie, the former chair and chief executive officer (CEO) of Hughes, had tried to create an electric vehicle start-up around repurposed laptop computer cells, without success.[29] For Matsushita and Toyota, the commodity cell strategy effectively reconciled their interests. Matsushita already produced commodity cells for consumer electronics and they were proven, while the purpose-built electric vehicle modules were not. If commodity cells were used to build a battery pack and it turned out that Toyota wasn't serious or the hybrid didn't sell, there would be little loss. If the Prius succeeded, on the other hand, Matsushita would have a vast new market.

POLITICS OF PATENT MONOPOLY

Another compelling reason for this approach was OBC's patent position. In the early 1990s, Matsushita began producing nickel-metal hydride commodity cells, and when it marketed them in the US, OBC claimed its patents had been infringed. The parties worked out an agreement in December 1992 licensing Matsushita to use OBC battery patents limited to the field of consumer electronics. What most worried OBC was Matsushita's May 1992 claim to have invented the world's first nickel-metal hydride rechargeable for electric vehicles. OBC lawyers would argue that the claim was false but that the battery-maker still hoped to license its intellectual property relating to batteries for electric vehicles to Matsushita. Over the next two years, the companies engaged in periodic negotiations without resolving this question.[30]

Matsushita would later claim that the 1992 agreement actually allowed it to develop electric vehicle batteries. Its lawyers would argue that the agreement gave the company the right to license all patents owned by OBC and Energy Conversion Devices (ECD) covering inventions for small batteries conceived or developed prior to December 31, 1992. Matsushita also claimed that OBC and ECD agreed not to assert their rights on licensed patents they owned covering inventions conceived prior to May 4, 1992, relating to batteries of any size and for any application that also employed Matsushita's AB_5 metal hydride electrode.[31] In effect, Matsushita argued that it could use small batteries (its euphemism for commodity cells) licensed from OBC without restrictions on applications, and moreover that a battery using a mix of Matsushita and OBC compounds was a novel thing for which it could claim legal ownership.

Stanford Ovshinsky hoped that OBC and ECD's differences with Matsushita could be resolved. He maintained friendly relations with Shosuke Kawauchi, a senior managing director who would rise to become executive vice president of Matsushita's battery division.[32] When GM and OBC joined forces in 1994, however, the prospects of an alliance with Matsushita became much less likely. Matsushita and Toyota faced the possibility that GM might use OBC's patent position to block the importation of electric vehicles equipped with nickel-metal hydride batteries into the US. The RAV4 EV was vulnerable owing to its large, purpose-built battery, which potentially put Matsushita in violation of the 1992 agreement. Still, it suited the Japanese partnership to have this vehicle in the limelight. As a low-volume compliance car, the RAV4 EV was bound to lose money, but it had value as a means of testing the US reaction to Toyota and Matsushita's plans for nickel-metal hydride battery technology for electric vehicles in the US market, as well as distracting attention from the Prius.

THE PRIUS PRINCIPLE

Here was another important quality of the hybrid battery electric format. In the Prius, Toyota and Matsushita also sought to foreclose the emergence of an automotive version of the consumer electronics market for disposable primary batteries, something that OBC seemed to be aiming for. The all-battery electric car was a potentially open system. If a battery pack did not last the lifetime of the vehicle, it would have to be replaced, and battery makers would reap the rewards. This could not happen with the Prius because the battery pack was bundled into the propulsion system, so it could not easily be removed, and it was supposed to last the life of the vehicle. In short, the Prius was designed as a closed system. In its value chain, there could be no competitive battery market as such, only a long-term strategic relationship between two giant companies. The Prius could be interpreted as a type of electric car that aged like an ICE car and could therefore be marketed like one.

This was the Prius principle. Putting this principle into practice, however, proved a lot more complicated than suggested by the ostensibly off-the-shelf approach of using commodity cells. By the summer of 1995, Matsushita had built a new battery pack. It was nowhere near as energetic and powerful as the Prius platform required, but Fujii's engineers were preoccupied by another problem. They worried about how to integrate the battery with the ICE and the planetary gear. Until a new battery pack was ready, engineers used a temporary nickel-cadmium pack to resolve the systems issues.[33] By November, all the components were assembled into a test car.[34]

Still, Matsushita struggled to build new battery packs. When these finally became available for prototype cars in early 1996, a host of new engineering problems arose, especially overheating. As Toyota and Matsushita grappled with these problems, the automaker started selling the RAV4 EV equipped with Panasonic nickel-metal hydride battery packs to Japanese municipal customers in September 1996. This compliance car became the first road-legal vehicle to use such a battery, but Toyota was no less wary of encouraging impressionable US air quality regulators with demonstrations of all-battery electric technology than its US counterparts. The automaker chose to conduct the first evaluation of the RAV4 EV outside Japan not in the US, where there was intensive media scrutiny, but on the isolated island of Jersey, a self-governing British Crown dependency technically separate from the UK.[35]

Then, in late 1995, Toyota's newly installed president Hiroshi Okuda, backed by chairperson Shoichiro Toyoda, moved up the Prius's production date by one year, to December 1997.[36] It was a bold decision, given Matsushita's inability to deliver the promised battery pack on time, and it implied high confidence in the capacity of Toyota and its partner to manage technological complexity. Over the next year, however, troubles with the battery pack deepened, threatening to undo the new deadline. In a painful lesson of industrial cross-training, engineers discovered that battery and pack design and cell production dynamically interacted. Most automakers wired batteries of cells in series, connecting unlike terminals of each successive cell (anode to cathode), a more efficient arrangement than parallel wiring (anode to anode, cathode to cathode) and one that produced higher voltage. However, series-linked batteries are only as good as the least reliable cell, and consequently are more prone to failure. To maximize the lifetime of a battery pack, it was essential to equalize the state of charge of its cells, a task requiring a sophisticated battery management technology then in early stages of development.[37]

Sudden failures of series-wired battery packs were common in the early mandate years, and there were many failure modes.[38] Through 1996, Toyota was stumped by persistent bugs in the Prius battery pack, and it fell to Fujii, appointed executive vice president of Panasonic EV Energy, the partnership's joint battery production venture, to solve the mystery. In commodity cell-making, quality control occurred at the back end. Freshly made cells were aged, and defective ones slowly discharged over time. Sometimes, however, proofed cells failed.[39] If just one of the 240 cells in the Prius battery pack was bad, the hybrid system would not work. Fujii found that lint and metal dust at Matsushita's Tsujido plant were often the cause of cell failure. He instituted a new quality control regime to reduce contamination, and if the revamped line was not exactly at the standard of the semiconductor clean room, Toyota and Matsushita had moved it a step in that direction.[40] Even when the quality of individual cells was improved in production, however, cells still sometimes developed idiosyncratic operating characteristics in a battery pack environment depending on their placement in the array, causing thermal and electrochemical imbalances that reduced the reliability and lifetime of the pack.[41]

FUTURE SHOCK

Clinton had introduced his technology plan with a homily on making a friend of change, using planned obsolescence in Silicon Valley to analogize the role of government in the neoliberal era. In the computer industry, designers planned new products "knowing they'll be obsolete within twelve to eighteen months."[42] The president promised to apply this principle in cutting bureaucracy, encouraging initiative and diversity of opinion, and collaborating with the private sector.

For all their political appeal, the popular tropes of Silicon Valley had limits in making sense of and informing change in the automobile field. Hardly anything in the imagined world of electronics and information technology was an appropriate guide in developing the advanced propulsion automobile. Moreover, deep-rooted stereotypes clouded analyses of the changes then underway in the global industrial complex. Established automakers, especially the US ones, were often analogized with the ponderous government bureaucracies that Clinton hoped to reform. In the received wisdom of the day, the auto industry was inflexible, hierarchical, and technologically conservative. In contrast, high technology was believed to be agile, full of small, smart start-up enterprises with horizontal organization and no middle managers to impede the process of inventing and innovating. Such generalizations obscured complex realities. After all, GM had created the Impact, the world's most advanced road automobile, albeit with massive assistance from AeroVironment and Hughes.

But there were grains of truth in these views. Venture capital generally ignored alternative automobile enterprises in the 1990s, partly because of the long lead time to commercial scale, but mainly because capital was preoccupied with the information technology revolution, where vast fortunes were being made, increasingly in dot-com. What little experience there was of the business of electric cars by the mid-1990s suggested that it was a difficult way to make money. In 1995, US Electricar went bankrupt after two years of rapid expansion, sharing the fate of Sebring-Vanguard and Petro-Electric before it.[43]

Yet public policy compelled automakers to adapt. For Toyota and Matsushita, the hybrid battery electric car was, in part, a means of solving the regulatory conundrums posed by the US market. The Japanese partnership also had to contend with GM and OBC, an alliance that ostensibly represented the competition in the electric vehicle space. But the Impact was a substantially different sort of enterprise, one whose partners held incommensurable interests.

6 BOUNDING BATTERY RISK

Bringing an electric car to production is in many ways similar to bringing any car to production, but with a few key differences: an electric motor, lots of wiring, lots of computers and electronics, and lots of battery.

—Robert Stempel, December 1994

Robert Stempel had some reason to view the year 1994 with optimism. In 1992, he had suffered the humiliation of becoming the first General Motors (GM) chief executive to be forced out of the post since William C. Durant in 1920.[1] With his discovery of Stanford Ovshinsky and the Ovonic Battery Company (OBC), however, Stempel found renewed purpose. In his much-diminished capacity as a GM consultant, Stempel agitated behind the scenes to keep the Impact alive and saw OBC as the means to achieve this. He helped arrange the secret test of the Ovonic Impact in January 1994. In March, he helped broker the deal that revived the Impact program and resulted in the creation of GM-Ovonic, the joint venture that was to manufacture some of the Impact's batteries. While GM had relegated the Impact program to test status, it had not terminated it, either, suggesting the automaker was intent on further developing the technology of the all-battery electric car.[2]

Stempel became a senior technical advisor of GM's electric vehicle program. With this nominal promotion, the former executive became an evangelist, joining Ovshinsky in preaching the virtues of the nickel-metal hydride rechargeable in the all-battery electric car. Stempel had earned respect in electric auto enthusiast circles and lent his still-considerable prestige to the cause. He was among the speakers at the twelfth Electric Vehicle Symposium in Anaheim in December, where he mingled with celebrities including the actor-environmentalist Ed Begley, Jr. Also present was John W. Adams, appointed president and chief executive officer of GM-Ovonic in September, who announced that the joint venture would deliver its first nickel-metal hydride batteries in one year.[3]

But Stempel also had doubts. He was not sure that Delco Propulsion Systems (DPS), the new division responsible for GM's electric vehicle technologies, properly appreciated OBC's battery. In a November letter to Ovshinsky, Stempel wrote that he had urged the division to promote the technology "for what it is: a life-of-the-car, high power battery."[4] However, DPS was responsible for managing a range of technologies including controllers, chargers, and motors, as well as batteries, both lead-acid and nickel-metal hydride, and although its purview included hardware developed for Impact, the division had a broader mission. DPS was designed to coordinate research and development relating to electric vehicles at AC Delco Systems (the former Delco Remy, so renamed in early 1994), Delco Electronics, and Allison Transmission with a view to selling core industrial content to other automakers, as well as companies in the conversion market.[5]

DPS was the product of a protracted reorganization of GM's parts empire that unfolded through the 1990s, initiated by executives who were uncertain and apprehensive about the future industrial-technological landscape in the ZEV mandate era and were also under pressure to reassess vertically integrated operations at a time when capital increasingly favored divestiture and cost-cutting. The division was formed in September 1994 as a part of GM's Automotive Components Group (ACG), itself created in 1991 to consolidate myriad parts operations at some 200 plants, and it was managed by J. Byron McCormick, the engineer who had been impressed by Ballard's fuel cell test in 1987 and who had since become a proponent of advanced propulsion technologies. The mission of DPS of supplying an emerging market in electric vehicles that GM was unwilling to directly stimulate with its own commercial-scale electric vehicles sent mixed signals that Ovshinsky and Stempel struggled to interpret. Having triggered unanticipated regulatory and industrial-technological change with the Impact, GM moved tentatively across unfamiliar terrain. Impact was then in a preproduction phase managed by Robert Purcell at AC Delco Systems, which was absorbed along with the rest of ACG into Delphi Automotive Systems in 1995. A close advisor of Jack Smith, Purcell wanted the first vehicles ready by the spring of 1995 and Adams predicted that GM-Ovonic could deliver the first nickel-metal hydride batteries by the end of that year.[6]

Adams's task, amidst this tumult, was to translate and ultimately manufacture OBC technology. From OBC's perspective, the first-generation Impact rechargeable was ready for production. Over time, however, Adams grew less confident, and Stempel and Ovshinsky chafed at the delays. In the meantime, OBC tried to foster other industrial alliances, part of a series of maneuvers launched by all the parties with a stake in the nickel-metal hydride battery to jockey for advantage. While GM and OBC were united in their wish to prevent Toyota and Matsushita from using the technology in the US market, OBC was not prepared to rule out cooperation with anyone and held out the hope that some arrangement could yet be made with the Japanese partnership. For their part, Toyota and Matsushita needed information on the nature of OBC's patent claims and GM's intentions in the nascent electric vehicle market. As the talks unfolded through 1995 and into 1996, they precipitated a legal clash between OBC, Matsushita, and Toyota that revealed the collective nature of innovation in this context and the difficulty of drawing boundaries around parochial commercial interests.

COST CONUNDRUMS

From its inception, the GM-Ovonic partnership was philosophically divided on the purpose of the joint venture regarding battery cost and industrial innovation more broadly. OBC believed that the primary means of meeting cost targets was manufacturing at scale, while GM saw cheap battery power as an outcome of research and development.[7] GM enlisted the staff of GM-Ovonic from AC Delco Systems, where Adams had served as chief engineer of manufacturing and had begun his career working with auxiliary lead-acid rechargeables for internal combustion engine (ICE) vehicles. This experience would prove of limited value in the Impact. Cells for auxiliary batteries were produced by batch processing, a start-and-stop method that tended to induce variations in cell quality. While such variations did not matter in auxiliary batteries for ICE cars, they mattered a great deal in large series-wired battery packs, whose cells had to work with 100 percent reliability. The best way to reduce cell variation was through continuous processing, a sophisticated and expensive industrial enterprise akin to process chemistry.[8]

But Adams planned to move cell production to continuous processing only after OBC met strict cost and performance targets. As the parties worked out the terms through 1994, what emerged was a to-do list not so much for GM-Ovonic, but for OBC. GM-Ovonic had only six full-time employees. On the other hand, OBC was responsible for all battery research, development, design, and engineering in order to meet ambitious performance goals set not by GM, but by the USABC.[9] The advanced battery consortium wanted the cost of battery power cut to $150 per kilowatt-hour at a time when OBC was charging around $6,000 per kilowatt-hour for prototype batteries, amounting to well over $100,000 per battery pack. OBC also had to boost the energy density of its cells from around 70 watt-hours per kilogram to the USABC target of around 100 watt-hours per kilogram.[10] And for the time being, OBC could not count on an economy of scale to accomplish this. Its materials science researchers would have to make breakthroughs at the company's Maplelawn sample build facility in Troy, and there was no guarantee that the USABC would be a consistent research patron. The consortium provided OBC with $18.5 million in April 1992 and another $5.5 million in August 1994, but it was much less generous thereafter.[11]

Integrating the battery into the Impact's electric drivetrain induced complications that slowed the translation process and lent validity to the argument of automakers that the terms of the mandate had to correspond with industrial-technological realities. Circumstances decreed that for the Impact, OBC and GM-Ovonic would have to move from prototype to preproduction batteries while simultaneously confronting systems problems arising in the integrated drive system. Continued testing revealed many problems at the systems level. The Impact's famous first run on an OBC pack occurred in a dry and cool Arizona winter, but it turned out that nickel-metal hydride chemistry was sensitive to warm weather. The pack heated up when it charged, and, in warm ambient conditions, when it discharged. Much of the trouble stemmed from poor cooling design. The Impact's T-shaped battery pack did not circulate air well, an oversight that was fixed relatively easily when GM engineers installed a blower. A more serious problem was that charged packs lost charge over time. Charge depletion was a materials issue, and OBC was still modifying the battery chemistry even as it prepared to ramp up assembly.[12]

These problems persisted into early 1995, causing cost targets to be lost in a welter of trade-offs. Issues with charge depletion prevented OBC from freezing battery chemistry, and without a standardized chemistry, it was impossible to fully understand cycle life. In turn, not knowing how a battery pack aged complicated the cost equation. By mid-1995, GM believed that it would be difficult for OBC's first batteries to achieve even $500 per kilowatt-hour, meaning that a 30 kilowatt-hour pack could be worth a princely $15,000. Automated production was predicted to cut that figure, but in 1995, production was still labor-intensive.[13]

Adams's spring 1995 deadline could not be met, yet GM had little interest in accelerating the pace of research and development. Over the course of the year, the car companies grew apprehensive as deepening support for the ZEV mandate in the northeastern states presented the prospect of major engineering overhauls to outfit compliance cars for the arduous demands of northern winters. Automakers lobbied fiercely to revise the mandate timeline, arguing that they could not deliver a suitable commercial electric vehicle battery before 2000. In the fall, the California Air Resources Board (CARB) convened a battery panel and negotiations began. Counterintuitively, car companies wanted to introduce Zero Emission Vehicles (ZEVs) ahead of schedule in exchange for reduced quotas. They wanted to build a total of 5,000 cars by late 1996 and early 1997 (well before the 1998 deadline) and 14,000 instead of 22,000 vehicles in 1998. Having long warned of the risks of rolling out an unfinished product, the auto industry's new position implied that it had solved its technological problems.[14]

But it had not. OBC developed an alloy that helped resolve the temperature issue, and air quality regulators came out of the panel under the impression that GM-Ovonic was closer than anyone else to manufacturing nickel-metal hydride batteries. On the cost question, however, GM-Ovonic was as divided as ever, and the auto industry's proposal to CARB only widened the divide. OBC president Subhash Dhar argued that an order for 20,000 packs could drive costs down first to $250 per kilowatt-hour and then as low as $180 per kilowatt-hour within a few years.[15] GM-Ovonic's goal for initial production capacity was 250 cells per day, the equivalent of one complete battery pack. In 1995, however, the joint venture was operating on a sample build basis, yielding only 7 cells per day, and it did not plan to reach 200 cells per day until mid-1997.[16]

MODERATING THE MANDATE

As Adams revised the GM-Ovonic timeline, air quality regulators made the first of a series of concessions to the car companies. Under John D. Dunlap, III, who in 1994 replaced Jaqueline E. Schafer as chair of the air resources board (Schafer replaced Jananne Sharpless in 1993), CARB modified the Low Emission Vehicle (LEV) regulations in March 1996. California eliminated the zero-emission percentage requirements from 1998 through 2002 and worked out memoranda of agreement with each of the seven largest automakers, committing them to produce up to a total of 3,750 vehicles for marketplace demonstration programs over the calendar years 1998, 1999, and 2000. Importantly, regulators began to make explicit technological recommendations. They gave multiple credits to car companies for using advanced power sources in electrics, identifying the nickel-metal hydride, sodium-nickel chloride, and lithium ion and polymer rechargeable batteries as the most promising candidates. In exchange, the automakers would have to meet stricter emissions standards and agree to continue the research and development of ZEVs. The 10 percent quota for 2003 remained unchanged.[17]

GM was the chief architect and beneficiary of the arrangement. On January 4, 1996, Jack Smith introduced the EV1, the production version of the Impact, at the Los Angeles Auto Show. It was first GM-branded product in history and was to be marketed by the corporation's Saturn division in twenty-four locations in the cities of Los Angeles, San Diego, Phoenix, and Tucson starting in the fall. However, it was not the OBC-enabled supercar that Ovshinsky, Stempel, and the enthusiasts had hoped for. GM decided to delay deploying the nickel-metal hydride battery until at least 1997. The Generation 1 version of the EV1 was instead equipped with a Delphi lead-acid pack that gave a range of between seventy to ninety miles, and GM said it would sell the car for around $35,000. The EV1 was to be part of GM Advanced Technology Vehicles (GMATV), a rebranding of GM's electric vehicle program. The GMATV program would be part of Kenneth Baker's research and development division in Troy and Robert Purcell would remain in charge of the overall effort. Almost lost in the media hype was Smith's announcement that GM would also launch a converted electric pickup truck for fleet use by utilities and government agencies in 1997.[18] A week later, at a Phoenix meeting of the Edison Electric Institute, an organization that represented investor-owned electric companies, Toyota announced that it would deliver seven converted RAV4 EVs equipped with lead-acid batteries for evaluation and trials with Southern California Edison.[19]

INDUSTRIAL TRADECRAFT

GM reaped a public relations dividend from early deployment. The company claimed credit as the first major automaker since World War I to market an electric car, a vehicle more technologically advanced in every respect than the RAV4 EV, its nominal competitor. Even better from GM's perspective was that the automaker did not have to produce large numbers of EV1s. Media coverage was equivocal. The *Los Angeles Times* correspondent Donald Nauss credited GM for its historic achievement but also chastised it for a "confusing public display of corporate schizophrenia."[20] Harry J. Pearce, the architect of the automaker's environmental policy and electric vehicle program, was promoted to vice chair and elected to the board of directors.[21]

Out of the spotlight, not all was well with the joint venture between GM and OBC. In December 1995, Stempel was appointed chair and executive director of OBC's parent Energy Conversion Devices (ECD) and as he took up these posts early in the new year, he reviewed the situation with Adams. GM-Ovonic needed $20 million to get to the point of production, and Delphi Automotive was getting cold feet. Stempel expected that the auto parts supplier would handle the integration, manufacturing, marketing, and sale of the OBC battery pack, but the company claimed that it was barely breaking even on sales of $6 billion and believed that electric vehicles were a money-losing proposition. Adams and Stempel discussed the possibility that Hughes or an outside investor like Honda or Peugeot might be interested in replacing Delphi.[22]

Another option was Matsushita. Over the course of 1995, OBC and the electronics giant quietly discussed the possibility of cooperating in manufacturing batteries. One obstacle was that the two companies used different metal hydride alloy chemistries with different industrial-technological implications. Matsushita was committed to AB_5 while OBC used AB_2, a vanadium-rich compound selected because it enabled the maximum possible hydrogen storage capacity. Vanadium is a relatively rare element that at that time was not expensive in the volume used in cells for consumer electronics. When employed in cells for electric vehicles, however, the volume of vanadium could potentially be much larger, implying significantly different economies at commercial scale.[23] Adams and Stempel believed that Matsushita would insist that any production collaboration had to utilize its AB_5 formula.[24] Such cooperation was unlikely because it suggested OBC would have to abandon not only its sunk investment in AB_2 research, development, and production but its associated patent claims to nickel metal hydride technology as well.

But Matsushita and Toyota had other reasons for engaging OBC. They needed information on the nature of OBC's patent claims and GM's intentions for electric vehicles, and they used the RAV4 EV to start conversations around these questions. Ovshinsky and Stempel hoped that these conversations meant that a production deal might yet be worked out with the Japanese alliance. GM encouraged the negotiations, likely to surveil the opposition. As the talks unfolded through 1995 and into 1996, they precipitated a series of legal actions that turned on the difference between a consumer electronics battery and an electric vehicle battery, a science controversy that in turn laid bare the US partnership's patent strategy to control nickel-metal hydride chemistry.

In the version of events told by OBC's lawyers, Matsushita informed OBC in February 1995 of its plans to develop the RAV4 EV battery, ostensibly to sort out licensing terms. In a May meeting in Tokyo, OBC averred that it had said it would license production for the US market, but only through GM-Ovonic, "consistent with its commitments to GM." It also claimed that Matsushita said it preferred to join GM-Ovonic outright. Talks were inconclusive. Then, in August, OBC learned that Matsushita had approached Ford, and negotiations between OBC and Matsushita broke down. They resumed in the fall at the request of GM, according to the OBC lawyers.[25]

However, when Matsushita's lawyers asked OBC in the summer of 1995 to specify the patents that Matsushita would be violating if its nickel-metal hydride batteries for electric cars found their way to US shores, there was silence.[26] At least, that was Matsushita's side of the story. With no answer from OBC, Toyota and Matsushita made a bold move. On January 22, 1996, Toyota announced that the demonstration of the RAV4 EV with Southern California Edison would employ not lead-acid but rather nickel-metal hydride rechargeables. Suddenly, the RAV4 EV had become a means of testing OBC's claim of having the exclusive right to use such batteries for electric vehicles in the US. Ovshinsky learned of the switch through Japanese media and reproached Shosuke Kawauchi in a February 5 letter. He was still interested in collaboration, he wrote, but licensing issues would have to be settled first. Until OBC and Matsushita sorted their differences, Ovshinsky wanted Toyota to delay its test.[27]

But Matsushita wanted to know the legal basis for denying its electric vehicle batteries access to the US market. On February 26, Chester Kamin, Ovshinsky's personal attorney, provided the answer. Matsushita was infringing on US Patent No. 5,348,822, or the "822." Unless Matsushita gave written assurance that it would stop importing such batteries into the US, Kamin warned, OBC would take legal action.[28] Days later, Matsushita filed to invalidate 822, moving to preempt what it claimed was an imminent lawsuit from OBC. Shortly thereafter, Toyota and its US affiliate joined as plaintiffs.[29] OBC's countersuit against Matsushita and Toyota laid bare the scope of the US company's claims. Patent 822 described a cathode made of nickel hydroxide and other materials informed by the same principle of compositional and structural disorder that informed the hydride materials comprising OBC's anode. Kamin noted that the patent had been filed on March 8, 1993, and was therefore not protected by the 1992 agreement.[30] OBC did not possess a Matsushita electric vehicle battery, had no direct evidence of infringement, and so had to work by inference. Matsushita product literature indicated that the performance characteristics of the company's consumer batteries and electric vehicle batteries were similar, suggesting, held OBC's legal team, that the batteries were made of similar materials. OBC also analyzed the cathode of a Matsushita consumer battery in a test that it claimed indicated the presence of substances encompassed by the 822 patents.[31]

For its part, Matsushita did not deny that it had infringed 822. Instead, it called attention to the hydride materials comprising the anode, and to its AB_5 alloy in particular. Matsushita's lawyers claimed that the company had been working on this substance with a view to applying it in electric vehicle batteries as early as 1990. Matsushita believed that the 1992 agreement allowed it to market a nickel-metal hydride battery of any size and for any application if it contained AB_5, a claim that OBC did not directly contest. Matsushita lawyers were also concerned about OBC's long delay in identifying 822. They were curious about what Kamin had meant in his February 26 letter when he indicated that 822 was only one of a number of "other violations" of OBC's rights. Matsushita suspected that the battery company had other patents that it was in no hurry to identify but that it might later try to assert rights to.[32] In effect, Matsushita accused OBC of being a patent troll.

SCIENCE AND THE FACTORY FLOOR

With no direct evidence of infringement, OBC lawyers made a claim for intellectual ownership of the technology of the nickel-metal hydride battery that was rooted in the science of its materials. In essence, it was an argument for linear innovation framed as a story of David and Goliath narrated in nationalist rhetoric. OBC, a small US company supported by the federal government, had conceived the fundamental technology of disordered materials to make the world's first commercially viable nickel-metal hydride rechargeable battery. But giant foreign corporations planned by "deceit" to "pre-emptively attack" OBC's rightful market. This conspiracy to "infiltrate" infringing batteries into the US without a license, held OBC lawyers, would cause irreparable harm to OBC, the only entity legally authorized to license or produce nickel-metal hydride battery technology for the electric vehicle market.[33]

As in other fields of science-based industry, however, innovation in this context was nonlinear. There was constant traffic in ideas from the battery laboratory to the factory floor and back again. OBC did not monopolize knowledge of industrial manufacturing and electric vehicle systems, all of which were pertinent to production-grade nickel-metal hydride chemistry. In these areas, Matsushita and Toyota possessed a great deal of expertise.[34] In 1996, OBC was still transitioning from materials research to component production and had not yet begun the process of scaling the nickel hydroxide cathode materials covered by 822 and incorporating them in complete cell designs.[35] OBC also struggled to master quality control in materials that it had moved to production, encountering difficulty with controlling the thickness and weight variability of sheets of negative electrode that it produced for licensees of its consumer cells.[36]

Moreover, the market for electric vehicles as OBC imagined it did not exist in 1996. Only weeks before OBC lawyers filed suit, the car companies compelled CARB to delay the mandate. The US industrial base for electric vehicle batteries was just beginning to be built up and progress was slow. In the 822 affair, Matsushita and Toyota accused OBC of obstructing research aimed at understanding electric cars in real-world conditions. OBC was not facing the prospect of imminent widespread use of an infringing battery, argued Matsushita and Toyota, and its opposition to the RAV4 EV pilot study risked delaying the introduction of electric cars in the US.[37] This claim could be interpreted as self-serving, but in fact no one party held all the technological cards, as Stempel suggested in a note to the media: "This was a difficult action for ECD to take since we are committed to support the developing electric vehicle market, and Toyota is a potential customer for some of our products."[38]

It was because of this interdependence that Matsushita and Toyota had initiated their lawsuit. They sought to loosen OBC's patent grip, but the stakes were much larger than the unprofitable RAV4 EV. Legal action kept the attention focused squarely on Toyota's compliance car while the Prius remained in the shadows. The lawsuit also allowed Matsushita to test reactions to its argument that its consumer and electric car batteries were essentially the same, a circumstance it and Toyota would have to address when the Prius was introduced in the US market and the nature of its battery pack became known. Perhaps the most important effect of the lawsuit was surveillance. Matsushita's suit and OBC's countersuit enabled the Japanese partnership to probe the operations of the US partnership and its intentions for the EV1.[39]

These legal actions cast into relief the ambitions of the parties in the emerging electric vehicle market, the potential conflict of interest between automaking and battery making, and the balance of power in collaborative research and development in this context. OBC lawyers claimed that the intimate synergy between power source and electric drivetrain technologies meant that it was imperative that replacement batteries be the same brand as the original equipment in a battery electric car lest users incur significant "switching costs" by using replacement batteries from other sources. Matsushita was the world's largest battery manufacturer, they noted, and if the corporation took a leading position in the electric vehicle battery market, its advantage could be insuperable.[40] In principle, the same logic applied to OBC. If it had not been apparent before, the 822 affair suggested that if automakers wanted to use the nickel-metal hydride rechargeable, they would have to deal with OBC.

HEDGING WITH HONDA

Ford and Chrysler had helped promote OBC through the USABC, but these automakers had no intention of deepening their involvement with the battery company if they could help it.[41] OBC did have other prospects besides the Big Three, Toyota, and Matsushita. Honda had been the first automaker to collaborate with OBC, as part of a project to develop an electric car for the US market that began in 1988, likely inspired by GM's Sunraycer. Honda's approach differed significantly from its competitors. The company planned a purpose-built all-battery electric car on a subcompact platform that could be ready for large-scale production if conditions warranted, and built and tested a series of prototypes through the 1990s.[42] OBC's relationship with Honda took on a different complexion with the formation of the Toyota-Matsushita and GM-Ovonic alliances. By mid-decade, it was not clear whether multiple nickel-metal hydride formats or a standard format would be commercialized for electric vehicles so Honda hedged its bets, testing Matsushita batteries while taking care to preserve relations with OBC.[43] Initially, the automaker seemed to favor the US company. In January 1996, Honda tested its first EV electric car, and in March, it purchased a 3 percent stake in OBC, an investment that Stempel believed made the battery company worth $300 million.[44]

OBC welcomed the deepening relationship with Honda but it posed fresh complications. The car company wanted a cell with greater capacity than OBC's standard GM cell module, which meant that OBC would have to add two more cells to its existing module can. More troublesome was Honda's suggestion that it would like the cells to conform to dimensions stipulated by the Society of Automotive Engineers (SAE), the professional and standards-making association of the automobile engineering establishment. The SAE standard electrode had a larger surface area than the OBC version, and conforming to this standard required the redesign not only of OBC's module can, cover, and lid, but of the electrode belt of the process roll itself.[45] Satisfying these requirements added to costs that were already high thanks to the chicken-and-egg production conundrum with GM. In contrast, Matsushita had already adopted the SAE standard, and the company's manufacturing connections with Toyota gave it a cost advantage. These factors ultimately led Honda to equip its EV Plus car with Matsushita batteries.[46]

7 FUEL CELLS, HYDROGEN, AND ENVIRONMENTAL POLITICS

Fuel cells, in my opinion, are one of the most promising long-term technologies.
—Alex J. Trotman, Ford chief executive officer, October 1997

The struggle for control of nickel-metal hydride electric vehicle battery technology was an important front in the technopolitics of green automobility, but far from the only one. As the 1990s progressed, fuel cells became increasingly prominent in the discourse of sustainable transportation. The Clinton administration favored fuel cells as a power source in the advanced hybrid configurations under study by the Partnership for a New Generation of Vehicles (PNGV). Partly this was due to the activism of two politically well-connected physicists in the early 1990s.[1] For Henry Kelly, a senior associate at the US Congress's Office of Technology Assessment, and Robert H. Williams, a senior research scientist at Princeton University's Center for Energy and Environmental Studies, energy independence and clean air were vital national interests that could not be enabled by internal combustion engine (ICE) technology. But an electric car powered by a fuel cell, they argued in an influential policy paper, would constitute a green vehicle that was as easy to refuel and as convenient to drive as an ICE automobile. At commercial scale, wrote Kelly and Williams, fuel cell electric propulsion would move energy conversion beyond the "age of fire" into the "age of electrochemistry." The technology, they argued, would allow the US to continue to use fossil fuels, but in a clean and efficient manner that would wean the country from oil imports in the process.[2] In February 1993, Kelly joined the

White House's Office of Science and Technology Policy (OSTP) as its assistant director for technology, where he became a leading architect of the PNGV.[3]

Policymakers began paying more attention to fuel cells at a time when the technology was being popularized by Daimler and Ballard Power Systems, the small Canadian engineering research firm whose proton exchange membrane design produced enough power give an electric vehicle performance comparable to the battery electric vehicles of the day. As always in the fuel cell field, the question of commercial feasibility turned largely on the question of fuel. Like all fuel cells, Ballard's membrane fuel cell worked best on preprocessed pure hydrogen, which yielded water as the so-called waste product. Storing pure hydrogen onboard an automobile poses certain engineering challenges because diatomic hydrogen diffuses easily through conventional steel. Since the 1970s, however, progress had been made in technologies of hydrogen storage, including metal hydrides. Moreover, in Daimler, Ballard had an industrial partner with the resources to develop or acquire all the other components of a hydrogen fuel cell electric car and the expertise to integrate them at commercial scale.

The much more difficult problem was how to supply the hydrogen. Proponents of the fuel cell electric vehicle had long hoped for development of fuel cells capable of operating on hydrogenous carbonaceous fuels, either directly, or indirectly, following the processing of these substances into a hydrogen-rich fuel gas. In the 1960s, large corporations including General Electric (GE), Esso, and Shell did preliminary research on carbonaceous fuel cells, but the engineering challenges of integrating such technology into an electric drivetrain were so great that such work was largely abandoned by the early 1970s. Most subsequent research in carbonaceous fuel cells focused on stationary units in the utility generation role.

The alternative was to produce large volumes of pure hydrogen, an approach that implied construction of a hydrogen economy. Futurists had been proposing variants of this sociotechnical regime since the 1960s but these schemes received industrial and governmental support only at the level of studies and white papers. The energy crisis of the 1970s had spurred modest research in hydrogen storage and fuel systems applied in fuel cell electric and converted ICE vehicles for demonstration purposes. With the return of cheap petroleum in the 1980s, hydrogen receded from the headlines.

In the environmentally conscious 1990s, however, the idea of a hydrogen economy gained new currency and influential adherents thanks to fuel cell futurism. As fuel cell advocacy burgeoned in this period, thanks in no small measure to Daimler and Ballard, it became associated with and breathed life into the hydrogen economy imaginary at a time when automakers were under increasing regulatory pressure. While the car companies had convinced the California Air Resources Board (CARB) to weaken the initial Zero Emission Vehicle (ZEV) mandate quotas, they still faced the 10 percent quota for 2003. There was also a chance, however remote, that carbon dioxide might become a regulated emission on a global scale if the industrial and industrializing nations were able to negotiate the treaty then being prepared for the Third Conference of the Parties to the United Nations Framework Convention on Climate Change (UNFCCC) in Kyoto, Japan, in December 1997. For car and oil companies eager to strike favorable bargains with environmental regulators, the hydrogen economy and the fuel cell electric car as the ultimate ZEV would become important promissory bargaining chips in the technopolitics of green energy and automobility.

THE BEST OR NOTHING

Like most automakers, Daimler was conservative in its choice of commercial propulsion systems but sought to cultivate an image of engineering virtuosity, in part by acquiring aviation and aerospace assets.[4] The ethos of technological progress as expressed in the motto of Daimler's Mercedes division ("the best or nothing") also informed the company's practices of researching and developing advanced propulsion technology. From the early 1970s, the automaker experimented with hydrogen ICE propulsion and metal hydride storage and, from the mid-1980s, with the molten salt battery. The sodium-metal chloride battery impressed CARB as a promising power source, and in the early 1990s, Daimler began demonstrating electric conversions of Mercedes production models equipped with this device at high-profile events, including the New York City Marathon of 1991 and the 1992 Olympic Games in Barcelona.[5]

As other automakers were discovering, however, commercializing advanced rechargeable batteries was an exceedingly difficult task. The sodium-nickel chloride battery represented a considerable improvement over sodium-sulfur battery technology, but it still had the intrinsic shortcomings of all high-temperature power sources. The device had to be heavily insulated, typically with stainless steel, and supplied with sophisticated thermal management systems to maintain its heat and keep its electrodes molten, all of which added weight and complexity.[6] Initial versions were expensive and not especially durable and gave an electric vehicle a modest range of around sixty miles.[7]

Daimler continued testing electrics equipped with sodium-nickel chloride batteries through the 1990s. Gradually, however, the automaker began to shift emphasis to the low-temperature membrane fuel cell. Daimler engineers were impressed by Ballard's Mark 5 stack, which had nearly thirty times the power density of the fuel cell used in the *Gemini* spacecraft.[8] In 1989, one year after concluding a demonstration of hydrogen ICE propulsion in Berlin, Daimler leased a Mark 5 for trials and in 1993 signed a four-year deal with Ballard to build a fuel cell electric demonstration vehicle. The partners would go on to build several such vehicles, known as NextCars or Necars, and use them to promote fuel cell power. The first of them appeared in April 1994 and resembled the 1960s-era Electrovan from General Motors (GM). Like its predecessor, the Necar I was based on a van chassis, and its cargo space was similarly occupied by propulsion equipment, indicating its function as a test bed. However, the Necar I was more powerful and polished than the Electrovan and also had the potential to use methanol, or so Daimler claimed.[9]

The mention of methanol suggested a major advance in capability, and the media began paying attention to the newsworthy novelty of an electric car that did not use a battery. The *Wall Street Journal* reporter Oscar Suris hailed the Necar I as a breakthrough. To be sure, there were skeptics. Suris quoted Bradford Bates, Ford's chief of alternative power sources, who held that there was a "tremendously long list of engineering issues" to be addressed in packaging fuel cells in automobile drivetrains. Nevertheless, Suris reported that Daimler's demonstration bolstered voices calling for expanding fuel cell research in the PNGV, and suggested that Detroit and Washington might have to respond, lest they relinquish "a potential technological edge" to a foreign competitor.[10] Indeed, the consortium launched new work on fuel cell systems shortly thereafter. Three months after the appearance of Necar I, the US Department of Energy (DOE) awarded contracts to Ford and Chrysler to develop hydrogen fuel cells and supported GM in a project to develop a methanol fuel cell.[11]

By December 1995, some California air quality officials believed that fuel cell electric propulsion was becoming a viable alternative to battery electric propulsion.[12] By then, Ballard had boosted fuel cell power density to around 570 watts per liter, well above the PNGV benchmark of 400 watts per liter, using hydrogen fuel.[13] In May 1996, Daimler-Benz displayed another fuel cell electric van in Berlin, now with space for six passengers, with the press lauding the vehicle as a technical and aesthetic advance.[14] Environmental groups were starting to pay attention. In August, the *Washington Post* published a letter by Jason Mark, a transportation analyst with the Union of Concerned Scientists (UCS), praising Daimler's efforts to develop the fuel cell electric vehicle as a viable solution for both smog and climate change.[15]

COUNTDOWN TO KYOTO

Daimler and Ballard's leadership in fuel cell electric technology intersected with Ford's efforts to cope in a global marketplace that was increasingly being defined and delimited by environmental regulation. Then the world's second-largest automaker, Ford had proportionately large ZEV commitments, yet it was among the least prepared of the major car companies to deliver its mandated quotas. The automaker had in effect ceded the race for the most advanced compliance car to GM and Toyota by committing to a problematic advanced battery that foreclosed the possibility of near-term deployment. Ford's Ecostar was a converted European Escort delivery van equipped with a sodium-sulfur battery, a device that had helped spark a revolution in the science of solid-state ionics but proved a failure as a practical power source. Ecostar's battery was so plagued by chronic overheating and fires that Ford withdrew the vehicle from public road testing in 1994.[16]

These problems gave Ford an excuse to delay action on manufacturing electric cars but by 1997, the company was in a quandary. It had no credibly green vehicle in commercial development at a time when the world's industrialized and industrializing nations were finalizing the terms of the greenhouse gas emissions treaty due to be signed in Kyoto in December. While Ford was part of the PNGV, progress in collaborative research and development was slow. The automaker's most important initiative in green car technology was the alternative fuel ICE vehicle. Ford had six types of natural gas-fueled vehicle in its 1998 model year and claimed it was planning to build hundreds of thousands of vehicles capable of operating on ethanol mixtures, but that effort was constrained by supply, infrastructure, and environmental issues. Toyota increased the pressure on its competitors when it announced in March that it would introduce a commercial hybrid electric vehicle at the end of 1997.[17]

The Japanese automaker had suddenly positioned itself to take leadership in the green automobile space. Shortly thereafter, Daimler and Ballard deepened their relationship, and it was in this context that Ford began to pay closer attention to events in the fuel cell field. On April 15, Daimler purchased 25 percent of Ballard and the two companies created a joint venture that in some ways resembled GM-Ovonic. The enterprise aimed to develop and produce fuel cell electric propulsion units, not complete fuel cell electric vehicles, and sell them to other automakers, as well as license their manufacture. Ferdinand Panik, a Daimler-Benz senior vice president, remarked that the joint venture was "open for business" with other automakers and hoped to produce 100,000 units, with the ambitious goal of making the ICE obsolete between 2005 and 2010.[18]

Ballard and mandate-exempt Daimler seemed to be positioning themselves to dominate the supply of a core component of an alternative ZEV for the US market. One week later, on Earth Day, Ford announced that it would build a prototype hydrogen fuel cell electric car with the support of the DOE under the auspices of the PNGV, for evaluation by 2000. Some environmentalists welcomed the move. Jason Mark opined that the fuel cell was "perhaps the most promising" sustainable power source.[19]

The fuel cell turn coincided with intensification of the campaign by US automakers to prevent policymakers from defining carbon dioxide as a pollutant. In July, this effort culminated in the Senate's unanimous passage of the Byrd-Hagel resolution, a measure that barred the US from signing the climate change treaty. The Clinton administration was blocked from ratifying the Kyoto Protocol, but Detroit continued to lobby Washington, now with a view to leveling the playing field in the environmentally regulated market.[20]

In October, Ford chair and CEO Alex J. Trotman delivered a speech at the National Press Club warning that iniquitous emission controls at the domestic and global levels risked eroding US industrial competitiveness. Trotman noted that he disagreed not so much with climate change science as with climate change policy. He criticized the mandate without mentioning it by name and made the familiar plea to allow market forces to determine which technologies the industry adopted. If the regulatory state insisting on intervening, Trotman remarked, it should help and not hurt industry. In the short term, argued the executive, the federal government could facilitate the greening of personal transportation by purchasing Ford alternative fuel vehicles for its own fleet and helping to set up the requisite fuel infrastructure. The solution in the long term, held Trotman, was to adopt advanced technology. He mentioned Ford's hydrogen fuel cell project, adding that he saw the fuel cell as the one of the most important technologies of the future.[21]

In December, a few days after the adoption of the Kyoto Protocol, Ford made a seemingly major commitment to fuel cell technology. In a ceremony in Stuttgart, Ford joined the Daimler and Ballard alliance, committing $420 million and taking a 15 percent stake in Ballard. The partners described their relationship as "technology-sharing." Each member would contribute a core competency: Ballard would supply fuel cells, Daimler would integrate them into fuel supply and control systems, and Ford would provide the electric drive components, including the computer, motor, and transaxel. Ford executives lavished praise on the technology. Trotman held that fuel cells were "one of the most important technologies for the early twenty-first century." William Clay Ford, Jr., the great-grandson of Henry Ford and then chair of the corporation's finance committee, believed that the fuel cell would give the automaker a "competitive edge." For years, said Ford, electric vehicles had been held back by limitations in battery capacity. Fuel cells, he claimed, had no such limitations.[22] The new alliance was notably vaguer in its objectives than the previous Daimler-Ballard arrangement and essentially emulated the function of the PNGV. No provisions were made to manufacture anything at scale, and each automaker aimed to build its own vehicle, with Daimler opting for methanol and Ford for hydrogen fuel.[23]

FIXING A FUEL

The decision of Daimler and Ford to select different fuel cell fuels had the effect of bolstering fuel cell discourse while giving a misleading impression of progress in the field. The auto industry's concurrent experiments with hydrogen and alcohol fuel cell systems served to elide completely different technologies into a single notional fuel cell. The resulting belief that the fuel cell was a kind of universal chemical energy converter, or even a power panacea, was a fallacy, but this imaginary performed important social work.

Car companies learned to highlight specific features of fuel cells calculated to appeal to particular audiences, without mentioning the various sociotechnical trade-offs. Environmentalists were attracted to the idea of a power source technology that produced water as waste, while the oil industry anticipated a vast new market for carbonaceous fuel cell fuels. Daimler was particularly adept at modulating this dramaturgy of expectations. The automaker kept its hydrogen Necars in circulation even as it began testing the methanol-powered Necar III, a converted Mercedes A-class subcompact and the first of the company's fuel cell demonstrators to be based on a passenger automobile.[24]

As a result, media pundits often erroneously assumed that low-temperature fuel cells could use a wide variety of fuels interchangeably.[25] On the spectrum of carbonaceous liquids, methanol and ethanol were long considered to be among the substances that, after pure hydrogen, were most readily and sustainably electro-oxidized. However, the technoscience of alcohol fuel cell propulsion had barely advanced since the early 1960s. Old lessons of the characteristics of these systems had to be relearned. Once again, engineers looked to reformer technology to crack hydrogen out of alcohol and create a hydrogen-rich fuel gas that fuel cells could process more easily than straight alcohol. But reformers added weight, complexity, and cost to the electric drivetrain. The alternative was to use carbonaceous fuel directly in a fuel cell, but that brought a corresponding increase in the complexity of the internal electrochemical reactions of the system. Chrysler fuel cell chief Christopher Borroni-Bird observed that direct alcohol fuel cells required more platinum than hydrogen fuel cells owing to alcohol's relatively low reactivity, and also suffered from crippling side reactions.[26]

In a sense, the Necar III was an acknowledgment of the standstill in the state of the art because it used a reformer that occupied most of the passenger and cargo space. Even as researchers struggled to develop alcohol fuel cell technology, PNGV planners began tackling the considerably more challenging gasoline fuel cell system. The process of reforming gasoline requires a much higher temperature than methanol (700 degrees Celsius as opposed to 200 to 300 degrees) and more energy to dissociate hydrogen from the hydrocarbon molecules. In turn, this creates thermal expansion and corrosion and degrades the reformer catalyst. More problems arose in the resulting fuel gas stream, which contained contaminants like carbon monoxide and sulfur that had to be stripped out, further complicating the design of the electric power train.[27]

The auto industry's seemingly abrupt transition from hydrogen to methanol to gasoline fuel cell systems dismayed environmentalists. In April, Jason Mark had applauded Ford for committing to hydrogen fuel cell power. Only months later, in October, he criticized the federal fuel cell program for "veering off course." Gasoline reforming technology made the fuel cell dirtier as well as more complex and costly and prevented the power source from realizing its full potential.[28] To be sure, the PNGV had not exactly changed course. From its inception, the consortium had supported concurrent research on several fuel cell systems, working simultaneously on hydrogen and alcohol from 1994 before adding gasoline and multifuel reforming to the agenda around 1996. At any rate, proponents of the gasoline fuel cell like Borroni-Bird argued that the gasoline infrastructure was ubiquitous and that fuel cell researchers had to adapt to that reality.[29]

HYDROGEN AND THE HILL

The technological challenges of the gasoline fuel cell proved too great to surmount in the near term. By the end of the 1990s, pundits, experts, and policymakers were looking to hydrogen fuel cell systems and infrastructure as a total solution to the problems of sustainable energy conversion in automobiles. The fuel cell turn Increasingly became a hydrogen fuel cell turn, aligning developers of the membrane fuel cell with hydrogen proponents in industry, universities, the federal science establishment, and Congress. After a hiatus through most of the 1980s, hydrogen discourse reemerged in mainstream energy and environmental circles at the end of the decade, thanks in part to the congressional activism of George E. Brown, Jr., a representative of California, and Spark M. Matsunaga, a senator of Hawaii, both Democrats. In 1989, Matsunaga introduced a bill that become the Spark M. Matsunaga Hydrogen Research, Development, and Demonstration Program Act. The legislation called only for $20 million but gave hydrogen new visibility in the federal research and development agenda. Among the act's provisions was the establishment of a Hydrogen Technical Advisory Panel (HTAP), a board of experts charged with guiding the energy secretary's conduct of federal hydrogen and hydrogen-related research and development. The HTAP was set up in 1992.[30]

Hydrogen received another boost when Robert S. Walker replaced Brown as chair of the House Science Committee in 1995. A Republican representative of

Pennsylvania, Walker served as chief deputy whip in the mid-1990s and was an important figure in the Republican triumph in the midterm congressional elections of 1994. He had close ties with the military-industrial and aerospace complexes and saw hydrogen as a means of enabling energy independence and abolishing antipollution regulations. In early 1995, Walker sponsored a bill to boost the DOE's annual hydrogen research budget by $100 million over the next three years, an initiative Brown initially opposed.[31] But hydrogen had bipartisan appeal and in 1996, Brown co-sponsored a bill with Walker that became the Hydrogen Future Act of 1996. Designed to study all aspects of the hydrogen economy, the act authorized $164.5 million over six years to demonstrate the technological feasibility of producing hydrogen from renewable energy and integrating these systems with fuel cells in industrial, residential, transportation, and utility applications.[32]

Air quality technopolitics and its increasing emphasis on fuel cell electric technology and associated hydrogen systems overlapped the interests of oil companies as well as automakers. The oil industry produced hydrogen as a byproduct of refining and increasingly used the gas to desulfurize refined petroleum products in response to increasingly stringent air quality regulations. Oil companies disliked the ZEV mandate and regulated technological change as much as automakers, and preferred the prospect of selling fuel cell fuels, even hydrogen, to a scenario where battery electric propulsion drove even a modest portion of the light duty automobile fleet.

These forces had the effect of embedding hydrogen discourse even more deeply in environmental and energy policymaking. What Daimler and Ballard had pioneered in fuel cell dramaturgy would be elaborated by most of the car companies subject to the mandate on a larger scale. In 1999, John Dunlap was replaced as the chair of CARB by Alan C. Lloyd, an air quality scientist, hydrogen enthusiast, and former HTAP chair. That same year, the state of California helped coordinate a group of oil and auto companies devoted to promoting the fuel cell electric car as their ZEV of choice. These developments coincided with the post–Cold War economic expansion, a boom driven by the information technology revolution and renewed faith in science and technology to reshape society for the better in almost every conceivable way. Hydrogen fuel cell advocacy was an important part of millenarian techno-utopianism, and it would inform the struggle between regulators and the car companies to determine the technology, industry, and infrastructure of the sustainable automobile.

8   KYOTO CARS

In the annals of car history, there are many examples of companies falling victim to rivals after missing a narrow window of opportunity, even while in a hurry. The environment was a critically important theme to carmakers and we couldn't afford to take second place.

—Satoshi Ogiso, Toyota managing officer, 2017

As interest in hydrogen and fuel cell electric propulsion grew, competition in clean car technology between General Motors (GM) and Toyota was reaching a climax. In March 1997, Toyota publicly announced that at the end of the year, it would introduce a commercial hybrid electric automobile equipped with a nickel-metal hydride rechargeable that would have double the fuel efficiency of conventional cars and produce 90 percent less effluent. The automaker did not specify the price or name of the vehicle, but suggested that it would be in the class of the Corolla subcompact and indicated that it had not yet decided whether to offer it for sale in the US.[1] At GM, the Ovonic Battery Company (OBC), and its parent Energy Conversion Devices (ECD), attention at the time was still focused on Toyota's RAV4 EV, the compliance car that the US camp believed it was competing against in the all-battery electric space and that Stanford Ovshinsky and Robert Stempel were confident would be no match for the Ovonic EV1. But OBC was growing anxious. Since early 1996, the company had been engaged in costly legal action with Matsushita and Toyota over Toyota's right to introduce electric cars equipped with advanced Matsushita nickel-metal hydride batteries into the US market, events discussed in chapter 6. Even more worrying for OBC was that GM was developing the Ovonic EV1 at an inexplicably leisurely pace. So concerned were Ovshinsky and Stempel that they began to more openly discuss the scenario of working with the Japanese alliance, probably with the intention of goading GM into taking bolder action.[2]

The spring of 1997 did bring some good news for OBC. The US District Court for the District of Delaware moved to dismiss Matsushita's lawsuit and validate the 822 patent. The ruling restricted the industrial giant's nickel-metal hydride technology to an earlier chemistry yielding 63 watt-hours per kilogram of capacity, meaning that the RAV4 EV would have less range than the OBC EV1, whose battery pack was then rated at about 70 watt-hours per kilogram.[3] For Stempel, the legal process was doubly heartening because it caused GM to announce that the first module of production-intent equipment was operational, contradicting Matsushita's claim that GM-Ovonic was a joint venture in name only. OBC used this information in its countersuit to argue that a US manufacturing base for electric vehicle batteries existed and was vital to the "overall EV strategy."[4]

Stempel hoped that competitive pressures accentuated by the court ruling would concentrate GM's mind on the OBC EV1. Harry Pearce had advised OBC and ECD in the case against Matsushita, and with legal victory all but certain, Stempel wrote to thank the vice chair for his help: "Now that the Matsushita EV battery market position has been defined (and limited), it is clear that GM can be in a dominant, controlling position with nickel-metal hydride electric vehicle batteries."[5]

What GM intended to do with this dominance was another question. The automaker was poised to trump the RAV4 EV with the OBC EV1 in the dramaturgy of high-technology supremacy, its chosen field of battle, and its leaders were satisfied that Toyota's limited-production compliance car posed no threat. But Toyota was poised to pioneer and monopolize an entirely new market niche with its commercial hybrid electric car. If GM leaders were unconcerned with the immediate economic ramifications of that project, they began to be aware that it could at least cause them some public relations problems, with the signing of the Kyoto climate treaty only months away.

Managers of GM's hybrid electric programs felt a more urgent need to demonstrate progress, but their room for maneuver was limited. The

Partnership for a New Generation of Vehicles (PNGV) administered the research and development of batteries for Detroit's experimental hybrid electric cars, and the consortium internalized the US auto industry's philosophy of mobilizing the best as the enemy of the good in order to delay action on mandated technological change. Where Toyota and Honda used first-generation nickel-metal hydride rechargeables in their first-generation commercial hybrid electrics, the PNGV interpreted the hybrid as another type of electric supercar. Moreover, consortium planners insisted on developing not one, but two systems: a power assist hybrid and a dual-mode hybrid. The former required only a battery of high power and modest energy, since the average depth of discharge would be comparatively shallow depending on whether the battery was coupled to a prime mover that responded relatively quickly (the internal combustion engine) or slowly (the fuel cell) to demands for power. Dual-mode hybrid electric drive, on the other hand, required a larger battery that was energetic as well as powerful and also was robust enough to repeatedly deeply discharge when providing periods of electric-only transport.[6]

The dual-mode hybrid electric car was similar to the all-battery electric car, requiring precisely the sort of superbattery that US automakers insisted was necessary for all-electric cars to be competitive but that they could not quickly develop. OBC was caught in a game of technological leapfrog, wedged in the uncomfortable space between GM and Toyota's tacit cooperation in suppressing the all-battery electric format and their nominal competition in commercial advanced propulsion technology, an arena that Detroit had effectively ceded. OBC bore the burden of cutting the costs of the nickel-metal hydride battery for the EV1 by means of materials science before GM would consider moving the car into production. On the other hand, GM's sudden interest in hybrid electric propulsion, following Toyota's revelation of its commercial hybrid electric, offered the possibility of another application of OBC technology.

But the PNGV was becoming skeptical about the nickel-metal hydride rechargeable in the hybrid electric application, an ill omen for OBC's future in the automobile industry. As the countdown to Kyoto progressed and the media limelight shifted to the hybrid electric car, the tensions and contradictions between GM and OBC were cast ever more starkly into relief.

CONSORTIA CONUNDRUMS

The quandary of OBC was that it had partnered with a corporation that believed its status as the world's largest automaker gave it the role of arbiter of the technological progress of the global industry. GM and its domestic peers agreed that rapid technological change was undesirable, and the doctrine, structure, and governance of the US automaking industry's research and development consortia reflected this view. These organizations investigated almost every conceivable form of automobile propulsion technology and judged it to very high standards of performance. With the assistance of the national developmental state, Detroit created three distinct entities for these purposes. The United States Advanced Battery Consortium (USABC) was responsible for batteries for all-electric cars, and the PNGV was responsible for all other kinds of alternative propulsion systems, including batteries for hybrid electric configurations. Both groups were formally part of the United States Council for Automotive Research (USCAR), the umbrella organization for collaborative research in the US auto industry, but the PNGV's hybrid battery programs were collectively managed by the USABC. While the agendas of the two consortiums in galvanic batteries substantially overlapped, the car companies effectively treated these organizations as separate entities. Moreover, the Big Three automakers did not decide their priorities in advanced propulsion technologies collectively; rather, they apportioned responsibilities among its members in specific fields.

In short, the architecture of the US auto industry's collaborative research and development complex inhibited coordinated rapid action on specific technologies. OBC's experience with its hybrid battery illustrated these dynamics. While the USABC had helped fund OBC battery technology for all-electric vehicles, the PNGV excluded the company from its hybrid electric battery program, a notable omission because the partnership did fund research in nickel-metal hydride rechargeables but chose to contract with Varta and Yardney, companies with less experience in the technology. Both of these programs failed to meet initial PNGV milestones.[7]

Nevertheless, OBC used its own resources to develop its own battery for hybrid electrics, which it demonstrated to leaders of GM's hybrid electric program in a March 1997 meeting. The GM team was impressed. It had been working with an Optima spiral-wound lead-acid rechargeable, a device that fell well short of the PNGV's performance standards, which had been devised by Ford. Ford called for 50 kilowatts of power from a 0.5 kilowatt-hour pack, a goal that the GM team felt was unrealistic.[8] The team agreed to have the GM contractor AeroVironment test the OBC technology and intimated tantalizing prospects. Larry Oswald, a senior member of the GM hybrid group, indicated that the automaker expected to sell over 50,000 hybrid electrics per year. But the GM team did not offer OBC any funding. Oswald suggested that OBC inquire at the USABC because he believed the consortium had surplus cash, $24 million out of a fund of $32 million for research on hybrid batteries left over as a result of the conclusion of Varta and Yardney's programs.[9] Oswald also recommended that OBC make funding inquiries with Harold Haskins, a Ford engineer who served as a PNGV team leader responsible for technical targets and who had designed the consortium's analytical tools for testing hybrid batteries.[10]

OBC was not optimistic that the USABC would cooperate, but it was heartened by the AeroVironment report on its hybrid electric battery. The device's energy density of 70 watt-hours per kilogram was around two and a half times that of the Optima lead-acid rechargeable, the best result, claimed OBC, of any battery that AeroVironment had tested for the PNGV.[11] Oswald was pleased, but he wanted OBC to manufacture a slightly smaller module at its own expense. If OBC could do the job, said Oswald, GM might purchase one or more complete hybrid electric battery packs for evaluation.[12] Haskins offered similar praise, noting that OBC's accomplishment was all the more impressive given that the company had funded all the work. But the Ford executive also echoed Oswald's criticism, holding that OBC's basic hybrid electric battery cell was too heavy for PNGV requirements. He wanted to increase the cell's power-to-energy ratio. If OBC continued to improve the battery, Haskins indicated that there was a chance the technology might be evaluated by the Idaho National Engineering and Environmental Laboratory as part of the cooperative agreement between the US Department of Energy (DOE), the PNGV, and the USABC. It was possible that the DOE might lend OBC some direct support.[13] In the meantime, Haskins recommended that OBC try to access PNGV resources through GM's Oswald.[14]

OBC had been given the runaround, and not only as a result of the consortia's circuitous chains of command. Auto industry planners were taking an increasingly dim view of the nickel-metal hydride battery in the hybrid electric role. The National Research Council (NRC)'s 1997 review held that without a materials breakthrough, the PNGV cost goal of $150 per kilowatt-hour could not be met with nickel-metal hydride chemistry. The NRC agreed with the PNGV that the lithium ion rechargeable was a more promising power source in the hybrid electric application. The PNGV had already awarded contracts for lithium rechargeables to the French battery maker Saft and the US nonprofit SRI International as part of its Phase I program.[15] The new enthusiasm for lithium power was a harbinger of an effort by GM and the US auto industry to wind down their investment in nickel-metal hydride battery technology and reduce OBC's influence in their affairs.

COUNTDOWN TO PHASEOUT

OBC's prospects in the EV1 were hardly better than in GM's hybrid electric car. The battery company had the USABC's support in developing the EV1 battery only until the end of 1997, and much work remained to be done. GM-Ovonic chief John Adams was seeking to cut the cost per pack from $60,000 to less than $9,000, which was still twice the consortium's original target, and had only months to accomplish this goal. GM had promised to build a new dedicated battery manufacturing plant in Dayton, Ohio, only if quick progress was made at GM-Ovonic's Maplelawn preproduction plant.[16]

With limited resources, Adams worked to mechanize a facility that was one-tenth the size of a commercial production plant. By mid-1997, the negative electrode assembly equipment was operational, as was some of the equipment for *formation*, the process of curing new cells by repeated charging and discharging so that microstructures of electrode surfaces are prepared for optimum operation. Formation is sensitive, time-consuming, and vital for reliable and durable cells. GM-Ovonic's formation process then took fourteen days, and Adams wanted to cut that to three.[17] Additional formation equipment had been ordered, but it would not arrive for six to eight weeks. Also, the positive electrode coating machine had to be debugged and started. Until this was done, workers had to assemble cells by hand. By August, capacity was two packs per week, and Stempel hoped that this could be increased to one a day and five per week by December.[18]

OBC's future hung in the balance. There was just enough good news to give hope. In April, Adams informed OBC president Subhash Dhar that road tests of vehicles using OBC batteries gave good results. Personnel at GM Advanced Technology Vehicles (GMATV) used Ovonic EV1s on the Lansing-Detroit commute, a round-trip journey of around 180 miles that Dhar argued was not possible in an electric car equipped with lead-acid batteries. One vehicle shown to GM's board of directors accumulated 18,000 miles.[19] OBC batteries were also being tested in Chevrolet's converted S-10 light truck, which competed with Ford's converted Ranger in the market for compliance cars.[20]

But Stempel wanted GM to take full advantage of OBC's "battery family," a lineup of four progressively more advanced power sources derived from the baseline 70–75 watt-hour per kilogram cell for the EV1. OBC also had a plan to develop a nickel-metal hydride auxiliary battery for starting and lighting internal combustion engine vehicles. Stempel urged GM research and development chief Kenneth Baker to think of OBC technology not simply as a battery but as a multielement energy storage system that could serve every possible automobile propulsion format. Baker himself had promoted the hybrid and fuel cell electric formats in his 1994 presentation to the GM President's Council, the automaker's main decision-making body, in justifying the restart of the Impact car project.[21]

These appeals bore no fruit, and Ovshinsky found GM's seeming disinterest in OBC technology incomprehensible. He saw potential applications of OBC batteries everywhere in the automaker's industrial empire, from the buses and locomotives the company then manufactured to forklift trucks, standby power, and industrial robots. At times, the inventor expressed impatience with and even contempt for what he interpreted as an antiscience ethos embedded in GM's corporate culture. After a visit to GM's research laboratories, he wrote to Stempel, lamenting the automaker's "minimal" knowledge of advanced materials. To Ovshinsky, the self-taught materials genius, it was "obvious that we can be a problem-solving resource to serve GM's plans for new products as well as improve their older technologies."[22]

A CAR FOR CLIMATE CHANGE

Stempel personally lobbied Harry Pearce to accelerate GM's electric car programs. In an August 13 meeting with the vice chair, he warned that GM risked being outmaneuvered by Toyota unless it took action. Over the course of 1997, Stempel had surveyed global industrial and environmental politics with growing concern. In June, he attended the Summit of the Eight in Denver, hosted by Admiral Richard H. Truly, director of the National Renewable Energy Laboratory, and had been taken aback by the blunt language of Secretary of Energy Federico Peña, the keynote speaker. Peña's assessment was that the Clinton administration had no policy going into Kyoto, a situation that was going to create a great deal of confusion in the treaty negotiations because the other industrial powers were expecting leadership from the US. The energy secretary believed that things could continue as they were or that the US could develop a comprehensive plan. On an ad hoc basis, Peña and Vice President Al Gore's office were mustering representatives of US industry, including Stempel, to help formulate such a position.[23]

In Stempel's view, Toyota was going to use the Kyoto summit to position itself to dominate the global automobile industry by offering its commercial hybrid electric car as the solution to the problem of climate change. The automaker's message would be that better fuel economy equated to lower emissions of carbon dioxide, a substance that was not then subject to controls. Indeed, noted Stempel, Japanese and German automakers were already working hard to lobby against specific reduction targets at Kyoto. In the technopolitics of smog, the local air quality regulator held most of the cards; in the technopolitics of climate change, much of the authority on what counted as green swung back to the automaker. In building the market for the sustainable automobile, Toyota would rely on consumer acceptance, not public policy targets. This was the challenge facing Detroit, argued Stempel. GM could still regain the initiative, he continued, if it altered its relationship with OBC. Stempel proposed to restructure OBC as a stand-alone company through an initial public offering, allowing the automaker to purchase OBC shares and gain direct access to its technology.[24]

This was not to be, but with Kyoto only months away, GM felt it prudent to maintain some level of support for OBC. In September, the automaker approved the Ovonic Product Development Program, providing $16.4 million for research to cut costs and increase the energy of batteries for hybrid and all-battery electric cars. In a letter to Baker, Stempel opined that this was not a lot of money, but it allowed OBC to maintain operations, and he hoped that it also signaled a commitment to support the company's product lineup. Stempel added that OBC could also help meet GM's requirements for hydrogen fuel cells. OBC had yet to develop such a technology, but it had an important component in metal hydride storage, a key enabler of the nickel-metal hydride battery.[25] Stempel's mention of the hydrogen fuel cell was a sign that the executive was aware that the all-battery electric format was in real jeopardy.

KYOTO CARS

By the end of the year, OBC batteries were installed in some thirty GM vehicles, mostly S-10 conversions as well as a handful of EV1s. At Solectria, James and Anita Rajan Worden were using OBC batteries to achieve important range milestones, including mileage records of 216, 249, and 373 miles on a single charge.[26] OBC had helped make GM the most progressive of the US automakers in alternative propulsion systems, and Ovshinsky's celebrity ascended accordingly.[27]

As impressive as these accomplishments were under the circumstances, they were completely overshadowed by the debut of the Toyota Prius as the world's first commercial passenger hybrid electric car. Toyota devoted the entire fall of 1997 to the rollout. Initial production started in September, and the automaker officially introduced the car at a special event in Tokyo in October, thereafter promoting preproduction models at a series of high-profile venues, including the Tokyo Auto Show and the site of the 1998 Winter Olympics at Nagano. Priuses were also on hand at the climate summit in Kyoto, where they shuttled officials and dignitaries between venues. On December 10, the day before the climate treaty was signed, Toyota held a ceremony for the Prius lineoff.[28]

The Prius was a triumph of social as well as physical engineering. The program illuminated the intimate industrial-technological relationship between Japan and the US, highlighting contrasting approaches in public policy. Jack Smith would later claim that Toyota had developed the Prius with the help of a public subsidy, a hint that trade sanctions might be forthcoming should the vehicle be exported to the US.

However, the US national developmental and regulatory state had also been involved, directly as well as indirectly, in virtually all the alternative propulsion vehicle projects under development for the US market. Energy and environmental policies forced most automakers to build cleaner and more efficient cars for this market, and the national developmental state supplied US automakers with science and technology resources to help them accomplish these goals. Ironically, some of these stimulus initiatives were modeled on those said to have been deployed by Japan's Ministry of International Trade and Industry (MITI), the planning agency that to many American pundits and policymakers was antithetical to the principles of capitalist fair play and emblematic of Japan's industrial rise. From the mid-1980s, US officials set aside their ideological assumptions and tried to emulate the imagined Japanese model in research consortiums such as Sematech (the DARPA-organized association of US semiconductor manufacturers), the USABC, and the PNGV.[29]

A double irony was that MITI had no more power to compel or control domestic automakers than the US research consortia did. Japanese car companies managed to maintain their independence from Japanese state planners and had initiated their electric car projects largely in response to regulatory developments in the US market. Like the US federal science and technology complex, MITI long supported the research and development of various aspects of electric vehicle technology. But the ministry followed Toyota's and Honda's lead in hybrid electric technology, only indirectly contributing research relevant to parts of the hybrid drivetrain, including the permanent magnet motor.[30] It was in the manner of the provision of financial assistance that the Japanese national developmental state differed from its US counterpart. Toyota initiated the Prius project without the direction or aid of MITI but the ministry subsequently subsidized cars purchased by Japanese corporations, a form of stimulus that American entrepreneurs of electric cars and electric car batteries had long asked of the US government.[31]

The way in which Toyota introduced the Prius also illustrated that the company had far grander imagination and ambition in technological dramaturgy, not to mention a willingness to coordinate with government in staging it, than its peers. Years later, the G21 member Satoshi Ogiso claimed that the Prius production deadline had been moved up one year precisely so that the launch would coincide with the signing of the climate treaty.[32] Toyota president Hiroshi Okuda made this decision nearly seven months before Kyoto was formally announced as the site of the Third Conference of the Parties of the UN Framework Convention on Climate Change (UNFCCC) in July 1996, and so had been privy to negotiations at the highest level.[33] Pundits were skeptical that hybrid electrics could make money, but if the Prius had been intended solely as green propaganda, it was propaganda on an unprecedented scale.

On December 23, a judge in the US District Court for the District of Delaware announced a verdict that GM, OBC, and ECD knew was coming: Matsushita's lawsuit was finally dismissed. Where nickel-metal hydride rechargeable technology for electric vehicles sold in the US market was concerned, Matsushita and Panasonic EV Energy (PEVE), as Matsushita's joint battery venture with Toyota had been known since January, were restricted to a battery with relatively low energy density called the MHI-BX, which had been covered under a previous agreement with ECD, and they could not use OBC technology to improve it. Over the course of the year, OBC had spent $3 million in scarce capital in defense of its patent rights, a victory that may have seemed pyrrhic in light of the Prius.[34]

9   ART OF THE POSSIBLE

Our practical experience with the EV1 gives us a significant advantage to build
from as we develop options. As you look at these vehicles today, remember that
they are not traditional show cars meant to pique a fantasy. They are a family of
advanced technology concept vehicle that demonstrate the art of the possible.

—Jack Smith, General Motors (GM) chair and chief executive officer, January 4,
1998

Under the terms of the Partnership for a New Generation of Vehicles (PNGV),
American automakers were not expected to introduce prototype hybrid
propulsion cars until 2004. However, the shock and surprise caused by the
unveiling of the Prius demanded an immediate response, so the Big Three
hastened to display the fruits of their work on alternative propulsion technology at the North American International Auto Show in Detroit in January
1998. Contemporary observers noted that the event compared unfavorably
with the pageantry of Toyota's Kyoto car.[1] Ford had its P2000, based on the
chassis of the midsize production Contour sedan but utilizing a lightweight
aluminum frame and an advanced diesel engine. Somewhat more ambitious
was Chrysler's ESX-2, a series hybrid built around a turbo-diesel engine and
a small lead-acid battery pack. This was a mild hybrid, or "mybrid," so-called
because it did not depend heavily on its battery electric system, which was
used only to run accessories and to assist in hard acceleration.[2]

General Motors (GM) mounted the largest and most elaborate display in
the form of a mock press conference. Robert Stempel, Stanford Ovshinsky,

and Ovshinsky's wife, Iris, sat in the audience, while the corporation's leaders, clad in green sweaters, waited in the wings for their cues. As scripted, the event situated the executives across from three preteens wearing business suits playing future GM executives, whose role was to quiz the grown-ups. Jack Smith opened the proceedings, driving an Ovonic EV1 through a curtain. Stepping out of the vehicle, Smith greeted the audience: "As I have said before, no car company will be able to thrive in the twenty-first century if it relies solely on internal combustion engines."[3]

The issue was climate change, or "global climate warming," as Smith put it, and solutions, he held, would come through innovation, not regulation. The EV1, said Smith, was essentially a platform to build experience in developing technological options. To illustrate this point, he gave the floor to research and development chief Kenneth Baker, who introduced the Ovonic EV1. Equipped with a nickel-metal hydride rechargeable, the car had a range of around 160 miles per charge, double the range of the lead-acid version of the EV1, which made the vehicle the undisputed performance leader in the all-battery electric class. It was doubtless a gratifying moment for the Ovonic Battery Company (OBC) and electric car enthusiasts.[4]

But GM also sought to balance energy efficiency, environmental responsibility, and consumer appeal, continued Baker, so the company had developed a range of alternative advanced propulsion vehicles. In addition to the all-electric Ovonic EV1, four other variants of the EV1 competed for attention. One had a three-cylinder ICE using compressed natural gas, while another utilized a methanol fuel cell electric system. Another version was a parallel hybrid equipped with a turbocharged Isuzu direct-injection diesel engine driving the rear wheels and an Ovonic battery–powered electric motor driving the front ones. Still another variant was a series hybrid equipped with Ovonic batteries that could be recharged from an external source, a format that had a much longer all-electric range than conventional hybrids. On paper, this so-called plug-in hybrid could drive about forty miles on its battery. GM quietly developed this variant with the help of Andrew A. Frank, a professor of mechanical and aerospace engineering at the University of California at Davis who had been building plug-in hybrid conversions with his students since the 1970s.[5]

The EV1 plug-in hybrid had intriguing potential, not least of which was the prospect of surpassing the Prius in capability. Baker himself favored fuel cell power, as did research and development chief Larry Burns. This propulsion technology gave excellent range (more than 300 miles), claimed Baker, and had the potential to be the best long-term solution. There was a question from one of the child executives. If the hybrid electric car looked like the next step and the fuel cell electric car was the future, when would they be available? GM, responded vice chair Harry Pearce, would have a production-ready hybrid ready by 2001 and a production-ready fuel cell electric by 2004 or sooner.[6]

The presentation represented the realization of the plan that Baker had presented to the GM President's Council in early 1994. The 1998 Detroit auto show marked the beginning of the end of the experiment with the all-electric automobile equipped with the nickel-metal hydride rechargeable and the start of a new phase in the technopolitics of clean cars. Smith's GM had echoed other automakers in acknowledging climate change, a discursive shift that coincided with the gradual disintegration of the Global Climate Coalition in the late 1990s. Some analysts ascribed the organization's demise to its declining reputation as the scientific evidence of climate change accumulated.[7]

An alternative explanation for the disappearance of the Global Climate Coalition is that it no longer served the purposes of the car and oil companies to act as "merchants of doubt" when they realized that the industrial and industrializing nations would or could not agree to regulate carbon dioxide.[8] With the failure of the US to ratify the Kyoto Protocol, recognizing climate change became a way for corporations to burnish their environmental credentials without risking regulatory consequences. Toyota had shown how this could be done in its campaign to market the Prius as an alternative to the all-electric vehicle.

In November 1998, the California Air Resources Board (CARB) acknowledged the efforts of automakers to develop advanced propulsion technology. Air quality regulators amended the Low Emission Vehicle (LEV) statute and created the Super Ultra-Low Emission Vehicle (SULEV) category, allowing automakers to earn partial Zero Emission Vehicle (ZEV) credit for vehicles that had super ultra-low emissions and all-electric capability, employed advanced componentry, or used inherently clean fuels. These measures implicitly recognized the carbonaceous fuel cell and hybrid electric cars as partial ZEVs and the hydrogen fuel cell electric car as an alternative ZEV. In principle, CARB now offered automakers the flexibility of configuring blends of these types of vehicles in meeting their ZEV mandate quotas. Air quality regulators also proposed awarding multiple credits to ZEVs with very long ranges, incentiv-izing automakers to invest in hydrogen fuel cell electric propulsion.[9]

Jack Smith referred to clean car dramaturgy as the "art of the possible," but executing this choreography in ways that plausibly satisfied the public interest was no trivial matter.[10] Air quality regulators showed that while they were amenable to renegotiating the terms of the mandate, they were not prepared to do away with it. The 2003 quotas still had to be met, and years of well-publicized demonstrations of ZEV technology by the auto industry had stoked expectations that electric cars of one kind or another would soon be on the market. Mindful of the risks of subverting a popular policy, the car companies worked to prepare public opinion for the termination of the all-battery electric project by embracing environmental discourse and intensifying the promotion of alternative propulsion technologies.

UNDOING THE KNOT

The 1998 Detroit auto show generated badly needed political capital for GM. The editorial board of the *New York Times* characterized Smith's per-formance as an "epiphany" and framed the event as a vindication of the Clinton administration's environmental policy.[11] Days after the press con-ference, Stempel attended a working meeting with the White House cli-mate change team in which Katie McGinty, Vice President Al Gore's lead environmental advisor, praised the automaker. In the Ovonic EV1, held McGinty, GM had finally accepted its environmental responsibilities and provided the White House with a model of how technological solutions to climate change could also be good business. Some of McGinty's staffers were skeptical, interpreting GM's auto show display not so much as a coherent response to climate change and Kyoto as a hasty riposte to the Prius. Stem-pel tried to explain to them that the initiatives that GM had introduced were the logical extension of the automaker's electric vehicle program.[12]

Privately, however, Stempel worried. GM-Ovonic was gradually boosting production but at a scale that made only marginal progress in cost reduc-tion. In 1998, only 130 battery packs were due to leave the Maplelawn facility, each worth $64,000. Production was set to go up to 165 packs in 1999, at a unit cost of $55,000, and to 210 packs in 2000, at a unit cost projected to be a still-exorbitant $46,000.[13] Tensions over the boundaries of research and manufacturing were as persistent as ever. When OBC made prototype batteries, it was in charge of the entire process, and workers made frequent changes to the chemistry at all points of production. In the EV1, OBC was a junior partner in a manufacturing enterprise, responsible for making the negative electrode. GM-Ovonic was responsible for everything else, including the cathode, cell formation, and component assembly. The overriding preoccupation of GM-Ovonic chief John Adams was quality control, and continuous processing could not easily be stopped once initiated. As one GM-Ovonic manufacturing engineer explained, "we want production to flow like water."[14] Adams was under orders to produce only first-generation nickel-metal hydride cells, and he was not about to interrupt the flow. To Ovshinsky's dismay, GM-Ovonic rejected a plan to manufacture OBC's more energetic third-generation formula, as well as a proposal to introduce calcium hydroxide in the cathode. This "minor change," Ovshinsky held, would improve high-temperature charge acceptance and compensate for the EV1's poor thermal management system. With cells rolling off the production line, however, it was too late to further alter the battery chemistry. There was more bad news in reports that GM was questioning OBC's claims for its technology. Ovshinsky believed that this would cause Toyota to tell its customers to use batteries made by Matsushita/Panasonic rather than GM-Ovonic.[15]

However, the open market in electric vehicle batteries that Ovshinsky imagined remained elusive. Matsushita's relationship with Toyota enabled it to monopolize the supply of nickel-metal hydride rechargeables for electric vehicles in Japan.[16] OBC had long hoped to cement an alliance with Honda, but the automaker equipped its EV Plus compliance car with Matsushita/Panasonic cells and did little to promote its own all-battery electric car program. In April 1997, Honda quietly rolled out a single EV Plus in a line-off ceremony at its Takanezawa plant, ultimately building around 300 of these vehicles in meeting its mandate quota before ending production in April 1999.[17] Honda planned to deliver half of the vehicles to individual customers and the rest to institutional and fleet clients including Budget rentals and the University of California at Riverside, all on three-year leases. The next phase, held American Honda president Robert Bienenfeld, was research. Like Toyota, Honda used its experimental limited-production all-battery electric car to help develop its commercial hybrid electric car, a vehicle that, like Prius, used a battery pack of Panasonic commodity cells. Honda would also use the EV Plus as the basis of its experimental FCX fuel cell electric vehicle.[18]

Automakers were gradually winding down their all-battery electric programs and GM-Ovonic faced a reckoning. In December 1998, General Motors
announced that it would introduce the Ovonic EV1 in 1999. By then, the
joint venture had been in low-volume production for nearly three years, and
while GM opened the new dedicated manufacturing facility in Dayton, engineers were still struggling to move the basic module from hand to automated
production. One of the things they discovered was that at low manufacturing volume, 60 per cent of costs came from materials. As OVC executives
tried to find ways of cutting costs that did not involve scaling production,
they pondered adopting materials used in consumer electronics batteries. It
was an indirect concession to Matsushita's dominance in the field of commercial nickel-metal hydride chemistry and an ironic acknowledgment of
the success of the industrial giant's commodity cell strategy.[19]

HYDROGEN FUTURISM REDUX

As the 1990s drew to a close, automakers sharpened the research and
rhetoric of hydrogen fuel cells. The tasks that engineers and policymakers
asked of the carbonaceous fuel cell electric car, a virtual mobile chemical
plant, were beyond the capabilities of the technoscience of electrochemical
energy conversion as it stood at the turn of the millennium. Methanol fuel
cells required up to thirty minutes to warm up, and the already-complex
process of gasoline reforming was complicated further by the oil industry's
determination to fight the low-sulfur standard for gasoline imposed by the
Environmental Protection Agency (EPA).[20] New studies suggested that off-
board systems producing pure hydrogen from natural gas or gasoline at
existing service stations could be a way to quickly build up an infrastructure
to serve the hydrogen fuel cell electric car.[21] GM was a relative latecomer to
fuel cell technology, but it would make up for its slow start. In March 1998,
J. Byron McCormick, GM's alternative propulsion chief, claimed that the
automaker's investment in fuel cells was at least as large as its commitments
to hybrid and battery electric propulsion.[22]

Daimler's May 1998 merger with Chrysler led the new transnational
corporation DaimlerChrysler to accelerate its work on hydrogen fuel cell
systems. Daimler had been exempt from the mandate owing to its small
market share in California, but DaimlerChrysler had a considerable sales
footprint in the state, and it needed a ZEV. The corporation had a potential

compliance car in the converted Mercedes A-class hatchback equipped with the sodium-nickel chloride battery, but it cancelled this project in favor of a pledge to develop a hydrogen-powered supercar. In March 1999 in Washington, DC, DaimlerChrysler rolled out the Necar IV, a concept car promoted as the first fuel cell electric vehicle to be driven on public roads in the US. The vehicle was based on the A-class platform and featured a liquid hydrogen system, technology that in principle afforded a relatively long range. The car could run for about 280 miles on a full tank, substantially farther than an Ovonic EV1, at least in theory. The EPA administrator Carol Browner hailed the technology as a step toward sustainable transportation.[23]

One month later, at the State Capitol in Sacramento, flanked by the Necar IV and Ford's P2000 hydrogen fuel cell electric automobile, air quality officials and California governor Gray Davis launched the California Fuel Cell Partnership (CaFCP). Here was yet another research consortium, comprising not only car companies but now oil companies as well. The fossil fuel industry was allying with CARB and the auto industry to test new technology, develop standards, and educate consumers.[24] Shortly afterward, GM and Toyota announced a five-year cooperative agreement to research fuel cells. It seemed, wrote *New York Times* journalist Andrew Pollack, as if fuel cells were "stepping out of the laboratory and onto the road."[25]

ELECTRIC BABYLON

Detroit's conceptual theater in non-all-battery electric advanced propulsion technologies continued at the North American International Auto Show in January 2000. Two years after the debut of the Prius, US automakers displayed their latest hybrid concepts, now featuring advanced turbo diesel engines mated to a variety of advanced power sources. Ford's aluminum-bodied Prodigy was a mybrid, equipped with a Saft nickel-metal hydride battery linked to a starter-alternator. Reputed to get seventy miles per gallon, the concept car featured restrained styling and looked like a possible candidate for production.[26]

For its part, GM unveiled a vehicle whose styling seemed calculated to leave the opposite impression. The Precept looked like a wheeled spacecraft. Harry Pearce addressed its finer points after delivering unwelcome news for enthusiasts of the all-battery electric car: the automaker was stopping production of the EV1. Only 600 customers had opted to lease the vehicle and GM, held the executive, was recognizing the shortcomings of the "pure electric" format. The EV1 had limited range and its battery pack took too long to recharge. On the other hand, GM was not cancelling the program outright and the cars would stay on the road for the time being. There was a possibility that production would be restarted, said Pearce, if costs could be cut and if demand materialized. In the meantime, GM was negotiating with Toyota about the possibility of jointly producing a hybrid electric car.[27]

One important consequence of the EV1, continued Pearce, was that the program had provided GM with useful experience relevant to other advanced propulsion technologies including hybrids and fuel cells. The automaker applied these lessons in Precept, a car in which no effort had been spared to reach the PNGV goal of 80 miles per gallon. GM engineers tried to eliminate every source of drag and used the lightest materials, including aluminum, not only for the frame but for the skin as well, a sophisticated achievement owing to the difficulty of stamping aluminum sheet. The Precept had two variants. One was a parallel hybrid electric equipped with an Ovonic battery that served as a backup for a lithium battery polymer pack that was not yet functional but promised even higher energy and power once fully developed. The other variant was equipped with hydrogen fuel cell electric drive supplied by two hydride storage vessels. Precept was not then operational but was expected to have a fuel efficiency of 100 miles per gallon.[28]

This complex object cost over $1 million and communicated a technopolitical claim similar to the one communicated by the EV1: the hybrid electric format was no more commercially feasible than the all-battery electric format. The *Economist* registered and reiterated this message in its review of the Precept, arguing that hybrid electrics did not make sense at a time of cheap oil. Toyota was heavily subsidizing the Prius, asserted the liberal newsmagazine, and the PNGV had spent hundreds of millions of dollars with little to show for it. GM was moving in the right direction with fuel cell electric propulsion, it argued. Unlike hybrids, held the *Economist*, fuel cells were not consuming vast quantities of taxpayer money and almost certainly represented the future of automobile propulsion.[29]

In fact, national developmental states had a hand in shaping all the advanced propulsion systems that automakers were then investigating, not least fuel cell technology. From Ovshinsky and Stempel's perspective, the *Economist* had made a more galling error. It omitted mention of the Precept's Ovonic battery pack, referring only to the concept car's nonfunctional lithium battery. Ovshinsky and Stempel expressed their annoyance in a letter to the editor that further rebuked the newsmagazine for failing to mention all-electric vehicles at all.[30]

The letter went unpublished, and its plea for recognition underscored how isolated OBC was becoming in the automobile world. Company president Subhash Dhar did his best to find a silver lining. Shortly after Pearce's announcement, Dhar announced that OBC's high-powered hybrid battery could also serve an auxiliary role in the conventional internal combustion-engine car.[31] With the all-battery electric car project frozen and no US commercial hybrid on the horizon, options for OBC in the automobile space were rapidly diminishing.

BIG OIL AND ECD

At the same time, ECD's longstanding ties to the oil industry were deepening, thanks in part to the hydrogen turn. In the late 1990s, ECD obtained funding from the National Institute of Standards and Technology's Advanced Technology Program for research on hydrogen storage materials, a project that also gained support from Shell. In spring 2000, ECD made an agreement with Texaco that resulted in the oil company purchasing 20 percent of ECD stock in a deal worth around $500 million, the largest single transaction in the technology company's history to that point. That fall, the two companies agreed to collaborate in fuel cell and hydrogen technology, enterprises that Texaco held would enhance energy diversity and environmental stewardship.[32]

These arrangements took on added significance in October when Texaco acquired GM's share of GM-Ovonic, creating Texaco Ovonic Battery. GM had liquidated its alliance with OBC, a move that unfolded in the context of the automaker's broader divestiture of its auto parts holdings. For six years, said Harry Pearce in a statement explaining the decision, GM and OBC had worked to cut the cost of nickel-metal hydride rechargeables for electric vehicles. In the future, GM would obtain such batteries from Texaco Ovonic Battery, which would inherit a plant in Kettering, Ohio, the only facility in the US dedicated exclusively to manufacturing them. Ovshinsky and Stempel may well have wondered what vehicle platforms these batteries would equip. In a terse statement, they noted the "great transition" confronting OBC, a company whose development owed much to its partnership with the world's largest automaker. Now, the battery maker's future would be determined in partnership with a "global energy leader."[33] The character of this new association quickly changed when Chevron acquired Texaco in the midst of the negotiations to phase out GM-Ovonic.[34]

These sudden developments represented a serious blow to Ovshinsky and Stempel's ambitions yet the partners still found reason to hope. Nearly twenty years earlier, OBC had been spun out of materials research in hydrogen storage that ECD conducted for the oil company ARCO. Neither of the Ovshinsky enterprises had real experience in fuel cell technology, but in principle, hydride compounds could be integrated into fuel cell systems and nickel-metal hydride batteries could be integrated into hybrid fuel cell electric drive systems. It was possible to believe that OBC and ECD technology could thrive in the transition to the hydrogen fuel cell electric vehicle.

INTERRING THE ALL-BATTERY ELECTRIC CAR

Years of negotiations between CARB and automakers on the technological identity of the ZEV produced regulatory concessions that incentivized the continued research into and development of all-battery electric cars but disincentivized their manufacture. By 2000, the auto industry's ZEV commitments in California had been reduced to only 1,800 units.[35] Regulators also incentivized the deployment of non-all-battery electric advanced propulsion cars by means of the partial ZEV rule, which in principle allowed automakers to use hybrids to meet some of their mandate commitments. In 2001, CARB phased in these requirements with the hope that they would encourage automakers to finally manufacture clean advanced propulsion automobiles.

However, quasi-planned US energy and environmental regulatory controls on the auto sector had far greater influence on the material practices of Japanese industry than domestic industry, and the resulting structural imbalances made the task of enforcing zero emission automobility all but impossible. All automakers welcomed CARB's tacit recognition of carbonaceous and hydrogen fuel cell electric propulsion as partial and pure zero emission technologies, respectively, but Honda and Toyota were best-positioned to take advantage of the partial ZEV rules thanks to their commercial hybrid electric car programs. Most other car companies at that time did not plan large-scale production of such vehicles. With the mandate set to take effect in January 2002, this would create new problems of public policy that would lead the national developmental state to deepen its involvement in the affairs of US automakers.

In the meantime, car companies launched a legal attack on the mandate that exploited the complexities of regulating hybrid electric technology in the US context. Hybrids were promoted as clean, fuel-efficient cars, and while California air quality was jointly regulated by the state and federal governments, regulating fuel efficiency was the sole prerogative of the federal government. In January 2002, DaimlerChrysler, GM, and Isuzu, a Japanese automaker in which GM held a large equity stake, filed a lawsuit in federal court charging that California's amendment of the mandate to encompass hybrids represented an effort to usurp federal powers to regulate fuel efficiency.[36] In June, a federal judge blocked CARB from enforcing the mandate for the 2003 and 2004 model years. A measure that air quality regulators had intended as an olive branch and devised as a means of helping ease automakers into the zero-emission era was being used against them.[37] It would not be the last time that opponents of forced technological change would attempt to take advantage of the jurisdictional divide between fuel efficiency and environmental regulation to inhibit California's ability to compel automakers to deploy clean cars.

Simultaneously, the administration of President George W. Bush moved to support hydrogen. In January 2002, the White House replaced the PNGV with FreedomCAR, a research and development consortium devoted to hydrogen fuel cell technology that it claimed would make fuel efficiency regulations obsolete. The accompanying Hydrogen Fuel Initiative committed $1.5 billion for infrastructure, the largest-ever such investment.[38] Hydrogen technology had some bipartisan support in Congress but divided opinion in environmental and business circles, not always along ideological lines. Al Gore had spoken approvingly of the hydrogen fuel cell during his presidential campaign, remarking that the technology yielded only "clean water" in its waste stream.[39] Influential green energy advocates like Amory Lovins long supported hydrogen, and representatives of the Natural Resources Defense Council and the Union of Concerned Scientists lobbied CARB to bolster the quota of hydrogen fuel cell vehicles. Other voices were more skeptical. Hydrogen research and development was nothing new, held the *New York Times* business section, and moreover would take years to yield practical results.[40] In hydrogen, wrote the *Wall Street Journal* reporter Jeffrey Ball, automakers had embarked on a "futuristic technological crusade."[41]

Hydrogen polarized public opinion, but after years of conflict with the auto industry, it held technocratic appeal for CARB under the leadership of its chair, Alan Lloyd. In April 2003, the air resources board deleted references to fuel efficiency in the mandate and expanded its credit system to allow automakers to meet all their ZEV obligations with a mix of hybrid electric and hydrogen fuel cell electric cars. In August, GM and DaimlerChrysler ended their lawsuits.[42]

The all-battery electric car had been neutralized, thanks in no small measure to GM's prodigious efforts in litigation and technopolitical theater. Ford had also played an important role in this project but the company took a less confrontational approach to air quality politics and expended relatively less energy in its resistance than its peers. The number two US automaker preferred to let GM, DaimlerChrysler, and Toyota vie for leadership in alternative propulsion technology while participating in the discourse of clean cars and limiting its investments in them. All automakers depended in varying degrees on external suppliers of original equipment relating to electric propulsion systems, but Ford tended to outsource more than the major players, including from vendors aligned with and even owned by its competitors. The company experimented with Ballard fuel cells and equipped the initial model of its converted Ranger light truck with lead-acid batteries made by Delphi, then still part of GM. For its 1999 electric Ranger, Ford chose Panasonic nickel-metal hydride batteries.[43]

Ford also had the Th!nk City, an all-battery electric car launched in 2000 by Jacques Nasser, Alex Trotman's style-conscious successor. The project targeted the environmental hipster and was based on the PIV (Personal Independent Vehicle), a two-seat city car clad in pastel-colored thermoplastic and powered initially by a nickel-cadmium battery pack. The automobile was designed by the Norwegian company Pivco and a few dozen of the PIV3 variant were demonstrated in California, where it was known as the CityBee. In November 1998, Pivco went bankrupt and was later acquired by Ford, which rebranded the latest CityBee model as the Th!nk City. The automaker then partnered with the US Department of Energy (DOE) in what it claimed was the largest urban electric vehicle demonstration in North America, eventually involving some 376 vehicles leased across the US and Canada beginning in 2001.[44]

By then, GM was well into the process of deactivating the EV1 and the triumvirate that had navigated the company through the 1990s was no more. John Smale, the executive who had led the boardroom rebellion against Stempel, relinquished the chair of the board of directors to Jack Smith in 1995. In May 2001, vice chair Harry Pearce retired from GM. In June 2000, Rick Wagoner, a Harvard-trained economist who had served as GM president and chief financial officer, succeeded Smith as chief executive officer. Smale, Smith, and Pearce had planned the demise of the EV1, but Wagoner was credited with (and later blamed for) officially cancelling the program. The unfortunate denouement of the saga of GM's all-battery electric car would earn Wagoner infamy in the annals of corporate mismanagement and became metonymic of GM's broader problems in the early twenty-first century.

All told, GM manufactured 1,117 EV-1s by hand, including several hundred fitted with Ovonic batteries. When leases expired, the automaker repossessed vehicles and destroyed almost all of them, sometimes ostentatiously. Having played perhaps the leading role in giving rise to the ZEV mandate by commissioning AeroVironment to build the Impact prototype and then further entrenching the mandate by creating production variants of the car and popularizing the all-battery electric format, GM seemed intent on sending an unequivocal message about where the authority in automobile technology was ultimately vested. The company generated a good deal of bad press and ill will in the process.[45] No single event would dominate the collective memory of electric car enthusiasts more than the death of the EV1, which would serve as a rallying cry for a new generation of activists.

In effect, GM created political cover for its peers, who were just as eager to suppress the all-battery electric car but had a somewhat better understanding of the consequences of the public relations fiasco in the making. Ford terminated the Th!nk program in 2004 and initially planned to scrap the vehicles but when enthusiasts brought environmental activists into the fray, Ford reconsidered and shipped 300 units to Norway, where they found a ready market in a society that embraced the city car concept. Toyota had employed the first-generation RAV4 EV essentially as a technopolitical decoy to distract attention from the Prius, so it was less concerned about the ideological risks of allowing a few all-battery electrics to remain in the public eye. The company followed industry practice in offering the vehicle mainly for lease, but it did sell several hundred units to American users.[46]

OVONIC BATTERY COMPANY CODA

Neutralizing the all-battery electric car involved much more than physically destroying automobiles. For GM, OBC was a serious threat, one it manipulated to restrict the ability of Toyota and Matsushita/Panasonic to use nickel-metal hydride technology. Having accomplished that task, the automaker sold its share of GM-Ovonic to an oil company, whose views of battery electric propulsion technology mirrored those of the car companies. Ovshinsky and Stempel's seeming lack of awareness of OBC and ECD's place on the industrial chessboard could be ascribed in part to a belief in meritocracy that traced to self-confidence born of initial successes enabled by public policy.

Another important factor informing Ovshinsky and Stempel's perspective was that circumstances increasingly compelled them to perceive innovation through the lens of national interests. According to the historians Lillian Hoddeson and Peter Garrett, Ovshinsky believed that ECD survived in the 1970s thanks largely to partnerships forged in Japan's consumer electronics industry.[47] Perhaps inevitably, however, Ovshinsky and Stempel's efforts to develop the market for electric vehicle batteries pulled them into the orbit of the US auto and oil industries. This national perspective only deepened after GM-Ovonic was liquidated and guided the resolution of unfinished business between OBC, Matsushita, and Toyota. It was only a matter of time before someone in the US opened up a first-generation Prius battery pack and discovered that it contained dozens of electronics commodity cells. In March 2001, OBC sued Matsushita, Toyota, and their Panasonic EV Energy (PEVE) joint venture on the grounds that the Prius battery infringed OBC patents. In announcing the suit, OBC also accused Matsushita of failing to pay royalties on the cells in Prius battery packs. Matsushita may well have reasoned that electronics cells ceased to exist as such when they were repurposed for use in electric vehicle systems.[48]

OBC immediately terminated the 1992 agreement and received support from ChevronTexaco. In October, Cobasys, the manufacturing joint venture between OBC and the oil giant, joined the action as coplaintiff, and in July 2004 the parties reached a cross-licensing agreement on nickel-metal hydride rechargeable technology. Matsushita and PEVE had to pay OBC and ECD a $10 million patent license fee relating to commodity cells while

Cobasys was awarded a \$20 million fee on patent licenses granted to Matsushita, PEVE, and Toyota relating to batteries for electric vehicles. In addition, the joint venture was to receive royalties through December 31, 2013, on certain batteries sold by Matsushita and PEVE in North America. On the other hand, the licenses secured by Matsushita, PEVE, and Toyota did not grant rights to sell certain batteries in certain transportation applications in North America until mid-2007 and to sell commercial quantities of certain batteries in certain transportation as well as stationary applications in this market until mid-2010. Cobasys intended to manufacture its own nickel-metal hydride batteries but it also won the right to distribute PEVE batteries using this chemistry in certain North American markets until mid-2010. Cobasys and PEVE also agreed to cooperate in developing the next generation of nickel-metal hydride batteries for hybrid electrics.[49]

ChevronTexaco now had a powerful say in how the Japanese group developed and marketed nickel-metal hydride rechargeables for electric vehicles in the US market. To a degree, PEVE was able to obviate the legal restrictions on innovating this technology. The company developed a prismatic cell pack for the Prius of improved energy and power by slimming and eliminating parts without substantially modifying battery chemistry.[50] Increasingly, however, Toyota and PEVE looked to replace the nickel-metal hydride rechargeable in the hybrid electric application with the lithium ion rechargeable, a power source technology that promised certain performance advantages and, perhaps even more importantly, was less constrained by patent claims.

Like GM, Chevron cooperated with OBC and ECD only insofar as it served its interests. Even as the US companies joined in legal action against the Japanese partnership, the oil giant appeared to be trying to "take over, neutralize, or destroy" ECD, as Chester Kamin, Ovshinsky's attorney, reported in 2004.[51] Ovshinsky remained haunted by the prospect that connections with erstwhile collaborators in Japan might have been preserved. In April 2001, the longtime ECD chemist Dave Strand reported that the Matsushita researcher Takeo Ohta, a personal friend of the Ovshinsky family, had fondly recalled the "good handshake" era of the early 1990s. Ohta confessed that he felt his employer had handled the situation badly and wished that relations between the two firms in the battery sector could be as amicable as they were in the realm of optical memory. Ohta held out hope that some sort of arrangement could still be reached.[52]

Ultimately, OBC's pact with GM ruled out substantive cooperation with Japanese industry on electric vehicle technology. Toyota had already decided on the hybrid electric format before relations between OBC and Matsushita soured from mid-1995, but the ensuing legal imbroglio validated the strategy to bundle commodity cells in the hybrid drivetrain. In the 2000s, OBC and ECD would promote their technologies in service of the hydrogen economy that many observers expected to take shape, and a new era of green car technopolitics would begin.

10 COMPUTERS ON WHEELS

There were two things that everybody knew about electric cars in the year 2002: one, they suck, and two, they're dead. My goal was to radically change the opinion of what an electric car was in the public mind.

—Martin Eberhard, cofounder of Tesla Motors, 2015

In late July 2003, an unusual event unfolded in west Los Angeles. A convoy of vehicles including RAV4 EVs and EV1s wended its way through the Hollywood Forever Cemetery, disembarking drivers who gathered at a "funeral" for the electric car. Among the mourners were actors Alexandra Paul and Ed Begley, Jr., and members of the AeroVironment team responsible for the original Impact including Alec Brooks, Alan Cocconi, Paul MacCready, and Wally Rippel.[1] The event was organized by the filmmaker Chris Paine to protest the auto industry's termination of its electric car programs. Paine, whose own EV1 was slated for recall in August, delivered one of the eulogies. The EV1 died before its time, he said, but there was a ray of hope. Even more capable electric cars were waiting in the wings, equipped with lithium ion battery technology that would enable a much longer range than the EV1. Yet they were unlikely to flourish, held Paine, because politicians and manufacturers had convinced California to scrap the Zero Emission Vehicle (ZEV) mandate, ending all research and development of electric vehicles.[2]

It would have been more accurate to say that automakers sought to suppress all-battery electric technology. The car companies continued researching and developing other kinds of advanced propulsion technologies.

Toyota and Honda were commercializing hybrid electrics, and Toyota and Matsushita's joint manufacturing enterprise (Panasonic EV Energy or PEVE) was commercializing battery systems for these vehicles. The Ovonic Battery Company (OBC), formerly partnered with General Motors (GM) and then involved in a joint manufacturing venture with ChevronTexaco, hoped to supply some part of the hybrid market. At the national laboratories, the US Department of Energy (DOE) continued its longstanding research in lithium ion chemistries for electric traction. Governments and automakers were also committing considerable resources to hydrogen fuel cell electric drive.

Many in the electric car enthusiast community saw these efforts as propaganda, or *greenwashing*, in the vernacular of the environmental movement, that obscured harsh facts. Automakers had produced all-battery electric cars of high quality and then snatched them from the hands of consumers and disposed of them. Mawkish though the mock funeral had been, it represented something of the depth of public feeling. It was widely believed that automakers, and even air quality regulators, had conspired to subvert the popular will. Alienated and angered, some enthusiasts, environmentalists, and ordinary motorists turned to activism, giving rise to a new phase in the appropriate transportation technology movement.

As often occurred in the history of the contemporary electric car, idealism informed material practice. Enthusiasts formed groups like Jumpstart Ford and DontCrush.com to save their cars from destruction. From these ad hoc responses emerged Plug In America, an advocacy group cofounded by a former GM employee named Chelsea Sexton, an EV1 marketer-turned-activist who would achieve fame as a protagonist in Paine's popular feature-length 2006 documentary *Who Killed the Electric Car?* Plug In America promoted the plug-in hybrid electric as the solution for the nation's energy and environmental problems, and the organization would wield significant influence in Washington, swaying former Central Intelligence Agency (CIA) director James Woolsey as well as an increasing number of politicians, including future president Barack Obama.[3]

For others, the only real electric car was one that drew its energy solely from the galvanic battery. The death of the EV1 and the other all-battery electrics inspired fresh grassroots initiatives in the format. As in the past, practitioners initially sought to use available technology, applying lead-acid rechargeables in converted and purpose-built platforms, but many dreamed of developing more sophisticated systems. Cocconi was among them, as was Martin Eberhard, a computer and electrical engineer and successful dot-com entrepreneur. In the late 1990s, Eberhard and his collaborator, Marc Tarpenning, developed the Rocket eBook, one of the first electronic book readers, and formed NuvoMedia to commercialize it. The enterprise failed, but at the peak of the tech bubble in 2000, Eberhard and Tarpenning were able to sell the technology to the media company Gemstar for $187 million.

Newly wealthy, Eberhard and Tarpenning had time to ponder the causes of the collapse of the electric vehicle revival. Hydrogen hype was one factor. Many enthusiasts blamed the pro-hydrogen Alan Lloyd and the California Air Resources Board (CARB), and Eberhard would become a trenchant critic of both.[4] Another factor, believed Eberhard, was that automakers had deliberately designed their electric vehicles to fail. He felt that most compliance cars were ugly and underpowered, with the worst of them being "heartless" conversions of existing production vehicles that nobody wanted to drive. An exception was the EV1, one of the best electric cars ever built, in Eberhard's opinion.[5]

To be sure, automakers never intended to produce the EV1 and the other all-battery electrics as consumer commodities. These vehicles disappeared as a result of corporate decisions to prevent them from coming to market in the first place. Automakers then devised the theory that it had been consumers who had rejected the electric car, ostensibly because the limited energy density of existing batteries could not provide performance on par with gasoline-fueled internal combustion engine (ICE) propulsion.

Nevertheless, Eberhard believed that public attitudes on all-battery electrics were misinformed and had to be changed. He and Tarpenning aimed to build and commercialize the electric supercar that mainstream automakers claimed could not be built and commercialized. The vehicle would utilize lithium commodity cells in the most energetic and powerful battery pack yet devised for an electric car, and it would be marketed not to enthusiasts or environmentalists, but to well-to-do skeptics who favored kinetic thrills and who were not averse to environmental virtuosity if it did not come at the price of acceleration, range, and style. The project would eventually enlist many of the pioneering enthusiast-experts involved with the Sunraycer, Impact, and EV1 including Brooks, Cocconi, and Rippel. As this group of talented outsiders searched for allies and built up a community of alternative expertise in the shadow of the automaking establishment, they drew heavily on resources and methods from the consumer electronics and information technology sectors. In the new millennium, these and other likeminded groups would bring about the reconceptualization of the all-battery electric car as a computer on wheels, an analogy that guided a new generation of experiments in electric propulsion that contributed important innovations in the material culture and practice of automobility in a context of profound sociotechnical change. Deindustrialization, the information technology boom and bust, the deregulation and marketization of electricity, and the rise of the politics of climate change would intersect with and inform these activities and help stimulate the revival of the electric car.

THE TAO OF SILICON

Eberhard and Tarpenning professed little knowledge of battery or automobile engineering, but the world of information technology seemed to offer a wealth of possible solutions. By the early 1970s, Silicon Valley had become the electronic workshop of the world, but in succeeding decades the region and the state of California began to deindustrialize.[6] Offshoring and outsourcing gave rise to the so-called fabless foundry, an enterprise concerned solely with the design of semiconductors and microprocessors that could be produced cheaply by enterprises in other countries, increasingly in Asia. Out of the decline of US manufacturing arose the idea that outsourced electronic components could be seamlessly integrated in novel configurations. The personal computer commodifier Michael Dell referred to this as "virtual integration," a management model that emphasized marketing and logistics over research and development and took the trend toward corporate vertical disintegration to its logical conclusion.[7]

American automaking was undergoing similar structural changes. In 1997, GM sold Hughes Aircraft to Raytheon and two years later it spun off Delphi Automotive Systems, ending 90 years of virtually integrated operations at the car giant. In 2000, Ford divested Visteon. As Eberhard researched the auto industry, he learned that its core competencies were shrinking. Car companies still built their own engines, but most of them outsourced many or most other auto parts to varying degrees. If an electric car could not easily be completely built from scratch, reasoned Eberhard, perhaps one could be designed around components acquired in the open market. The technologies that most interested him were the induction motor and the rechargeable lithium ion cell, the latter being a key enabler of mobile consumer electronics. Eberhard and Tarpenning calculated that a pack of such cells would have sufficiently high energy density to give an induction motor–equipped car unprecedented acceleration and range.[8]

Not all of these ideas were new. Car companies had experimented with induction motors since the 1960s, and Toyota, Honda, and Matsushita pioneered the adaptation of the commodity cell for the hybrid electric format. The novelty of what Eberhard and Tarpenning envisaged lay in combining these systems and utilizing lithium chemistry. As the Japanese manufacturers discovered, however, using commodity cells to build battery packs was anything but an off-the-shelf solution. Resolving the problems connected with cell quality control, pack management, and systems integration had tested the resources of these large and experienced industrial enterprises.

Still, Silicon Valley investors were perennially looking for the next big thing and conditions favored diversification into transportation. Progress in information technology is often linked with Moore's Law, the miniaturization trend in semiconductor production named for Gordon Moore, the cofounder of Fairchild Semiconductor and Intel, who observed a correlation between falling costs and increased transistor density in silicon chips in 1965.[9] Semiconductor scientists, engineers, and manufacturers, as well as historians of science and technology, perceived this trend to be in a permanent state of crisis that they understood in terms of the anticipated physical barriers to scaling.[10]

Long before semiconductor manufacturers encountered the atomic limits of miniaturization, however, they faced the socioeconomic specter of overproduction. Integrated circuits were first used in military applications, and in 1965 Moore predicted that the scaled integrated circuit would someday enable a cheap home computer. However, he did not specify a timeline and was uncertain about nearer-term civilian applications.[11] The first non-military markets for integrated circuits turned out to be in rudimentary consumer electronics like electronic wristwatches and pocket calculators. By the late 1970s, Moore worried that scaling was outstripping the capacity of this market to absorb commodity microchip production and anticipated that the next major applications of chips lay in homes and automobiles.[12]

Integrated circuits and microprocessors did start to be introduced into ICE cars by way of engine control units in this period. At the same time, microchips began to be applied in the first personal computers, but personal

computing was not commodified until the emergence of the Wintel monopoly of Microsoft Windows and Intel microprocessors around the end of
the Cold War, and it was with that momentous development that the economic conundrum of Moore's Law was resolved, at least temporarily. The
proliferation of personal computing power made vast fortunes and in turn
enabled the construction of the internet and internet commerce, which
absorbed increasing volumes of surplus capital by the late 1990s. Speculation and overcapacity burst the dot-com bubble in 2000 and 2001, wiping
out $5 trillion in paper wealth, and in the recession that followed the terror
attacks of 9/11 there was apprehension about the direction of the information technology revolution.[13]

Eberhard and Tarpenning did not explicitly frame the all-battery electric
car as a solution to the economic crisis of Moore's Law, although some
observers would later make this suggestion. What the pair did do was lead
the way in arguing that an important avenue of growth for information
technology lay in electric automobility.[14] In making this case, the entrepreneurs drew on the communitarian energy of start-up culture, engaging
collaborators who contributed capital, expertise, and technology.

TAMING THE LITHIUM COMMODITY CELL

Eberhard and Tarpenning perceived that their immediate engineering
problem related to adapting and managing the chemical energy of the lithium rechargeable battery for the automobile application. The nickel-metal
hydride battery employed an aqueous or water-based electrolyte and would
not easily burn. In contrast, most lithium ion formulas were based on highly
combustible materials, including organic electrolytes and metal oxides. The
chief obstacle to a commercial lithium rechargeable had been the lack of a
safe anode. In the 1980s, researchers commonly used test anodes of metallic lithium, a material that caused lithium ions to plate unevenly on the
anode, forming "dendrites," or treelike encrustations of lithium that with
repeated charging and discharging could bridge the electrodes and induce
a short circuit, and possibly an explosion. In the mid-1980s, a safety breakthrough came when a number of researchers, including Asahi Kasei's Akira
Yoshino, established that a carbonaceous anode would enable relatively
unproblematic reversible lithium intercalation. Sony's Energytec division
successfully adopted this idea, mating a graphitic anode with a lithium

cobalt oxide cathode in what became the world's first commercial lithium ion rechargeable battery, which appeared on markets in 1991. Nevertheless, the materials of the lithium cobalt oxide system and other lithium ion chemistries still comprised a highly volatile mix. Producing and packaging them in commodity cells called for painstaking process quality control.[15]

It was crucial to understand how the failure modes of lithium rechargeables related to the duty cycle of the appliance served by the power source. This was not a simple matter, in good measure because mobile electronics design was largely alienated from power source design, particularly in the US. Lithium rechargeables could be ignited by a number of factors, including overcharge, overdischarge, and short circuits, and had to be equipped with numerous safety features, including current interrupters and gas vent mechanisms. American manufacturers of mobile devices did not produce the batteries for these applications, and problems of systems integration abounded. Designers of mobile electronics and computers tended to undersize battery cavities for expected performance or otherwise mismatched them with power-source form factors.[16]

The results could be explosive. If a cell in a lithium battery ignited, it could trigger an uncontrollable chain reaction called "thermal runaway," a sudden release of chemical energy that could spread to other cells and cause a fire that, fed by the oxygen in the metal oxides, could not easily be extinguished. The potential for trouble was greater in lithium batteries for electric vehicles because of the larger quantity of combustible materials involved.

Here, Eberhard saw another advantage of the cylindrical commodity cell. Its smaller size, he reasoned, gave a safety as well as an economic edge over the larger prismatic cell. If a prismatic cell ignited, there was a greater likelihood that the resulting fire might engulf the rest of the pack, and then the vehicle. If a commodity cell ignited, on the other hand, the resulting small fire might be more easily contained if the battery pack had safety controls.[17]

In 2001, Eberhard and Tarpenning started sketching designs for a such a system, but they quickly realized that the commodity cell approach was more complicated than they had anticipated. The economics of the consumer electronics industry shaped lithium commodity cells in ways that made them difficult to adapt for electric vehicle applications. Pressure to cut costs came at the expense of quality. The lithium cell sector had rapidly become a highly competitive, low-margin industry dominated by a few firms based mainly in Japan. From around 2000, these companies began to offshore manufacturing to South Korea and China in operations that industry insiders observed were initially characterized by extensive bugs and high cell scrap rates.[18]

These problems illuminated the asymmetrical coproduction of materials, lithium batteries, and their applications. In the era of mobile computing, manufacturers assumed that consumers would throw away and replace old handheld devices long before their aging batteries became a problem. Hardly any research was devoted to battery reliability and safety.[19] The designers of cells were guided primarily by the imperatives of microprocessor scaling. The designers of mobile devices were constantly introducing faster processors that generated more heat and required more power. For some cell designers, an efficient solution was to make room for more reactive material in the cell casing by thinning the separator, a crucial safety device. Separators are polypropylene- or polyethylene-based polymer membranes that insulate electrodes and inhibit short circuits caused by dendrites, while offering minimal resistance to ionic transport. Separator micropores expand and cut off charge current in the event of a heat spike and are the last line of defense against thermal runaway if failure is not sudden.[20]

However, separator membranes had not been specifically designed for use in lithium ion batteries. The basic chemical formula dated to the late 1960s and had been developed for use in filtration equipment and breathable garments, including surgical gowns and recreational clothing, where it was popularized as the Gore-Tex brand.[21] When thinned in a lithium rechargeable, separators often failed to provide sufficient insulation in the integrated battery appliance, a factor in a spate of fires involving mobile devices through the 2000s and 2010s.[22] Clustering lithium commodity cells of uneven build quality and reduced margins of separator safety in a series-wired battery pack for an electric vehicle posed real hazards, especially if the pack's management electronics did not properly control the rate or state of charge of each cell. Charging cells too fast could induce dendrites, short circuits, and thermal runaway.[23]

ECHOES OF IMPACT

Designing and building a safe lithium ion battery pack entailed prodigious sociotechnical challenges, but there were many other obstacles on the road to the electric supercar. Eberhard needed to acquire the rest of an electric propulsion system, and his search for ideas and technology brought him in late 2002 to southern California and AC Propulsion (ACP), the small but influential company linked to AeroVironment. ACP had been founded in February 1992 in the Los Angeles county suburb of San Dimas by Wally Rippel and Alan Cocconi, the engineers who had developed much of the propulsion system for the original Impact as consultants for AeroVironment. The pair quickly tired of the tedium of working with Hughes and GM to turn the idiosyncratic Impact into a manufacturable commodity. Cocconi had been especially annoyed by the automaker's decision to equip the Impact with an inductive paddle charger and an off-board charging station, technologies that ensured a kind of proprietary control over the car and that Cocconi believed detracted from vehicle performance and posed serious complications for infrastructure as well. In response, he built his own integrated electric propulsion system, installed it in a converted Honda CRX, and convinced Rippel and Hughes's Paul Carosa to join him in developing the system as a seed technology for the auto industry.[24]

Free from direct corporate oversight, ACP could indulge its creative imagination, but within the confines of the paradoxical reality that the auto industry, its main customer, was hostile to electric propulsion technology. Consequently, generating revenue was a constant problem. After nine months of working without pay, Rippel returned to AeroVironment in the fall of 1992.[25] Over the course of the 1990s, ACP gradually developed a business model around a symbiotic relationship with mainstream automakers in the market for compliance cars. Using production facilities in China, the company supplied drive systems to Honda (and later Volkswagen) for packaging in converted production models. For the automakers, this was a quick and relatively inexpensive way to gain experience in advanced propulsion technology and also meet their mandate commitments. For ACP, the arrangement provided funds to continue its experiments.[26]

Cocconi's signature project was the tzero, a car based on the Impact formula. It was a two-seater that employed a lightweight structure and advanced electronics to wring the best possible performance out of lead-acid batteries. The vehicle had a primitive tubular frame and a fiberglass body wrapped around an integrated powertrain that combined an alternating current (AC) induction motor, regenerative braking, a lead-acid battery pack and management system, and a miniaturized charger. On paper, the initial lead-acid-powered variant of the tzero offered performance comparable to the second-generation EV1 equipped with Panasonic lead-acid batteries, a heavier and more polished vehicle. Cocconi claimed that the tzero could accelerate to 60 miles per hour in about four seconds and had a useful range of between 80 and 100 miles. To provide greater range, ACP devised a gasoline-electric generator to maintain the charge of the tzero's battery and packaged it in a trailer. This towable generator, or range extender, turned an all-battery electric car into a kind of hybrid electric car, but without the costly modifications that would have spoiled a perfectly good all-battery electric, recalled Cocconi. Range extenders became popular among electric vehicle enthusiasts in the late 1990s and early 2000s.[27]

The tzero debuted at the Los Angeles Auto Show in 1997, and its prospects received a boost when Alec Brooks joined ACP in 1999. Over the course of the 1990s, Brooks organized a team at AeroVironment that provided research and testing services and equipment in support of GM's EV1 and hybrid electric programs. This represented important business for Aero-Vironment at a time when its other ventures were not doing as well, but by the end of the decade, Brooks was growing restless. When he received an offer to manage ACP's tzero small-volume production program, he jumped ship. For Brooks, ACP offered a more exciting vision of the future than GM, along with all the headaches that came with a start-up. Where the engineer had overseen a group of around eighty people at AeroVironment, he now performed an array of tasks as a jack-of-all-trades in a company of six or seven people. Brooks set up a website, a database, and a parts-numbering system, and he also contributed to the tzero's handling development. Like Rippel, Brooks found the work intellectually stimulating but financially unrewarding. With ACP unable to pay him a regular wage, Brooks returned to AeroVironment in 2003.[28]

Brooks was on his way out of ACP when Eberhard began making inquiries there. Looking past the tzero's unpolished exterior, the dot-com entrepreneur was impressed. Eberhard commissioned the company to build him a copy, invested $500,000, and discussed his ideas for a lithium commodity cell pack with Cocconi and Tom Gage, an engineer and former race car mechanic and Chrysler employee who joined ACP in the mid-1990s and rose to become its president and chief executive officer (CEO). Cocconi was also convinced of the potential of lithium power, thanks in part to experiments he was conducting with remotely piloted solar-powered aircraft equipped with laptop cells.[29] Eberhard maintained that Cocconi and

Gage informed him that ACP could provide him with a tzero equipped with a lithium ion battery pack, and Eberhard invested more money. But by late 2003, the lithium tzero project was starved for cash because the market in components for compliance cars, ACP's primary source of revenue, had evaporated with the neutralization of the ZEV mandate.[30] The company could not complete Eberhard's lithium tzero, and the entrepreneur mulled the possibility of purchasing ACP outright before resolving to form his own electric vehicle start-up.[31]

A POWER PLANT ON WHEELS

The efforts of ACP to survive in this period fostered another project that had far-reaching implications both for automobility and electricity as energy conversion systems. This was bidirectional electric vehicle power and vehicle-to-grid, a sociotechnical imaginary that purported to situate the electric car as a decentralized power plant that could supply power to the grid. The concept originated in a collaboration between ACP and a group of environmental policy analysts that emerged around 2001 at the intersection of the crisis in mandated zero emission automobility and a crisis in California electricity precipitated by deregulation and marketization.

Regulated electricity was a legacy of the response to the collapse of the financial infrastructure that underpinned the electricity infrastructure during the Great Depression. Following the stock market crash of 1929, highly leveraged holding companies that owned utilities and other interests, exemplified by the empire of Edison protégé Samuel Insull, went broke. In response, New Deal regulators restricted ownership of electric utilities to public and private entities that specifically produced electricity, restructuring the industry mainly around intrastate and a few interstate holding companies. The resulting regulated vertically integrated utilities provided a variety of supply services that together served as much to order and stabilize the system as to meet consumer demand. In principle, the most profitable of these services was peaking power, delivered when demand was highest. Regulated utilities spread the high cost of peaking power across the much less lucrative stabilization services, including spinning reserve, or reserve power generation, and frequency regulation, the balance of load and generation that maintained the system at 60 Hertz. In this system, pricing did not reflect real-time shifts in supply and demand as in other consumer commodity markets, largely because, for most of the twentieth century, meter technology was only capable of measuring total electricity consumption. Had real-time metering then been possible, consumers would have faced steeply higher prices in high-demand periods and paid a premium for convenience. In regulated electricity, consumers paid a relatively low price averaged across the integrated high- and low-cost bundles of services, a system that socialized the rhythms of American industrial and middle-class domestic life.

From the late 1970s, planners made a series of efforts to dismantle this system as part of a broader public policy shift to seek efficiency by fostering competition in regulated infrastructure enterprises, including commercial aviation, trucking, and the gas utilities.[32] This process was guided by the doctrine of marginal cost pricing, an economic theory that prescribed pricing reflecting the extra cost of an additional unit of output. In the 1980s, proponents of marginal cost pricing pointed to the seemingly successful deregulation of the airlines, an initiative pioneered by the Cornell University economist and Carter administration advisor Alfred E. Kahn. An influential ideologue of deregulation, Kahn asserted that the deterioration in the quality of airline service following deregulation was more than compensated for by the decline in prices for consumers.[33] However, he gave relatively little consideration to what marketization might mean in the context of electricity.[34] In the late 1980s, one observer imagined that deregulation would be organizationally expressed in electricity as it had in other industries: the unbundling and shedding of "nonessential" services, the eliminating of cross-subsidization, and the commodification of core competencies.[35]

Kahn would later profess doubts about the assumed benefits of deregulated electricity, especially consumer choice.[36] In California, planners recognized that the marketization of electricity could destabilize the system, so they developed a hybrid approach that in essence represented the regulation of deregulation. In 1998, the state legislature created the California Independent System Operator (CAISO), a nonprofit organization that reported to the Federal Energy Regulatory Commission (FERC) and was designed to manage the emerging market. Generation was deregulated and separated from distribution, which remained regulated, as no group had an interest in duplicating transmission. Some 40 percent of installed capacity was sold to newly created independent power producers from which the three major utilities were compelled to purchase power auctioned on a day-to-day, hour-to-hour basis. Retail prices were capped, but wholesale prices were not.[37] In California's partially deregulated, disintegrated system, independent producers dominated peaking power as the most profitable service.[38] The less lucrative grid stabilization functions were spun off as so-called ancillary services that CAISO contractually compelled producers to supply.[39]

These interventions produced a highly unstable sociotechnical regime in which the public bore all the risk. In 2000, environmental and sociotechnical factors converged to cause disequilibrium. Hot weather stoked demand, traditionally met with peaking plant, typically gas turbines, built expressly for this purpose and used only a few times a year. It took time and money to add peaking capacity and deregulation disincentivized new construction. Moreover, nearly 20 percent of the state's generating capacity was idled, ostensibly for maintenance. As a result, producers operating peaking capacity kept it in operation longer to meet rising demand, causing a wholesale price spike of 800 percent in May, exacerbated by market manipulation. Independent producers profited, but Southern California Edison nearly failed while Pacific Gas and Electric went bankrupt and consumers suffered from rolling blackouts in early 2001.[40]

With prospects of a market in electric vehicles fading, ACP saw an opportunity in California's electricity crisis to market its technology. The company looked to the ideas of Willett Kempton, an anthropologist and environmental policy analyst at the University of Delaware, and Steven Letendre, a professor of management studies at Green Mountain College. In an academic article published in 1997, Kempton and Letendre argued that bidirectional electric cars could be pressed into service as mobile power plants in ways that would satisfy all interests. For the utilities, such a system promised a cheap and fast way of satisfying rising demand. Ratepayers would benefit from cheaper and more reliable electricity services. And individual private owners of electric cars would be empowered as entrepreneurs, selling electricity as a means of paying off their investments. Kempton and Letendre presented their case as a syllogism: because the aggregate generation capacity of the US light duty vehicle fleet was more than sixteen times that of stationary generation plant and because the average light vehicle was used only 4 percent of the time, a fleet of electric cars could constitute an important resource for electric utilities, even if it was only a fraction of the size of the ICE vehicle fleet.[41]

Cocconi and ACP originally developed bidirectional power technology not for electric utility applications, but like much else in their design philosophy, as a means of mitigating the limitations of the battery technology of the day. Bidirectionality was built into the company's second-generation AC-150 powertrain as an emergency feature to enable electric cars to transfer charge between each other at a time when battery capacity was still relatively limited and the possibility of running out of charge and becoming stranded were very real.[42] With California plagued by skyrocketing electricity prices and rolling blackouts, Gage contacted Kempton, who confirmed he had not patented the concept of the electric vehicle as a bidirectional power plant.[43] Then ACP launched a project to demonstrate what Gage dubbed "vehicle-to-grid," a task managed by Brooks.[44] Working with Gage and Kempton and consulting CAISO, Brooks set out to understand how marketized electricity was managed and how the electric car might function as a utility resource.[45]

Kempton and Letendre initially envisaged electric cars serving the peaking market, but that market was controlled by independent producers. That left the ancillary services as the only other electricity markets. Theorists including Kempton believed that frequency regulation was the most attractive of these because it constituted about 80 percent of CAISO expenditures on ancillary services, and it was to this market that ACP looked.[46] With funding from CARB and help from the National Renewable Energy Laboratory, the team staged an experiment. It installed an AC-150 drive into a converted Volkswagen Beetle, where it functioned as an integrated onboard charger that converted AC power to DC for use in the battery and operated the process in reverse. The team demonstrated bidirectional power flow using wireless power dispatch commands simulated from CAISO historical data, not linked in real time because the power capacity of a single vehicle was too small to be accepted by the operator's energy management system.[47]

The team assumed that all the basic technologies of vehicle-to-grid were at hand and had only to be applied in novel ways. Nevertheless, the experiment seemed to suggest that vehicle-to-grid did require some technological innovation. A crucial component was the aggregator, a notional sociotechnical entity that mediated between the grid operator and connected vehicles, tracking available electrics and representing them as a unified source of controllable capacity. Brooks noted that an aggregator required sophisticated software, technology that was beyond the scope of the experiment.[48]

The aggregator also had sweeping institutional-organizational implications. Vehicle-to-grid was predicated on interconnecting automaking and

electricity-making, completely different businesses governed by different regulatory bodies.[49] Brooks held that the chief obstacle to this ambitious interdisciplinary project, one that spanned transportation, the environment, and
energy, was the lack of an institutional champion. If utilities had an interest
in the scheme in principle, automakers did not, so that champion would
have to be a regulatory agency.[50] To serve as a practical solution to problems
caused by grid deregulation and marketization, the car-grid energy conversion imaginary paradoxically required further layers of governmentality.

INDUCING INNOVATION

Meanwhile, Eberhard and Tarpenning were preparing their start-up. On
July 1, 2003, they founded a company they called Tesla Motors, after Nikola
Tesla, the famed inventor of the induction motor, headquartered in San
Carlos, California. Before setting up a proper laboratory, Eberhard performed experiments in propagating catastrophic failures of lithium cells at
his home in Woodside. Far from prying eyes at his isolated estate off Skyline
Boulevard, he could bury and detonate batteries without attracting attention. Eberhard quickly ruled out some of the lesser-known manufacturers
like Thunder Sky, whose products were popular among hobbyists but that
he came to consider inherently unsafe.[51]

Convincing suppliers to do business with Tesla Motors was difficult. They
had little incentive to work with an outsider like Eberhard, whose initial
production plans promised unprofitably low volumes. Cell suppliers also
faced the prospect of legal liability if things went wrong. Lithium cells were
classified as a hazardous good, and Eberhard was proposing to use them in
one of the most hazardous applications. The main reason why enthusiasts
had not hitherto used lithium cells in electric cars was because cell manufacturers hitherto refused to sell them for this purpose. Moreover, virtually
all lithium cell manufacturers were based in Asia and operated in exclusive
supply arrangements. After engaging a number of companies and testing a
variety of cells, Eberhart and his team came to favor Sanyo by late 2003.[52]

Emulating Stanford Ovshinsky, Eberhard had to journey to Japan to
forge a personal relationship with this prospective partner. One of the cross-
cultural lessons that Eberhard learned was that in the Japanese industrial
context, sales decisions were often made not by corporate sales staff, as in
the US, but by plant managers. In Eberhard's recollection, the Sanyo plant

manager was not easily convinced and initially indicated that the company had no interest in an automotive market because margins were too thin.[53] Indeed, such a market then hardly existed at all outside the Toyota-Panasonic alliance around the Prius hybrid. However, Sanyo had ambitions. Panasonic was the world's largest producer of nickel-metal hydride cells for hybrid electric cars, but Sanyo became the world's largest overall producer of nickel-metal hydride commodity cells when it purchased Toshiba's nickel-metal hydride business in 2000. In electric vehicle applications of such cells, Sanyo was in a distant second place. It had a cooperative development program with Honda and was negotiating a production program with Ford to supply a battery pack for the Escape hybrid sport utility vehicle (SUV).[54]

These efforts were on a relatively small scale and involved a nonflammable battery formula that had been tried and tested in automobile applications. What Eberhard was proposing was unprecedented. He was bargaining for bulk sales of the lithium-cobalt oxide 18650 cell, so-named because it was 18 millimeters wide and 65 millimeters long, a dimensional standard set by Sony to suit the requirements of its camcorder. Over the years, the 18650 became the standard battery form factor for lithium rechargeables in notebook computers and other mobile devices and was manufactured by a number of companies in a variety of chemistries. Eberhard invoked an argument of scale, asking Sanyo managers to consider how many 18650 cells the average user consumed in a lifetime. Such cells were integrated into their applications, so most consumers never directly encountered them in the retail market. If one user purchased several notebooks and perhaps a camcorder over the course of a lifetime, that could constitute anywhere from ten to thirty cells. But one user of just one of the cars that Eberhard was proposing would consume up to eight thousand cells. One such car required as many cells as nearly 2,000 notebook computers, and 1,000 such cars would have as many cells as two million notebooks. Here was an opportunity, Eberhard told Sanyo, to vastly grow its business. Sanyo paid attention and allowed Eberhard to demonstrate his ideas for battery pack management and safety systems.[55]

With this opportunity, Eberhard and Tarpenning refined their plans. The idea now was that Tesla Motors would license the tzero powertrain from ACP, build its own lithium ion battery pack, and package everything in the chassis and frame of an attractive sports car that Eberhard hoped to develop in collaboration with Lotus. The famed British marque offered design services and was also one of the few companies that manufactured cars for competitors. To build the pack, Tesla hired the Stanford University graduate Jeffrey Brian (JB) Straubel, an electronics engineer who had also been intrigued by the prospect of applying lithium commodity cell power to electric drive. Straubel, too, had patronized ACP, an indication of the intimacy of the California electric car enthusiast scene.

The signs from ACP were encouraging. While the company failed to deliver a car for Eberhard, it used his money to build a battery pack of lithium 18650 cells for the tzero. The vehicle proved a resounding success, earning the best score in the 2003 Michelin Challenge Bibendum, the annual competition in sustainable automobility. For Cocconi, it was a high point in a distinguished career. The upgraded tzero was one of the most efficient cars in terms of energy use per mile, he averred, and would also outperform Ferraris and Lamborghinis on the drag strip. Eberhard borrowed the car for three months to evaluate performance, generate interest, and attract investment, subjecting the vehicle to test drives in the winding mountain roads between San Carlos and his home on Skyline Boulevard that Gage described as "untold in number and severity."[56]

PROJECT DARK STAR

Eberhard code-named the project Dark Star in homage to John Carpenter's 1974 eponymous cult science fiction film.[57] Eberhard dreamed of one day building Tesla's own manufacturing plant but in the meantime he needed help. In November 2003, he accompanied Gage to the Los Angeles Auto Show, where he met Lotus's legendary project engineer Roger Becker and made a pitch. Eberhard wanted Lotus to help him adapt the two-seat Elise for a car that would become known as the Roadster, and he wanted the British company to manufacture it as well. Becker agreed. Roadster would not be designed wholly from the ground up, but the vehicle was much more complex than a conversion. Tesla licensed the Elise chassis technology and then worked with Lotus to reengineer the aluminum frame to suit the electric propulsion system and redesign the spartan passenger compartment. The arrangement offered Tesla access to the Lotus supply chain and production plant at Hethel.[58]

Before collaboration could begin, Eberhard and Tarpenning needed to raise millions of dollars to build a prototype. Sand Hill Road, the wellspring of Silicon Valley venture capital, was only a short drive from San Carlos, but the early years of the new millennium were a far cry from the roaring 1990s. The bursting of the dot-com bubble signaled a new recession, exacerbated by the 9/11 terror attacks and wars in Afghanistan and Iraq in 2002 and 2003. With venture capital cautious, Eberhard and Tarpenning made the fateful decision to bring aboard an investor on a path not dissimilar from their own. Elon Musk, a South African immigrant, had also managed to make and keep a fortune in the tech boom. As a cofounder of the software startups Zip2 and PayPal, Musk cashed out before the bust and was looking for new opportunities, and not in dot-com, a sector awash in overcapacity. Like Eberhard and Tarpenning, Musk believed that the next big thing would be in transportation, although his visions were more exotic. In 2002, Musk founded SpaceX with the intention of radically cutting the cost of rocketry and a long-term goal of founding colonies on Mars.

In the short term, automobility seemed to promise more lucrative returns. Musk was fascinated by powerful automobiles, as much for their value as markers of wealth and status as for their kinetic potential. With the proceeds from the sale of Zip2, he purchased a million-dollar McLaren F1, one of only sixty-two in the world, he claimed to CNN at the time. That interest drew him into Silicon Valley's electric vehicle enthusiast culture. Musk had links to ACP through Straubel, a friend who had introduced him to the tzero. Introduced in turn to Eberhard and Tarpenning by Gage, Musk became the angel investor that Tesla Motors had been looking for. He committed $6.5 million, becoming chair of the company and its single largest shareholder in one stroke.[59]

ON THE ROADSTER

Tesla Motors developed along lines of the Silicon Valley start-up, embracing an ever-changing cast of players socialized to consultancy and the job-hopping common in California's advanced technology sector. The business plan followed the method developed by Eberhard with his borrowed tzero. In the absence of a commercial product, the company would market the promise of the electric supercar through a series of prototypes intended to advance the state of the art, generate excitement, and stimulate further investment in research and development.

Synthesizing the technology of ACP, Tesla, and Lotus proved more complicated than the principle of virtual integration assumed. In 2006, Tesla
hired Rippel on the basis of his familiarity with ACP's technology, and once
again the veteran engineer found himself involved in a major electric car
project. But Rippel was struck by what he characterized as the company's
lack of "engineering understanding." As planned, the Roadster was much
heavier than the tzero, and from Rippel's perspective, the solution was a
larger and more capable motor. However, the company believed that it would
be simpler to acquire a robust two-speed transmission. The plan nearly came
to grief when the transmission could not be synchronized with the high-
torque motor and became locked in second gear. Straubel later asked Rippel
to design and build a new motor, a project that became the basis of the
motor for the sedan that Tesla was planning as its next project.[60] There was
also trouble with Roadster's battery pack. Tesla assumed that pack assembly
could be outsourced, but efforts to set up a plant in Thailand were abandoned
when it became clear that that end of the operation required sophisticated
climate controls and specialized personnel.[61]

The first Roadster prototype was rolled out in early 2005 and was presented at Burning Man, the counterculture arts festival held in the Black
Rock Desert of northwestern Nevada in August.[62] By 2006, two more cars
were ready, trimmed in company colors: one black, the other red. Engineering Prototype 1 attracted the first serious investors, including Google's
Larry Page and Sergey Brin, venture capitalist Nick Pritzker, and JP Morgan.
Tesla took the vehicles on an extended summer tour of California, displaying them at a series of high-profile events that garnered media attention
and celebrity interest. Tesla also planned to supply a lithium ion battery
pack for the updated City car under development by a revived Th!nk, purchased in 2006 by a Norwegian consortium, but the US company was over-
extended and had to drop this scheme.[63]

In June, *New York Times* automobile journalist Matthew Wald introduced
Tesla to the national audience. Ten years after the launch of the EV1, wrote
Wald, a new electric car was about to come to market. The Roadster was
a hot rod that could also go more than 250 miles on one charge, much
farther than the GM car. Wald's tone was positive, but there were a few
discordant notes. Previous electrics had been stimulated by environmental
imperatives, stated the journalist, and cleanliness and efficiency were the

two qualities such cars possessed that ICE cars did not. Neither quality, Wald admitted, was likely to motivate those willing to pay the $85,000 to $100,000 that Tesla Motors was asking for a Roadster.

It was this article that publicly introduced Eberhard's business plan. Mobile phones, refrigerators, and color television had all started life as high-end goods, not commodities. So it would also be, held Eberhard, with the first commercial electric. Tesla would begin taking orders immediately for 4,000 to 5,000 Roadsters, the first of which would be produced in mid-2007. With the proceeds, said Eberhard, Tesla would bootstrap to a larger, more mainstream vehicle.[64] Musk reiterated the plan in an essay posted to Tesla's website in August.[65]

Musk and Eberhard agreed on business strategy. But their personal relations quickly deteriorated, ostensibly over the question of production costs, precipitating a struggle for control that unfolded throughout 2007. Some board members felt that Eberhard did not understand the economics of manufacturing and that his cost estimates were incorrect. This view was promoted by the Chicago-based investment firm Valor Equity, whose man in Tesla, the engineer Tim Watkins, made common cause with Musk. To be sure, Musk himself tended not to let thrift dictate design. He encouraged Eberhard to make a series of stylistic changes that added cost and complexity to the prototypes, including replacing the fiberglass body of the original Elise with one made of carbon fiber, which Musk judged more attractive but proved difficult to paint.[66]

Years later, Musk would claim that Tesla had been founded on the false premise that a commercial electric car could be built around components drawn from the open market, a fallacy that he claimed nearly destroyed the company before it even got off the ground. However, Eberhard seems to have understood virtual integration as a temporary expedient until Tesla could develop its own manufacturing capabilities. At the same time, Musk admitted that nobody in Tesla knew what they were doing in the early days.[67] When the bills came due for the Roadster, Musk weighed his options. By late 2007, Tesla's board forced Eberhard out of the company he helped found. Musk assumed control and threw his personal wealth and restless energy into an enterprise that would become synonymous with him. In pursuing the electric supercar, Musk would refine the techniques of virtual integration and enlist a legion of enthusiasts motivated by the dream of zero emission automobility, as well as the public resources that governments would make available to help realize it.

As the fledgling Tesla Motors struggled to realize the Roadster, the fortunes of ACP, the company that had played a key enabling role in Tesla's rise, took a new turn. In one sense, ACP became a casualty of Tesla's own troubled gestation. Always on the margins of financial viability, ACP was desperate for cash at a time when Eberhard licensed its technologies, but the company hardly benefited from the arrangement. Licensing was based on patents, not production, and ACP earned revenue for services rendered only in the initial stages of integrating the Roadster into the Elise platform. Tesla paid hardly any royalties to ACP, a consequence, according to Cocconi, of alterations that Tesla's lawyers made to ACP patents. By then, ACP was largely owned by Chinese interests, who purchased Cocconi's stake.[68] The trend in the transfer of advanced US propulsion technology to Asian industry would accelerate in coming years as US automaking faced its greatest reckoning to date.

11 MOTOR CITY TWILIGHT

We made a bad decision. Being known as the technology laggard is not conducive to selling automobiles.
—Robert A. Lutz, General Motors vice chair, March 11, 2007

In his time in the limelight, Martin Eberhard caused something of a stir in the automaking establishment. In mid-2006, the engineer received an invitation from GM executive Robert A. Lutz to pay him a visit at the company's headquarters in Detroit. A multilingual polymath who served in the Marine Corps, Lutz had spent most of his life in the auto industry, where he earned a reputation as an automobile enthusiast with a penchant for performance vehicles and the nickname "Maximum Bob." As the head of Chrysler's Global Product Development in the late 1980s, Lutz was one of the minds behind the Dodge Viper sports car. In 2001, GM president and chief executive Rick Wagoner recruited Lutz as vice chair for product development, charged with revitalizing automobile design in a company that, since the dismissal of the engineer-executive Robert Stempel in 1992, had been run by executives trained in finance, management, and law.[1] Eberhard could not help but notice the chrome-plated V-16 engine mounted in Lutz's office.[2]

The sum and substance of the meeting was a kind of corporate confessional. Lutz told Eberhard that Tesla's Roadster inspired him to launch a project for a commercial hybrid electric car called the Volt. The success of the Prius was an even more compelling factor in this decision. Many analysts predicted that Toyota's hybrid electric would lose money, but the company claimed that the line turned profitable in 2001.[3] Ironically, some of this success was enabled by US stimulus policies. Among the provisions of the Bush administration's Energy Policy Act, passed in August 2005, were incentives for alternative automobile propulsion systems. These measures were devised primarily with the fuel cell electric car in mind, but there was also a tax credit of up to $7,500 for conventional hybrid passenger vehicles that applied to up to a total of 60,000 units until December 31, 2010.[4] Toyota reached this threshold by mid-2006 and had sold a cumulative total of over 600,000 units worldwide by year's end, more than half of them in North America.[5]

The Prius was a vindication of the vision of Victor Wouk, who died in 2005 and became known in some quarters as the "father" of contemporary hybrid electric technology.[6] For GM, the Prius was becoming a symbol of corporate myopia. For years, the automaker led the industry in the fight against regulated technological change, not anticipating that governments might one day offer subventions for building cars that met social policy goals. In the post-Gulf War economic boom, Detroit largely abandoned the relatively balanced product lineup that air quality and fuel efficiency regulations had compelled it to adopt in the 1970s and 1980s and reverted to the classic formula of large, lucrative, and gasoline-thirsty vehicles. In the 1990s, US car companies enjoyed a decade of prosperity, but by the mid-2000s, economic conditions were changing. After nearly a quarter-century of relative stability, oil and gasoline prices began a precipitous rise in late 2004, affecting sales of the biggest and most profitable vehicles.[7] Only a few years earlier, Wagoner had argued that cheap gasoline undermined the rationale for a conventional hybrid electric car. Suddenly, fuel economy was an important metric of vehicle performance. Moreover, vertical disintegration, which had been promoted as a means of sharing risk and unleashing latent efficiencies, had destabilizing effects in an industry where suppliers of original equipment remained closely interconnected. In October 2005, Delphi Automotive Systems filed for bankruptcy, leaving GM liable for billions of dollars in pension, health, and life insurance payments that the terms of the 1999 divestiture obliged the automaker to shoulder in the event its former subsidiary failed. Some observers expected that GM itself would shortly follow its primary parts supplier into insolvency. Over the course of 2005, GM lost $10.6 billion, halving its stock value and reducing its debt to junk status. The company remained the world's largest automaker by sales volume, but its market capitalization plummeted to eighth place.[8]

By mid-decade, Toyota was on the verge of replacing GM as the world's dominant automaker, and pressure on the US company to develop a product that could compete with the Prius became irresistible. Still, both Lutz and Wagoner perceived the Volt's value primarily in terms of its symbolic power. In a 2006 interview with *Motor Trend* magazine, Wagoner remarked that he regretted terminating the EV1 because the decision detracted from GM's image, not profitability. The key to future profits, suggested Wagoner, was China, whose industrial revolution was stoking demand for Western products of all kinds.[9] Lutz similarly felt that the damage caused by GM's failure to develop electric cars was mainly to corporate reputation.[10]

The belief that technological prestige somehow informed market dominance had guided GM's historical approach to electric vehicle technology. With the Volt, the automaker planned to apply this logic on a much larger scale than it had previously attempted but still well short of commercial volume. The idea was to surpass Toyota in technological capability. The Volt was a dual-mode hybrid equipped with a large and powerful lithium battery that afforded a much longer all-electric range than conventional hybrids were capable of. Defined by the California Air Resources Board (CARB) as a type of extended-range electric vehicle (EREV), the car could also be plugged in and recharged and was more akin to an all-battery electric than a conventional hybrid electric. GM had experience in all of these formats. Had the company not terminated the EV1 plug-in hybrid along with the rest of the EV1 program, mused research and development chief Larry Burns to the press, it could have developed a product like the Volt a decade earlier.[11]

By the mid-2000s, conditions for bold new projects were much less auspicious than they had been in the mid-1990s. Like the EV1, the Volt was planned on the eve of a recession, and US automakers commanded relatively fewer resources than they had in the booming 1990s. The Big Three failed to achieve the broader Partnership for a New Generation of Vehicles (PNGV) goal of modernizing the industrial base for advanced propulsion technologies and moreover had also outsourced much of their conventional manufacturing capacity. The core content of electric cars had come to be dominated by Toyota, Matsushita (renamed the Panasonic Corporation in 2008), and other Asian enterprises, and US automakers including GM would have to look to them for help with their own belated green car programs.

Detroit's ability to sustain these programs while weathering rising energy costs and the deepening economic crisis in turn required additional resources that only the federal government could provide. By mid-decade, the US Department of Energy (DOE) began to redirect its efforts in alternative propulsion technoscience away from hydrogen and toward the plug-in hybrid and its lithium rechargeable battery. As the global economy slid into recession in 2007, however, the federal government prepared to intervene on a much larger scale for a contingency that would have been unthinkable in prior years. American automakers would find themselves on the verge of financial ruin and the full resources of the national developmental state would be mobilized to rescue them.

HYBRIDS HACK HISTORY

The Prius was a crucial impetus in the electric vehicle revival, and not only as an industrial-technological benchmark for and challenge to mainstream automaking. As at other critical junctures in the age of auto electric, outsiders played an important role in reinterpreting and repurposing commercial technology. Electric purists snubbed the Prius but other enthusiasts saw it as the platform for a conversion that could approach the potential of the all-battery electric car. In the early 2000s, they began modifying Priuses for plug-in capability, with an important group coalescing around Andrew Frank, the University of California engineering professor who had helped GM develop the EV1 plug-in hybrid and become known as the originator of this format. Frank inspired the creation of the non-profit California Cars Initiative (Cal-Cars), founded in 2002 by the activist-entrepreneur Felix Kramer as an open-source collaboratory that enlisted local talent including the electrical engineer Ronald Gremban, a veteran of the Great Electric Car Race of 1968 and Sebring-Vanguard, who served as CalCars's lead technical advisor. Enthusiasts disassembled Priuses and installed plug sockets and bigger batteries and modified the electronics. One problem was that the Prius's computer kept the battery at around 60 percent of charge, a feature designed to preserve the lifetime of the power source but that also inhibited all-electric operation. Another enthusiast, a former EV1 user named Greg Hanssen who operated his own Prius conversion business, devised a solution. He reprogrammed the computer into thinking that the battery was always nearly full, enabling deep discharges that gave a much longer all-electric range.[12]

The hack illuminated the technopolitical calculus that had informed the creation of Prius. The first conversion prototype by CalCars used lead-acid

batteries, giving an all-electric range of around 10 miles and a fuel efficiency of around 100 miles per gallon. Activist-engineers wanted to use
nickel-metal hydride batteries and were skeptical of Toyota's claim that
such batteries cost too much to be practical in the plug-in format, on the
order of seven or eight times the target of $150 per kilowatt-hour specified
by the United States Advanced Battery Consortium (USABC).[13]

But Toyota had neither the means nor the motive to cut the cost of the
Prius battery. The Prius was not just an advanced automobile. It was also
a closed sociotechnical system whose architecture was designed to balance
and protect the interests of two parties in battery electric propulsion technology. Toyota built and sold the cars, and Panasonic built and sold Toyota
the batteries for those cars. By neutralizing battery replacement costs and
an open market in batteries, the Prius mitigated much of the risk that the
temporal mismatch posed to each partner. If battery costs were too high,
Toyota would suffer, and if they were too low, Panasonic would pay the
price. In the Prius system, battery costs were insulated from market forces
and calibrated and stabilized to benefit both parties. In effect, Toyota and
Panasonic, along with Cobasys, functioned as a sort of cartel of nickel-metal
hydride power for electric vehicles.

This closed system further reinforced the growing preference for lithium
power among proponents of electric cars. Chevrolet planned the Volt's competitive advantage around a large pack of lithium ion pouch cells (a kind
of soft-pack prismatic cell that was lighter and cheaper than hard-pack prismatic cells) slated to be built by GM in a plant that was to be the first
such manufacturing facility in the US operated solely by an automaker.[14]
GM displayed the Volt concept car at the 2007 Detroit Auto Show, eliciting
criticism that the car was yet another exercise in public relations. Wagoner
responded that the plug-in hybrid electric was a "breakthrough idea" whose
time, he added equivocally, was "moving to be right."[15]

What all parties could agree on was that the Volt was a highly sophisticated technology with capabilities that in some ways surpassed the Prius.
Where the second-generation Prius had a 1.3 kilowatt-hour battery, the
Volt had a 16 kilowatt-hour battery, and where the Prius could operate in
all-electric mode for only a short distance, the Volt had an all-electric range
of about 40 miles. Thereafter, the Volt would switch modes and operate as
a series hybrid, using an ICE to drive a primary electric motor for another
350 miles and power a secondary electric motor that kept the battery at

a minimum state of charge. The Volt could also operate in series-parallel mode, with the ICE mechanically assisting both electric motors in driving the wheels as necessary. Homeowners could recharge the battery using the standard 120-volt household outlet.[16]

This strategy of competing in the hybrid electric space through bigger and better battery technology was reinforced by federal lawmakers, who had come to see the plug-in hybrid as an important element of public policy. In December 2007, Congress passed the Energy Independence and Security Act, authorizing the DOE to disburse $360 million in grants for cost-shared projects supporting plug-in hybrids from 2008 through 2012. The act also supported a national electric education program (named after Frank) and made provisions to extend loans for producing advanced batteries. All of this was part of a much larger initiative called the Advanced Technology Vehicles Manufacturing Incentive Program. This authorized $25 billion in loans to help US automakers and their suppliers develop and produce cars that met the federal Bin 5 Tier II emission standard and that were 25 percent more fuel efficient than similar vehicles from the 2005 model year.[17]

Like the car companies, lawmakers viewed the problem of the sustainable automobile largely through the lens of the technical parameters of the power source. With the success of the Prius, it became a matter of national pride that US commercial hybrid electrics be superior in every respect. Lawmakers therefore defined the plug-in hybrid as a vehicle equipped with a battery of at least 4 kilowatt-hours in capacity, more than three times the size of the Prius power pack.[18]

In effect, the US declared a battle of the hybrid electric batteries. In the Energy Independence and Security Act, the federal government elaborated the sociotechnical rationale that had informed Volt, signaling that it regarded the battery as industrial core content and the weapon of choice in the green automobile wars. Impressive in relative historical terms, the scheme was inhibited by some of the traditional bureaucratic shortcomings of quasi-planned stimulus. As in the past, industrial borrowers of federal funds had to be financially viable, a requirement that greatly complicated the program as the recession deepened. Crucially, planners did not consider the economic ramifications of the temporal mismatch that had partly informed the development of the first-generation Prius. Larger batteries were more expensive and accentuated the economic conflict of interest between the automaker and the batterymaker. However, the federal scheme made no real provision to coordinate the actions of enterprises that made cars and enterprises that made batteries.

Toyota quickly responded. Days after the unveiling of the Volt concept car, the automaker announced that it would develop a plug-in variant of the Prius, with some media hinting that the company was already secretly at work on the technology. In July, Toyota announced that it would deploy a prototype plug-in electric fleet for testing and evaluation by the University of California at its Berkeley and Irvine campuses.[19]

RECESSION AND REGULATION

These events led California air quality regulators to once again revise their definitions of and incentives for Zero Emission Vehicles (ZEVs). In the spring of 2008, CARB developed an equivalency formula around a range-based credit system that featured two new categories known as type IV and type V, defined as fast-refueling vehicles with a range of more than 200 and 300 miles, respectively. Regulators derived these qualities from the technology of fuel cell electric drive, classifying fuel cell and all-battery electric propulsion as "gold" ZEVs and giving the former a higher credit rating owing to recent range improvements in test equipment. For automakers with large sales volume, the zero emission gold quota for 2012–2014 was 25,000 type IV ZEVs. Car companies also had the option of meeting at least 30 percent of the gold quota with 7,500 type IV fuel cell electrics and the balance with 58,000 enhanced advanced technology partial ZEVs. The latter was the "silver" standard, and it denoted a new category of vehicle using "zero emission fuel," meaning hydrogen in an ICE vehicle or electricity in a plug-in hybrid.[20]

These tortuous calculations illustrated CARB's changing understanding of clean automobile technology over time. Under intense pressure from automakers, air quality regulators dropped their single-minded focus on emissions and considered other metrics of performance relevant to industrial competitive advantage, allowing the industry to get zero emission credit from hybrid and fuel cell electrics and delay the rollout of all-battery electric cars. But US automakers did not have time to take advantage of this compromise to diversify their fleets. Weighed down by a business model that had suddenly become burdensome in rapidly changing economic conditions, they lost tens of billions of dollars. In 2008, Toyota made history when it replaced GM as the world's largest automaker by sales volume, a title that the US giant had held since 1931.[21] The political climate increasingly favored federal intervention, initially with a view to reviving the competitive powers of US automaking by means of advanced technology. As Senator Barack Obama campaigned for the presidency in the summer, he outlined a goal of placing one million electric vehicles on American roads by 2015.

By the end of 2008, Wagoner and the other Detroit chiefs were asking for direct federal aid. After a sortie to Washington in private jets generated bad press, the auto executives regrouped and arrived for the December round of congressional bailout hearings in advanced propulsion vehicles. Wagoner chose to drive a Volt prototype and delivered a message that he hoped would resonate with lawmakers. The future of the US automobile industry was advanced technology, held the executive, and it was in the national interest to support it: "It would be a shame for the US to fall out of that race because the technology development in almost all cases is done in the market where the company is domiciled."[22]

Wagoner's appeal expressed the ambivalence of the US automaking establishment toward the national developmental state and the premise that national innovation systems were the basis of national economic growth.[23] Collaborative research and development had a decidedly mixed record. It was one factor in the successful campaign by the US semiconductor sector to reclaim global leadership from Japanese manufacturers in the 1980s and 1990s, but far from the only or even the most important one.[24] For their part, US automakers rejected not only the products of public-private research but the practice of national industrial development as well, joining manufacturers of all nations in outsourcing and offshoring production.[25]

As the economic crisis worsened, policymakers momentarily set aside their belief in the curative powers of advanced technology. In April 2009, Chrysler declared bankruptcy, followed by GM in June, precipitating a massive federal intervention configured not by energy, science, and technology bureaucrats, but by the lawyers of the US Department of Justice and the economists of the US Department of the Treasury. The federal government held an asset auction and spent $60 billion in Treasury loans to create a publicly owned entity as the sole bidder. Much the same was done for Chrysler. For the first time in peacetime, a substantial portion of the automobile industry was effectively nationalized.

REBUILDING THE RUST BELT

The restructured General Motors Company (GM) emerged in July shorn of four brands and thousands of employees, including Wagoner, who was forced to resign by the White House as part of the bailout deal. Through the Treasury's Troubled Asset Relief Program, the federal government became the automaker's largest shareholder. Policymakers framed the bailout as an investment that would yield real and public policy dividends. The federal government planned to buy back its stake after GM was restored to financial health through a mix of collaborative science and technology, updated regulations, and new loan guarantees and tax breaks. President Obama assured GM and Chrysler that private enterprise would retain the prerogative to decide what models to produce.[26]

But the federal government had determined that it was in the national interest to support commercialization of the plug-in hybrid electric passenger car. In May 2009, the Obama administration announced plans for a new national average light duty fleet mileage rule that aimed to increase the standard from around 25 to 35.5 miles per gallon by 2016.[27] In June, the DOE advanced low-cost loans to Ford, Nissan, and Tesla Motors under the Advanced Technology Vehicles Manufacturing Incentive Program, aid for which GM and Chrysler were deemed ineligible because they were in bankruptcy protection.[28] The federal government also provided $2.4 billion to support the construction of a national lithium battery manufacturing complex and granted tax credits for electric cars of all kinds.[29]

President Obama took stock of these efforts in his State of the Union address of 2011. With the economy in recovery, said the president, it was time to pay attention to stiffening foreign competition, especially from India and China. Economic growth was this generation's "Sputnik moment," and with more investment in science, technology, engineering, and math as well as the proper incentives, the US could be the first country to deploy one million electric vehicles by 2015.[30] It was a clear signal that the federal government wanted automakers to transition from emergency stimulus to normal market operations as soon as their competitive powers were enhanced by advanced science and technology.

However, federal efforts to build a domestic electric vehicle battery manufacturing base yielded mixed results both because US industry depended heavily on foreign industry and because the Obama administration's stimulus plan was optimized for supply, not demand, in a recessionary period. Chevrolet's experience was instructive. The company followed standard industry practice in outsourcing the cells for the Volt's battery pack and considered two lithium chemistries that were cheaper than the lithium cobalt oxide formula used by Tesla. One was lithium iron phosphate, a compound that John Goodenough helped pioneer and that was under development by A123 Systems, a start-up founded in 2001 by MIT graduate and materials scientist Yet-Ming Chiang and based in Watertown, Massachusetts. In May 2008, A123 received a $12.5 million USABC grant to develop cells for the plug-in hybrid electric car, an application for which the lithium-iron phosphate system was believed to be well suited.[31] Utilizing iron at nanoscale, which imparted this plentiful and cheap element with useful catalytic properties, lithium-iron phosphate chemistry had higher power than lithium-cobalt oxide chemistry at the cost of lower energy capacity in a package that was comparatively long-lived and safe owing to iron's relatively low reactivity.[32] In August 2009, A123 received major federal assistance in the form of a $249 million grant to build a cell manufacturing plant in Livonia, Michigan. Despite these major investments by the national developmental state, the company failed to secure the Volt contract, although it would supply cells to an enterprise founded by the renowned automobile designer Henrik Fisker to build a luxury plug-in hybrid sedan called the Karma, a project that would receive a $529 million loan from the DOE in 2010.[33]

Chevrolet instead chose cells of lithium-manganese oxide manufactured by the South Korean giant LG Chem, which licensed components developed by Argonne National Laboratory. For the automaker, this option represented the path of least resistance. Lithium-manganese oxide chemistry was not quite as powerful as lithium-iron phosphate and was less durable, but it had greater energy capacity and reasonably good safety characteristics. Another factor that favored LG Chem over A123 was experience in manufacturing at scale.[34] In February 2010, the DOE awarded LG Chem $150 million to build a state-of-the-art cell manufacturing facility in Holland, Michigan, with the capacity to supply 60,000 vehicles a year. In 2012, however, it was cheaper to import the initial batch of cells from LG Chem plants in South Korea. The Holland plant remained idle for months. Workers drawing federally subsidized salaries had little to do, and the resulting scandal embarrassed the DOE and triggered an internal audit.[35]

Indeed, global interconnectivity in advanced manufacturing meant that US national developmental stimulus often benefited foreign industry.

Nissan-Renault used its $1.4 billion loan to renovate its plant at Smyrna, Tennessee, to produce cells and battery packs for an all-electric car called the Leaf. A pet project of chief executive officer (CEO) Carlos Ghosn, the Leaf was planned as the first commercial all-electric car in the post-EV1 era. The vehicle initially had a 24-kilowatt-hour pack of lithium-manganese oxide pouch cells, a chemistry with an international lineage.[36] Nissan developed the formula in a joint venture with the electronics giant NEC called the Automotive Energy Supply Corporation (AESC), set up in May 2008.[37] This entity resembled Panasonic EV Energy (renamed Primearth EV Energy in 2010), but with some important differences. Toyota's dealings with Panasonic were restricted to batteries for hybrids, and the automaker had only a minority stake in its joint venture.

In contrast, Nissan held the majority stake in AESC, which produced cells and batteries for all-electrics as well as hybrids and hence exposed the automaker to the economic risks of the temporal mismatch. The joint venture had large production facilities in Japan and Europe in addition to the US and aimed to dominate what some observers thought would be the burgeoning business of battery swapping through a supply arrangement with the Palo Alto startup Better Place. Other observers warned that Nissan risked overproduction at a time when the economy was still in recession.[38]

MORE COMPLEX THAN ICE

Low demand for a new and untried product was far from the only issue confronting the US electric car project. Building a domestic lithium cell complex posed prodigious technological and organizational problems and presented battery engineers with a double challenge: they had to integrate the science of solid-state electrochemistry at the level of process chemistry and simultaneously adapt it to standards of hybrid electric propulsion being worked out by US automakers. As the Volt demonstrated, cell manufacturing depended on accessing science and technology through transnational joint ventures. Managing such enterprises was difficult enough when they involved conventional automobile technology but was even more complicated with advanced electric automobile technology. Federal electric car stimulus exacerbated this complexity because it was an emergency measure that set aside the tasks of building up a post-secondary knowledge base and even of coordinating manufacturing in favor of building manufacturing capacity with available technology as quickly as possible.

These dynamics were illustrated in a $200 million Obama administration project to develop lithium cell and battery production at Johnson Controls, then the world's largest producer of lead-acid car batteries. The Milwaukee-based auto parts supplier wanted to compete with Panasonic and Sanyo in supplying nickel-metal hydride batteries for conventional hybrid electrics but had been shut out of that market. With the emergence of the plug-in hybrid electric came opportunities to supply automakers with lithium cells and battery packs for new-build vehicles, and, potentially, battery packs in a battery replacement market as well.[39] In 2005, Johnson Controls part-nered with the French battery giant Saft and opened a $4 million laboratory in Milwaukee to explore lithium rechargeable technology in the hybrid electric format.[40] In early 2009, the partnership contracted with Ford to supply the battery pack for the Fusion Energi, Ford's answer to the Volt. Where Chevrolet designed and built its own battery pack around LG Chem cells, Ford adhered to its practice of outsourcing as much electric propul-sion technology as possible.[41] The five-year deal called for Johnson Controls to assemble complete battery packs using cells initially produced in France and, later, at US facilities. With characteristic caution, Ford committed to producing only a few thousand units of the Fusion Energi per year.[42]

The experiences of one materials scientist at Johnson Controls revealed the engineering challenges in this context. When the project began in 2009, Jack Johnson had some twelve years of experience at the company. He held bachelor of science degrees in technology and mechanical engineering tech-nology. In the automobile industry, recalled Johnson, "a lot of people just don't get an electrochemistry degree." His job was to scale laboratory pro-duction from five or six cells a day to three a minute and accomplish this in compliance with automotive manufacturing and quality standards includ-ing ISO/TS 16949 and Ford's Advanced Product Quality Planning. A protocol for producing advanced batteries did not exist then, so one had to be devel-oped from scratch. In consumer electronics, batteries were designed to last three to four years at most, a lifetime entailing some 400 charge-discharge cycles. Batteries for electric vehicles, however, had to last up to ten years and 2,500 cycles.[43]

Fabrication required exacting quality control and sophisticated produc-tion tools and instruments. Whereas the producers of electronics commodity cells sought to weed out defective cells at the end of the manufacturing pro-cess to keep costs low, the makers of advanced batteries for electric vehicles sought to prevent defects and contamination. Production began when electrode materials were mixed into a slurry and coated, simultaneously and uniformly, onto both sides of a current-collecting foil. This meter-wide sheet was then floated in a full flotation drier that transported the web between rollers 30 meters apart. Improvisation was sometimes necessary. Borrowing ideas from high-speed newsprinting technology, Johnson Controls built a $20 million machine to run sheets of foil, containing the process within an extremely dry cleanroom to prevent moisture and particles more than half a micron in diameter from contaminating the product. Sheets of anode, cathode, and separator polymer were then brought together on a high-speed winder.

Johnson held that it was the most complex machine he had ever worked on. The winder pulled the sheets together through dozens of points of adjustment that were monitored and controlled with vision technology to a tolerance of 300 to 400 micrometers. The next step was cell assembly. Cells were cut out of the composite sheet with lasers and placed into battery casings, filled with electrolyte under vacuum pressure, and laser-welded shut. Gas chromatographs were used to run helium leak tests to ensure that the cells were hermetically sealed. Finally, thousands of finished cells underwent formation cycling and testing to eliminate those with manufacturing defects.[44]

These processes revealed the degree to which practices of manufacturing electric vehicle technology had become science-based. An important consequence of the dramatic increase in battery performance and knowledge of materials morphology was reciprocal improvements in instrument technology.[45] In lead-acid batteries, measurements to within a tenth of a volt were long considered acceptable because this did not represent much energy density. In a lithium ion battery, however, a tenth of a volt represents very significant energy density, with important implications for the formation process. As a result, potentiostats and chargers had to be capable of accuracy to within thousandths of a volt. In the wake of the near-collapse of the US auto industry, a fledgling advanced auto technology sector was emerging that overlapped the traditional heavy industrial complex and yet was importantly distinct from it. Controlling the manufacturing quality of electric vehicle battery cells involved hundreds of thousands of line items of failure mode and effects analysis and entailed manufacturing complexity that Johnson believed was greater than for ICE technology.[46]

12 ELECTRIC CARS AND THE BUSINESS OF PUBLIC POLICY

The overarching purpose of Tesla Motors is to help expedite the move from a mine-and-burn hydrocarbon economy towards a solar electric economy, which I believe to be the primary, but not exclusive, sustainable solution. Critical to making that happen is an electric car without compromises, which is why the Tesla Roadster is designed to beat a gasoline sports car like a Porsche or Ferrari.

—Elon Musk, cofounder of Tesla Motors, August 2, 2006

Policing the quality control of cell manufacturing was perhaps the only thing Tesla Motors did not have to worry about. In the Roadster, the company faced a host of engineering and financial challenges exacerbated by the schism between Martin Eberhard and Elon Musk. In the summer of 2006, each entrepreneur posted an essay on the company website that expressed contrasting visions of risk and reward. Eberhard was preoccupied by Tesla Motors's collaboration with Lotus and spoke of the myriad design, stylistic, and engineering problems absorbing the two companies, a task he admitted was further complicated by the fact that none of Tesla's principals were automobile engineers.[1] Musk took a more expansive view in what amounted to a policy paper couched in marketing rhetoric that appealed directly to the customer. Tesla was building an electric sports car that could compete with a Ferrari, he wrote, in order to create a carbon-free world. On paper, the Roadster would be more efficient than any other car on the road, Musk held, but it was still only a sports car and of little consequence by itself. The project's real utility, he maintained, was as a stepping stone to the affordable commercial electric car, and from there, a solar economy, Tesla's ultimate objective.[2]

Musk did not invent the concept of virtuous consumption as public policy, but under his leadership, Tesla Motors became the first automaker to build its entire brand around this principle. Musk refined Eberhard's marketing ideas into a holistic corporate aesthetic that conceived of Tesla as a community of participants and the customer as an agent in a project whose moral imperative required the simultaneous gratification and rejection of the self.[3] A Tesla car was not a normal consumer product, and normal consumer expectations did not apply. The technology was designed to deliver an unparalleled physical and psychological sensation, with the understanding that its hardware and software were experimental and might not work as planned. When that happened, users became informal engineers whose experiences informed the product development process.[4]

In essence, the Tesla aesthetic and its premise of democratized knowledge-making constituted a form of lifestyle activism that articulated the neoliberal precept of socially conscious capitalism.[5] Tesla wedded the ideas of consumer agency and empowerment to the dramaturgy of the large-scale automobile demonstration. Where mainstream automakers pioneered such demonstrations as a means of proving the economic infeasibility of the Zero Emission Vehicle (ZEV), Tesla used them as a means of sustaining its operations at a time when the company had no commercial products. In its early years, Tesla was essentially an enterprise in engineering research and development not dissimilar from AeroVironment and AC Propulsion (ACP). The company would frame trials of what were essentially production prototypes as launches of new products, generating media attention and new investments that were plowed back into research and development.

Over time, Tesla became instrumentalized as a private-sector solution to public policy problems. The company's business plan intersected with the preference for the large-battery electric car among California air quality regulators and federal lawmakers. As Tesla built up its engineering base and started producing automobiles, it also began selling propulsion componentry to established car companies, allowing them to outsource their zero emission obligations. Tesla fulfilled a similar function in California's complicated air quality control regime, which allowed companies to bank, trade, and sell credits earned for producing ZEVs to other automakers. Because Tesla produced only ZEVs, it could bank and sell all its credits, which became a major source of income for the automaker at a time when the company struggled to make a profit. In turn, Tesla represented a vast new market for components, especially battery cells, made by Asia-based enterprises.

These relationships illustrate the sometimes-paradoxical interdependencies wrought by quasi-planned energy, environmental, and industrial-technological policies. Tesla derived crucial support both from mainstream automakers and from the regulatory and developmental state, but Musk frequently placed himself in opposition to these groups, reminding consumers at every opportunity that fossil fuel interests had killed the electric car and chastising regulators for impeding the operation of the free market. The entrepreneur worked assiduously to craft a persona in the vein of other Silicon Valley luminaries as an independent and eccentric outsider speaking truth to power, deflecting criticism through assertions of green moral rectitude. Detractors would often be accused of impeding Tesla's efforts to save the environment. Yet over time, Musk and Tesla became increasingly reliant on and embedded in the regulatory apparatus and the industrial and financial establishments. Enthusiast ardor, crowdfunding, and celebrity fandom coalesced with finance capital, public stimulus, and Asian core content in an outsider enterprise that would eventually challenge the material and business practices of the global car industry.

PROMOTING PROTOTYPES

Tesla Motors's early years were marked by cut-and-try engineering, managerial and financial dysfunction, and ever-increasing media exposure. Slated to debut in the summer of 2008, the Roadster underwent a protracted design overhaul that complicated work on the production prototype. The two-speed transmission had failed, so a one-speed transmission made by BorgWarner was substituted. The motor had to be modified, as did ill-fitting body panels, and there was a constant struggle to trim the weight, the great enemy of the electric car. The Roadster had a lightweight aluminum chassis and carbon-fiber body, but its pack of 6,800 lithium-cobalt oxide laptop cells, sourced from Sanyo, weighed half a ton. In some ways, the work resembled what General Motors (GM) had done in reinterpreting AeroVironment's Impact, particularly in electronics. Jeffrey Brian (JB) Straubel worked to turn ACP's analog controller into a more robust digital device capable of smoothly handling the powerful battery pack.[6]

Tesla eschewed conventional advertising in favor of celebrity influence marketing, enabled by social media. The first Roadsters shipped in June 2008 and the $109,000 cars were snapped up by wealthy enthusiasts and luminaries including Dustin Hoffman, George Clooney, and Arnold Schwarzenegger. But promoting what were in essence production prototypes risked running afoul of conservative automobile pundits. A review of the Roadster by the BBC show *Top Gear* in December triggered one of the more sensational disputes in the saga of the contemporary electric car. The program started well enough. In a set-piece race with a conventional Elise, the television personality Jeremy Clarkson rocketed a Roadster from a standstill to 60 miles an hour in less than four seconds. Clarkson was impressed and reported more good news in the car's cheap operating costs and impressive 200-mile range.

Then, with Clarkson behind the wheel, the Roadster suddenly seemed to shut down and was filmed being wheeled back to the garage. It looked as if the car had run out of charge, but in fact BBC producers staged the scene as a projection of what would happen if they had continued their punishing test regimen. Using data supplied by Tesla, *Top Gear* calculated that under such conditions, the Roadster would have a range of only 55 miles. Clarkson also claimed the Roadster would take 16 hours to charge. A second driving sequence showed him remarking that the engine was overheating and sapping power and he was going to pull over to let it cool down.[7] Tesla Motors responded that neither of the two cars used in the test had fallen below 20 percent charge, no car had been immobilized as a result of overheating, and users with fast charging could recharge in around 3.5 hours.[8]

A charitable interpretation was that the BBC had taken creative license. To be sure, some of Clarkson's observations were grounded in the voluminous research in battery electric cars that had accrued by 2008, as well as common sense. Technical troubles could be expected to arise in any automobile subjected to hard driving. Racing used stored energy faster and shortened the range of all types of automobile propulsion systems, a correlation that the pioneers of battery electric distance driving understood only too well.[9] From the perspective of environmental science, Clarkson's claim that electrics were only as clean as the primary energy conversion systems that supplied their electricity was hardly controversial.[10]

On balance, the review was equivocal. The Roadster was "an astonishing technical achievement" and the "first electric car that you might actually want to buy," opined Clarkson. But in the real world, he concluded, the car did not seem to work. The television host seemed to suggest that the Roadster would require further development if it were going to best its conventional counterparts in every respect.

Musk saw things differently. He asserted that the BBC had cost his company hundreds of orders and millions of dollars in lost revenue, and he later sued the network. For Tesla's growing legion of supporters, Clarkson and *Top Gear* became only the latest examples of the automobile establishment's disingenuity and perfidy.[11]

WHITE STAR AND STIMULUS

Even as Tesla Motors coped with the Roadster, it planned White Star, the code name of a large luxury sedan later known as the Model S. Where the Roadster was a craft-built, heavily modified conversion, the Model S was designed from the ground up as the ultimate electric supercar. The new vehicle used an integrated propulsion system that mated an induction motor to a variable-speed single-gear transmission drawing power from the largest battery pack ever constructed for an electric car.

White Star would have been an ambitious project in the best of circumstances, but 2007 was a year of turmoil for Tesla. Eberhard and Straubel recruited Alec Brooks to serve as White Star's chief engineer, and once again Brooks found himself involved in a pathbreaking electric car enterprise alongside Wally Rippel. Rippel had designed an alternative motor for the Roadster that informed the design of the motor used in the White Star sedan. But Musk did not want Brooks to manage White Star, and both he and Rippel were sidelined and retained only as technical consultants. By early 2008, Tesla had spent nearly $140 million on the Roadster, about five times the original budget, and the company was nearly broke. In February, Rippel was laid off and Brooks left the company shortly afterward.[12] With Tesla on the verge of bankruptcy, Musk threw in much of his personal fortune and borrowed money from SpaceX after securing approval from the National Aeronautics and Space Administration (NASA), the chief benefactor of his privately held rocket company.[13]

At this point, Tesla staged an astonishing recovery. In the summer of 2009, the company was awarded a $465 million loan under the advanced technology vehicles program of the US Department of Energy (DOE).[14] The intervention was notable because the government made such aid contingent on financial viability, a provision that excluded GM and Chrysler. Tesla was hardly in better shape, but its rescue by the Obama administration caused some in the auto industry to pay attention. In April 2010, Daimler invested $50 million in Tesla, and in May, Toyota invested another $50 million. In June, Musk took Tesla public and raised $226 million. The *Wired* journalist Chuck Squatriglia observed that investors did not seem to mind that the start-up had yet to earn any money.[15]

Like ACP, Tesla held value for the established automakers as a source of componentry for ZEVs, a far cheaper way for them to meet mandate quotas than building core competency in all-battery electric propulsion technology from scratch. Backed by public money, Tesla would supply battery packs and electric powertrains for a number of compliance cars, including the Smart Car, the Mercedes A and B class electrics, and the RAV4 EV.[16] Through Tesla, the US taxpayer was effectively subsidizing the US environmental obligations of foreign car companies, a relationship that would have important unintended consequences on the growth of Musk's business and the shape of the electric automobile market.

GLOBALIZED CORE CONTENT

Federal stimulus was transforming Tesla into a key tool of US industrial, energy, and environmental policy. The Model S itself was the product of a global network of regulatory and industrial-technological linkages. As part of its arrangement with Toyota, Tesla was able to purchase on preferential terms part of the giant plant in Fremont that Toyota jointly operated with GM since 1984 and which had been shuttered in April 2010.[17] It was here that Tesla planned to assemble the Model S, largely out of Japanese core material content.[18] The Model S was built substantially of aluminum, and fabricating aluminum car parts required advanced techniques and technologies. Aluminum has one-third the malleability of steel and is much more difficult to shape as a result. If not properly pressed, aluminum sheet can break, and with 60 percent of the hardness of steel, it is also easily scratched. Properly stamping aluminum sheet required the production of sophisticated dies to exacting specifications, which in turn required craft skills. Tesla found these skills in Fuji Technica, a die maker that had gone into decline as Japanese automakers moved production to South Korea and China. The Model S put

Japanese die craftsmen back to work and made Fuji Technica a silent partner in Tesla's ambitious plans.[19]

Tesla's most important collaborator was Panasonic, the source of the cells for the Model S battery pack. In 2009, the electronics giant acquired Sanyo, the supplier of the lithium-cobalt oxide commodity cells used in the Roadster battery pack. This cell chemistry gave the Roadster very high performance, but it was too expensive for mass-marketed cars. For the Model S, Tesla and Panasonic selected another lithium formula based on lithium nickel-cobalt-aluminum oxide (NCA), a variant that was cheaper than lithium-cobalt oxide and could store more energy, at the cost of safety. Indeed, lithium-nickel oxides had some of the highest energy densities of all the lithium compounds, but they also were among the least stable. Early test cells often met a fiery doom. In the 1980s, Goodenough had experimented with and abandoned lithium-nickel oxides, and Saft's work in this field as a Partnership for a New Generation of Vehicles (PNGV) contractor met with similarly poor results in the 1990s.[20] In 1998, the DOE launched a program to help manufacturers understand the failure modes of lithium-nickel oxide. Researchers found that it was possible to stabilize the compound with a variety of substances, including cobalt, cobalt and manganese, and cobalt and aluminum.[21]

Over the next decade, this program became an important part of the federal advanced battery initiative and influenced Panasonic in its efforts to adapt NCA.[22] In 2009, Tesla and Panasonic began negotiations on an alliance that deepened the following year, when the electronics giant invested $30 million in the automaker. In April 2010, Naoto Noguchi, president of Panasonic's energy division, presented Straubel with the first NCA production cell from a new plant in Osaka.[23] In 2011, Panasonic entered into a four-year agreement with Tesla to produce cells sufficient for 80,000 vehicles.[24]

BATTERIES AND THE BUSINESS OF PUBLIC POLICY

Public policy support for electric supercars bound electric car start-ups, the established automakers, and their cell suppliers in complex relationships that reflected the asymmetrical durability of battery and electric motor. The commercial large-battery electric car represented a vast market for cells, and Tesla built the biggest battery packs ever constructed, ranging from 60 to 80 to 100 kilowatt-hours of capacity. Panasonic stood to make windfall profits from the Model S and its successors, as well as from the powertrains that Tesla built for Daimler and Toyota.

But dealing with Tesla also posed risks. Panasonic's arrangements with the upstart automaker were less clearly delineated and secure than those forged with Toyota for hybrid batteries in Panasonic and later Primearth EV Energy. In the Prius system, there was no pressure to cut battery costs because the hybrid electric battery pack was relatively small and its cost approached parity relative to other major components, achieving a balance in the interests of the automaker and cell maker. But in contemporary all-battery electric cars, the battery pack was relatively large and typically constituted the single most valuable component of the vehicle. As the history of the EV1 demonstrated, it was in the interest of the maker of the electric car to exert pressure on the maker of battery cells to cut costs. In the late 2000s through the 2010s, the goal of the promoters of all-battery electrics was to slash the cost of lithium battery power from around $400 per kilowatt-hour to below $100 per kilowatt-hour, the assumed cost of gasoline power at the time.

This objective placed the interests of the cell maker and automaker in conflict, and federal policymakers hoped that subsidies could resolve it. In October 2008, Congress passed the Energy Improvement and Extension Act, offering tax rebates for passenger electric vehicles. The baseline credit was $2,500, but the measure allowed $417 to be added for each kilowatt-hour of capacity in excess of 4 kilowatt-hours, up to a maximum of $7,500 for vehicles weighing less than 10,000 pounds. The full rebate applied only to the 250,000th unit of all types sold in the US per company, after which the rebate was phased out over the period of a year, falling to 50 percent of the full credit in the first six months and 25 percent in the final six months of the program before being eliminated altogether.[25] The American Recovery and Reinvestment Act of 2009 reduced the vehicle limit to 200,000 units.[26]

Cell makers raced to scale production before automakers exhausted their electric car tax credits. In this period, industry generally treated battery costs as a trade secret. Moreover, the cost question was further complicated by the tendency of actors to conflate cell costs with pack costs. With their sophisticated control systems and precision assembly requirements, battery packs had per-kilowatt-hour costs considerably greater than cells.[27] Sunsetting subsidies, concurrent research, development, and production, the slow economic recovery, and the unresolved question of what party would pay the cost of replacement battery power further destabilized the relations between the makers of cells and the makers of electric cars.

Another subsidy that incentivized automakers to make larger-battery electric vehicles was the zero emission credit program that the California Air Resources Board (CARB) set up at the inception of the mandate.[28] Designed to motivate established automakers, the rules allowed companies to bank credits from excess production of ZEVs and trade and sell them to other established automakers, which could use them to meet their own mandate commitments.[29] By the late 2000s, however, a score of small companies that exclusively produced all-battery or plug-in hybrid electrics emerged alongside Tesla, including Coda, Fisker, Wheego, and Th!nk. CARB wanted such enterprises to survive and put as many electric cars on the road as possible, so it allowed them to bank all their zero emission credits and sell them to mainstream automakers. As CARB's Tom Cackette recalled, credit sales became an important source of income for Tesla, generating hundreds of millions of dollars and helping the company weather several "near-death experiences."[30]

Credit sales were another way of socializing the mandate commitments of the automaking establishment and the emerging business of all-electric cars. Tesla emerged as the chief beneficiary of this system, largely because the rules favored supercars, a consequence of the agreement between automakers and regulators to define the ideal ZEV around the qualities of fuel cell electric propulsion.[31] CARB awarded credits on the basis of the emissions, range, and recharging or refueling capability of five classes of vehicle. Each credit was worth $5,000 (the cost of the fine that mainstream automakers paid for one credit of noncompliance), and automakers could earn up to seven credits per car depending on the characteristics of the car. The most valuable was the type V fast-refueling car, with a range of more than 300 miles, which was worth seven credits and $35,000 in zero emission exchange value. This class was notional and derived from the theoretical qualities of fuel cell electric technology. In 2013, the most capable production all-battery electric car was a version of Tesla's Model S that had a range of around 265 miles and took between twenty to seventy-five minutes to charge, and thus was not "fast-refueling." It rated type III, worth four credits and $20,000 in exchange value.[32]

In the summer of 2013, however, Tesla demonstrated a battery-swapping technology that in principle turned the Model S from a four-credit type III into a seven-credit type V ZEV.[33] But the company built only one battery-swapping station, located at an isolated truck stop on the western edge of California's Central Valley on Interstate 5, a site equidistant from San Francisco and Los Angeles along a major route frequently traveled by new Tesla owners. Some observers believed that Tesla was taking advantage of a loophole in the air quality incentive structure.[34] Battery swapping implied very complicated logistics and economics because from the perspective of all parties, the replacement battery had to mirror the state of entropy of the original. If the replacement battery was inferior to the original, the user lost value. If it was in better shape, the user gained at the expense of the automaker or battery provider. Only weeks before Tesla demonstrated its battery-swapping technology, Better Place, the battery swap start-up supplied by the Nissan-NEC joint venture, went bankrupt.[35]

Battery swapping proved a footnote in Tesla Motors's overall business strategy. The company instead decided to invest in a national network of proprietary fast-charging stations as its primary range-extending infrastructure, a project that started in 2012.[36] But the foundation of Musk's long-term plans for growth was cheap battery power in massive quantities. To get it, Tesla and Panasonic planned Gigafactory, a manufacturing facility so-called because it was designed to produce billions of kilowatt-hours of energy storage capacity, sufficient to equip 500,000 electric cars per year.[37] The partners sought to cut costs by integrating all aspects of manufacturing, from cell production to cell bundling in battery packs, in one building, an ambitious enterprise that, like all Tesla's major ventures, depended on public subsidies. In 2014, Musk bargained with the state of Nevada to erect the plant near Reno in a deal that included $1.3 billion in tax breaks and credits.[38] Building and operating this enormous installation represented an unparalleled feat in the history of high-technology industry that paved new industrial-technological ground and, like so much else in the unfolding age of auto electric, effectively constituted a vast experiment. While the production space of Gigafactory was vertically integrated, the partnership that operated it was not, and the initial halting efforts of Tesla and Panasonic to apply technologies that had never been used at such a scale accentuated the tension in their respective interests.

THE TESLA WAY

The Model S was an instant sensation in the automobile world. Manufactured in much larger numbers than the Roadster, although far short of commercial scale, the car served the same role as its predecessor as a plat-form for marketing social influence, but at a new level of sophistication.[39] In November 2012, *Motor Trend* named the Model S the 2013 car of the year before it even shipped, the first time the award went to a car not pow-ered by an internal combustion engine (ICE).[40] To *Consumer Reports*, the Model S was a "car from another planet" and the best-built as well, scoring 99 out of 100 points, the agency's highest-ever rating.[41] In August 2015, the magazine gave the P85 variant a rating of 103 out of 100, a gesture at Musk's decision to customize the car's stereo volume with a dial that went to 11, a reference to the rock mockumentary *This Is Spinal Tap*.[42] Customer fandom burgeoned. Model S owners used the supercharger network to set new records in coast-to-coast "cannonballing" and promoted sales through the company's customer referral program.[43] In exchange for prizes, first adopters used the incentive of six months of free supercharging to con-vince friends to buy in. Would-be owners placed deposits to get on a wait-ing list for the next supercars off the production line, producing a pool of revenue that Tesla Motors put toward developing new products. The most important of these was the Model 3 sedan, planned as an affordable electric car intended to launch Tesla from the luxury niche into the mass market.[44]

As with any new consumer commodity, real-world use of the Model S revealed a host of problems. Many issues related to the sorts of routine fit-and-finish bugs and road damage that afflicted all automobiles to vary-ing degrees. Two years after delivering a glowing review of the Model S, *Consumer Reports* withdrew the recommendation, noting while the vehicle offered unparalleled performance, its overall reliability was poor. For such situations, Tesla relied on gold-plated service that typically left a favorable impression on users and resulted in high customer satisfaction.[45]

However, the proliferation of the Model S placed it in new envirotechnical contexts that sometimes highlighted the limitations of the battery electric propulsion unit itself. In one notable incident, a test drive by *New York Times* reporter John Broder in February 2013 revealed a deficit between what the Model S computer believed battery capacity to be and its actual capacity in winter weather. Broder's goal was to make a trip from Newark, Delaware, to Milford, Connecticut, using the supercharging network that Tesla was then building on I-95. It was a journey of 200 miles, comfortably within the car's estimated range of 265 miles, but cold weather adversely affected battery per-formance. Broder reported that the battery could not seem to hold a charge, and the car shut down in Branford, Connecticut, 18 miles short of the goal.[46]

Broder's reportage provoked a contretemps that recalled the *Top Gear* incident. Around the time Tesla was preparing to go public in mid-2010, its lawyers had asked the BBC to pull rebroadcast of the offending episode and retract some of its claims. When the broadcaster refused, Tesla sued for libel in 2011. British courts dismissed the suit on the grounds that range was relative and contextual and at any rate, it was impossible to demonstrate that *Top Gear* had caused the claimant harm.[47] Nevertheless, the case became important in Musk's response to Broder. It prefaced an essay that the entrepreneur posted to the Tesla home page that accused the journalist of sabotage. The car's datalink showed that Broder had repeatedly driven around the Milford supercharger rather than immediately plugging into it, and Musk interpreted this as a deliberate attempt to drain the battery. The journalist responded that there was a simple explanation. On the night in question, said Broder, he had been unable to locate the dimly lit charger and was circling the parking lot in an effort to find it.[48]

Another problem was the safety of large lithium batteries. Failures of Model S battery packs were rare, but when thermal runaway did occur, the resulting fires usually destroyed the whole car. One such fire in 2013 was associated with a 6 percent drop in the value of Tesla shares.[49] Musk responded that gasoline was a dangerous good and that there were incomparably more fires involving gasoline-fueled ICE cars than all-battery electric cars. These points were indisputable in and of themselves, but they did not address the relative catastrophic failure rate in the two fleets or the fact that the lithium rechargeable battery was also classified as a dangerous good. A string of incidents involving lithium batteries on aircraft through the 2010s prompted the International Civil Aviation Organization and the Federal Aviation Administration (FAA) to strictly regulate these devices both as components of aviation power systems and as freight on passenger and cargo planes.[50] Moreover, while the failure modes of gasoline-fueled ICE propulsion were well understood, the failure modes of lithium electric propulsion were not, especially in cases where fires started spontaneously in cars at low speed or at rest.[51]

In principle, the battery pack was amenable to engineering fixes. Making money from all-battery electric cars was a much more difficult proposition, however. In 2012, the cell supplier A123 went bankrupt, and Coda and Fisker failed the next year. Tesla Motors itself was consistently unprofitable despite its unprecedented achievements. Battery power remained expensive, and the case that Tesla and Nissan made for user ownership of all-battery electric cars was cloudy, in good measure because the useful lifetime of battery packs was then unclear.[52] These uncertainties in turn warped the aftermarket. Lower-end all-battery electrics depreciated faster than ICE vehicles and began to be available at bargain prices. A 2012 Nissan Leaf that sold for around $36,000 retail was worth a little over $8,000 on average three years later.[53] On the other hand, the Model S retained most of its value, thanks to a generous subsidized resale guarantee designed to encourage customers to trade in their cars for new, upgraded ones after three years.[54] In effect, this provision was a kind of lease, the traditional method of marketing electric cars.

In summer 2014, the Organization of the Petroleum Exporting Countries (OPEC) triggered a petroleum price war, and in October, Toyota and Daimler sold most of their stakes in Tesla and discontinued their production arrangements.[55] From the perspective of the established automakers, the lesson seemed to be that the economic viability of the all-battery electric car was questionable when oil was cheap and incentives were exhausted. But mainstream automakers did not reckon with macroeconomic trends that were unfolding in the wake of the Obama administration's stimulus program. Tesla lost money and repeatedly failed to meet production targets, but the Model S served its function as the stepping-stone to the Model 3. Backstopped by government, the dream of the affordable electric supercar was increasingly perceived by investors as plausible. Between the end of 2010 and the end of 2015, Tesla's share price increased nearly ninefold, and in the second half of the decade, the company's market capitalization would swell to new heights as part of the largest financial bubble in history.[56] An enterprise that began as a start-up would accrue an influence vastly disproportionate to its industrial capacity as an automaker and assume a central importance in US and global economic affairs.

13   SILICON VALLEY TAKES CHARGE

Tesla Motors almost certainly represents the most extreme test of the limits and capabilities of the Silicon Valley model of innovation.
—Jon Gertner, journalist, Fastcompany.com, April 2012

The Model 3 was the riskiest of Tesla Motors's ventures, both in terms of industrial engineering and finance. On the surface, the car resembled a smaller version of the Model S, but it was a substantially new design that used an advanced new permanent magnet motor powered by a battery pack built around a new purpose-built cell. Tesla and Panasonic had to learn how to scale new core content while transitioning from the luxury to the commodity business model. Production had to be coordinated with the remainder of the available federal tax credits while meeting private capital's increasingly insistent demands for profit, a maneuver that had practically no precedent in civilian industry. Tesla needed the publicity and revenue generated by its premium cars, and yet sales of such vehicles meant that fewer tax credits were available for Model 3, a product calculated to appeal to ordinary drivers socialized to the standards of modern internal combustion engine (ICE) performance. In the Model 3, Tesla and Panasonic engaged in concurrent research, development, and manufacturing, a common practice in military industry where cost was no object.[1]

But cost was central to the success of the Model 3, and Tesla Motors's ability to access public resources had limits. To save money, the company retrenched its premium customer service while retaining as much of the influence-marketing model as possible. In 2016, Tesla ended its resale value guarantee and unlimited free supercharging.[2] At the same time, Tesla promoted a host of upgrades, many in the form of downloaded applications that kept the company in the media spotlight, if not always for good news.[3] Perhaps the best-known of these features was autopilot, a system that became implicated in a number of fatal crashes and was viewed by some analysts as a distraction from the task of mastering the economics of battery power.[4]

Tesla's updatable automobile systems helped reinforce ideas of a distinctly Silicon Valley mode of automobile innovation.[5] Many observers interpreted the contemporary electric car as the next phase of the information technology revolution. Some engineers believed that the control requirements of advanced vehicles in themselves were driving the frontiers of innovation in electronics.[6] Microchip manufacturers had long looked to the auto industry as an important avenue of growth and predicted windfall profits in the emerging market for plug-in electrics.[7] In early 2016, Musk declared that Tesla Motors would deliver 100,000 to 200,000 Model 3s in the second half of 2017, a promise that made history. Droves of customers made $1,000 deposits, which raised $400 million in cash over the course of the year. In April, Tesla's stock rose to $312 per share, giving the company a market capitalization of $51 billion and making it the most valuable American automaker, displacing General Motors (GM), an enterprise that produced more than 220 times more vehicles.[8] As Tesla Motors and the investment community would learn, however, commercializing the Model 3 required resources that only globalized public policy could furnish.

THE MODEL 3 AND POST-FORDISM

Tesla's outsized valuation was based almost entirely on the anticipated success of the Model 3, defined by the investment community as the production of 5,000 cars per week. This meant that Tesla Motors (which officially changed its name to Tesla in late February of 2017) had boost its output by more than a factor of four, at a time when it was still struggling to build its older types. The new Model S and X frequently required rework after leaving the assembly line.[9] The Model 3 itself meant "production hell," as Musk put it, and as occurred so often in the past, solutions were improvised. Workers hastily set up a new general assembly line under a giant tent in the parking lot of the Fremont plant. By July 2018, the company had reached its production target, an impressive achievement, although opinion was divided on what it meant for the future.[10] Much of the media attention focused on the ad hoc assembly line, the subject of much derision by traditionalists.[11]

The more serious manufacturing problems involved core content. Out in the Nevada desert, away from the spotlight, not all was well at Gigafactory. With the Model S, Panasonic adapted cells from its 18650 commodity line and invested in Tesla but remained aloof from the automaker's manufacturing operations. With the Model 3, however, Panasonic was not only an investor but a colocated collaborator, and the company struggled to scale a cell using a revised nickel-cobalt-aluminum (NCA) formula in a new form factor known as the 2170.[12] Whistleblowers spoke of confusion on the factory floor. Under pressure to produce huge volumes of cells, Panasonic lost quality control. Insiders reported that chemical mixers became contaminated and scrap cells piled up, amounting to nearly a half million cells per day at a time when the battery maker was sending Tesla three million cells per day. Tesla's initial attempts to build battery packs had not gone well either. Thanks to a programming error, claimed insiders, the robot that bundled cells into modules was puncturing them during assembly, and the damaged cells were finding their way into production packs. Employees were reported to have disabled elements of the system that tracked battery parts, violating a standard protocol of quality control in the automaking industry. Some media sources suggested that the problems could be traced to Musk himself, who was said to have personally deactivated certain automated processes on the grounds that they took too long.[13]

The build quality of the first Model 3s compared unfavorably to other all-electrics like the Chevy Bolt and the BMW i3. An audit performed by UBS revealed numerous issues, including sketchy spot welds, missing bolts, loose tolerances, and shoddy fit and finish.[14] A similar judgment came from *Consumer Reports*. In August 2018, the agency gave the Model 3 a positive, if qualified, review. Only a few months later, it withdrew the recommendation on grounds of declining reliability and put the car in the company of the plebeian Chrysler 300.[15]

FINANCING THE PEOPLE'S ELECTRIC

Tesla's transition from luxury to commodity production created a host of short-term financial problems for the company. Desperate for revenue, the automaker prioritized production of the more profitable $55,000 version of the Model 3 over the $35,000 version, giving preferential access to owners of the Model S and bumping customers from the queue who had placed a deposit for the cheaper variant.[16] In May, Musk announced a $78,000 version and averred that building the baseline model would bankrupt the company, raising alarms in the business press that Tesla might be giving up on its hopes for the mass market.[17] As the high-value models rolled off the line and Tesla edged closer to its 200,000th production unit, these units consumed the company's reserve of available $7,500 federal tax credits for customers who most needed assistance. In July 2018, the company exceeded the threshold, and the credit halved to $3,750.[18]

In effect, Tesla used federal subsidies to treat the Model 3 as a kind of low-end Model S in the hope that affluent consumers would shoulder the expense of the all-electric learning curve and buy time for the company to cut battery costs. Some studies suggested that most purchasers of electric vehicles and claimants of electric vehicle tax credits were wealthier taxpayers making more than $100,000 a year.[19] In 2018, the media tended to interpret the Model 3 as a luxury model, and when framed in this way, it was the best-selling car in the US. For years, Tesla identified its Model S variants primarily by the size of their optional battery packs, an explicit marker of cost and status, but it dropped this approach with the Model 3. For months after the car's formal launch in 2016, the company did not specify the energy capacity of the battery pack options, information that was crucial for investors trying to get a sense of production costs. These were estimated to be between $190 and $200 per kilowatt-hour by early 2018, but Tesla did not disclose the actual costs. Only in August 2017, a month after the first Model 3s rolled off the production line and with a $1.5 billion Tesla bond issuance at stake, did Musk reveal that the baseline and long-range versions of the car would have 50- and 75-kilowatt-hour packs, respectively.[20]

In this instance, the investment community forced Musk's hand. Nevertheless, Tesla continued to regard explicit references to the relationship between battery energy capacity and cost as troublesome to its marketing. The company discarded the practice of citing pack capacity as a pricing instrument, referring instead only to classes of performance that emphasized range and speed for all its products.[21] In early January 2019, Musk tweeted that Tesla was discontinuing the 75D battery pack in the Model S and X, a tacit admission that the Model 3 sat at the low end of a luxury lineup.[22] The same month, Tesla removed references to the standard battery pack of the baseline Model 3 from its website, further alarming customers and analysts.[23]

SHAPING THE MARKET

As Tesla struggled to deliver the Model 3 through 2018, Musk's social media profile grew increasingly erratic, culminating in an August 7 tweet claiming that the entrepreneur had secured funds to take Tesla private. Musk had long mulled this strategy as a means of shielding against public scrutiny and punishing short sellers of the company's stock. Rumors circulated that the entrepreneur had the support of the Saudi Public Investment Fund. The petrostate's sovereign wealth fund had indeed built up a 4.9 percent stake in Tesla, just below the 5 percent threshold that required public disclosure, becoming the fifth-largest investor in the company in the process. Journalists Arash Massoudi and Richard Waters of the *Financial Times* reported this story on August 7, the day of Musk's infamous tweet. On January 28, 2019, Massoudi and Waters claimed that Musk sent out his electronic missive minutes after their story aired. If the *Financial Times* was to be believed, the Tesla chief's actions had been triggered by a quick scan of the day's news.[24]

These moves puzzled analysts. For the holders of one-third of Tesla's shares, Musk was proposing a buyout at $420 a share, an offer worth $24 billion. Musk needed to control costs, so holders of the remaining stock were expected to roll over into the new entity. What the entrepreneur did not explain was what stockholders stood to gain in this scheme. When companies go private, investors lose the ability to buy and sell at short notice and typically demand premium capital as compensation for their loss in liquidity.[25] Observers held that Musk's justifications for privatization were unclear. The executive averred that he wanted to penalize short sellers, but Wall Street had actually inflated Tesla's outsized share price. Musk believed that concentrated private money would cloak Tesla's corporate affairs from the world, but analysts argued that such a regime was equally likely to bring about much more intrusive corporate governance.[26] Nor was it clear how much of Tesla's diverse shareholding base could be accommodated in a private entity. Many of its investors, especially the large institutional ones, held Tesla stock in funds that could only own publicly traded stocks. Many small individual investors were similarly barred from investing in private firms.[27]

Some observers suspected Musk had been motivated by the pressure of Tesla's $11 billion debt load. The company required continual cash infusions, and debt had the advantage over new issues of equity in that it prevented dilution and preserved the grip of the chief executive officer (CEO) as the largest single shareholder, at least in the short term. The problem was that the scheme robbed Peter to pay Paul. Nearly $2 billion of that debt was convertible, meaning that it could be repaid in cash or in stock if the stock reached certain predetermined levels. In that event, a bondholder become an equity holder. If the stock was below these figures when the debts came due, the bond had to be repaid in cash.[28]

Analysts suspected Musk of promoting Tesla's stock to avoid an impending cash crunch, even at the risk of diluting ownership. If the move had been intended to manipulate the market, it backfired dramatically. Tesla actually had not secured funding, and as this became clear, the stock fell, erasing $15 billion in market capitalization. Musk was accused of market-fixing, and the Securities and Exchange Commission (SEC) launched an investigation. Following what some observers characterized as an unusually swift inquiry, Musk and the company were each fined $20 million. The executive also had to relinquish the post of chair of the board for two years as well as agree that company lawyers would exercise oversight over his use of Twitter.[29]

These penalties were relative trifles, although the new spate of securities actions did add to Tesla's already considerable legal problems. The company faced numerous lawsuits on a host of claim ranging from automobile quality and safety to deposit theft and vendor nonpayment.[30] The sanction that seemed to concern Musk most was the restriction on his use of Twitter, a punishment that he framed as a violation of his First Amendment rights in a television interview that he gave with Lesley Stahl on *60 Minutes* in December 2018.[31]

In principle, the Twitter restriction inhibited Musk's ability to shape the market, and yet it also indirectly empowered the entrepreneur because it further crystallized his personal association with the Tesla brand. The attempt at oversight cast into relief Tesla's identity as an exemplary promissory organization guided by a single charismatic individual whose pledges for future performance had implications for the global marketplace. Jay Clayton, the Trump administration's appointee as SEC chair, underscored this message in an unusual address following the settlement that noted the importance of "skills and support of certain individuals" to corporate success.[32] In effect, the federal government acknowledged Tesla as a national asset, and Musk's stature rose correspondingly. As the leading champion of all-battery electric automobility, Musk challenged the fossil fuel order. As an engineer, he stood up to the financial speculators. Now he was taking on an arm of the federal government as a tribune on behalf of free speech. Only days after the settlement, Musk felt sufficiently comfortable to taunt his antagonist publicly as the "Shortseller Enrichment Commission."[33]

Whatever his shortcomings as a manager, Musk had a keen grasp of public policy as well as the popular mood. If more evidence were needed of the establishment's senescence, it was furnished by the demise of an automobile once held out as the key to its future. In November 2018, GM announced that it was canceling its Volt program.[34] Chevrolet had lost its gamble to use advanced technology to compete with the Prius. Consumers seemed to prefer an affordable hybrid that possessed modest all-electric range, but very high gasoline mileage and relatively low emissions, over a more expensive hybrid offering a much greater all-electric capability and zero emissions when in that mode.[35]

In the late 2010s, however, Tesla was also in trouble. The company historically had high staff turnover, and the troubled launch of the Model 3 precipitated the departure of dozens of employees and managers, including Tesla's chief of global supply management and chief accounting officer.[36] In March 2019, Tesla finally unveiled the $35,000 version of the Model 3, but slack demand and a decision to lay off staff and close stores caused the company's stock to lose a third of its value in the first half of the year. In an attempt to bolster the customer base, Tesla began offering the car for lease in April.[37] In July, JB Straubel stepped down as chief technology officer, leaving Musk as the last of Tesla's original cofounders. Over sixteen years, Straubel developed many of Tesla's technologies, and his departure was interpreted by some as the end of an era for a company that had done more than any other to revive the all-battery electric car.[38]

GLOBAL ELECTRIC

Despite its travails, Tesla was buoyed by policy incentives and foreign demand. Its cars were popular in Europe, especially Scandinavia, which benefited from a range of subsidies.[39] The company also earned revenue from carbon emission credits.[40]

But Tesla's real object of desire was China. In 2018, the automaker opened Gigafactory 3 in Shanghai to manufacture the Model 3.[41] Where the China-based facilities of US enterprises like Apple produced for export, Tesla's new plant was geared for the domestic market, mirroring the approach that Japanese and German manufacturers had long taken in the US as a means of circumventing protectionism.[42] Gigafactory 3 was planned at a time when the partnership between Tesla and Panasonic frayed under the pressures of the Model 3 program, an opportunity that the automaker took to attempt to diversify cell supply, influence cell design, and move in the direction of vertical integration.[43] Tesla began to use cells produced by LG and the China-based CATL in addition to Panasonic cells, and promoted a new larger cylindrical cell form factor that it encouraged its suppliers to develop.[44]

Tesla's increasing assertiveness in the cell market unfolded as part of a dramatic new chapter in the electric vehicle renaissance. What capitalist quasi-planning helped create was being massively expanded by communist central planning. Over the course of the 2000s and 2010s, China became the world's largest market for automobiles, including electrics. Stimulus policies helped foster a multitude of Chinese companies in the field of electric propulsion, and the Shenzhen-headquartered BYD became a contender for the title of the world's largest producer of electric vehicles.[45] Thanks to incentives that encouraged foreign automakers to use Chinese-made batteries in electric cars sold in the Chinese market, BYD also became the first producer of electrics to vertically integrate battery manufacturing from cell to pack.[46]

These significant milestones in the new age of auto electric were rooted in China's industrial revolution, a sociotechnical movement that also served to support Tesla's ambitions. Chinese government policy aimed to ensure that by 2025, 20 percent of vehicles sold in the country would be electrics of one type or another, representing seven million units per year, and the industrial infrastructural aspects of this project were suitably ambitious. As conceived by Premier Li Keqiang, the "Made in China 2025" plan aimed to source 70 percent of high-technology goods from domestic manufacturers. These policies pulled important elements of the global electric vehicle supply chain into the orbit of China's industrial complex. In 2013, China's largest auto parts supplier acquired the bankrupt A123, a recipient of hundreds of millions of dollars of US aid and a former star of the Obama administration's stimulus program.[47] In 2019, Nissan and NEC sold their majority stake in the Automotive Energy Supply Corporation (AESC) to the Shanghai-based Envision Group.[48]

Musk evinced aspirations to develop Tesla as a vertically integrated enterprise, but the practices and economics of developing and manufacturing battery cells at scale were utterly dissimilar from those of the other components of electric cars including software, electronics, motors, and vehicle chassis and body shells. There were a host of sociotechnical reasons for this. Commercialization in the power source field took far longer than in the other fields relevant to the electric car, with progress measured in increments over decades. And non-nuclear power source technoscience (apart from photovoltaics) had historically been neglected in US education and industry. To be sure, the national developmental state supported pathbreaking research in battery materials by figures like Stanford Ovshinsky, M. Stanley Whittingham, John Goodenough, and Michael Thackeray, and it also made prodigious if belated efforts to build a battery manufacturing complex.

Yet US industry and capital were reluctant to exploit those achievements and sometimes even suppressed them, as the history of the Ovonic Battery Company (OBC) and Energy Conversion Devices (ECD) demonstrated. William O. Baker's post-Sputnik call for US society to embrace advanced materials only gradually impacted US power source technoscience, largely because there was little demand in this field outside military aerospace until around the last quarter of the century. Efforts to develop advanced power sources in the US also lacked the crucial industrial complementarity provided by consumer electronics, a sector that declined relative to its Asian counterpart from the 1970s. In the US, industrial applications of advanced materials instead occurred largely in the fields of aerospace and semiconductors, sectors that the national developmental state protected and nurtured as strategic assets. The effect of these dynamics on the culture of US engineering, as Wally Rippel and Jack Johnson suggested, was that relatively few students viewed electrochemistry, solid state ionics, and power source technoscience more broadly as the basis of viable careers.

The training, skills, and infrastructure to produce the advanced battery cell, the beating heart of the electric car, were instead developed mainly in Europe and even moreso in Asia. Tesla's worldview reflected the asymmetrical global development of high-technology industries relevant to electric automobility. The company was founded on the principle of virtual

integration, the hallmark axiom of the vertically disintegrated US information technology sector, a formula that enabled the automaker to rapidly
seize market share and that in turn made it an attractive market for the core
material content of Japanese and South Korean industry. In its China operations, Tesla seemed destined to fulfil a similar role for Chinese industry.

Tesla's turn to China coincided with the ascension of the company's
stock (not to mention Musk's personal wealth) to dizzying new heights,
underscoring the drastic changes that automaking underwent in the first
two decades of the new millennium.[49] Tesla confounded most of the conventional notions of business management and success as articulated by
Alfred D. Chandler, the eminent scholar of business history. The company
did have a multiunit structure in its solar panel, energy storage, and auto
divisions, but it was not built around a stable managerial class, nor did
Musk show much interest in grooming one. In the first two decades of its
existence, Tesla was operationally dysfunctional, plagued by low workforce
morale and high turnover. Its products were stylish and sophisticated but of
variable quality. The company was widely hailed as a leader in innovation,
but most of its basic technologies had been invented by others. Tesla had
relatively little core intellectual property to protect or trade, a fact that Musk
acknowledged and attempted to make a virtue of by making the automaker's
patents publicly available in the name of the open-source movement.[50]

Yet Tesla not only survived but thrived. For some observers, the company was exemplary of the *postindustrial*, an expression coined by Harvard sociologist Daniel Bell to describe the shift from enterprises of heavy
manufacturing to enterprises of digital electronics and software.[51] Some saw
in Tesla a new national champion in the vein of Intel, Microsoft, and Apple,
able to offer worthy competition to Toyota, if not fully replace GM.[52] In the
early 1990s, the business writers James Collins and Jerry Porras authored
Built to Last, an influential book claiming that vision and organizational
excellence were the keys to corporate success.[53] The book provoked a strong
response in the management studies community, led by the business studies professor Phil Rosenzweig, who argued that most management concepts
like leadership and corporate culture were nebulous and difficult to define.
In a dynamic economy characterized by rapid technological and structural
change, held Rosenzweig, success was contextual and could not be ascribed
to solely to strategy or vision.[54]

Silicon Valley supplied Tesla with business metaphors and models, but the salient contextual factors in the company's rise were the decline of US automaking, the privatization of public policy, financial speculation, and US and Western industrial integration with Asia. Tesla was part of an infrastructure of American manufacturing that was at once reinforced and complicated by collaboration with China, a relationship that helped make the automaker worth more on paper than Ford and GM combined, and Musk one of the world's wealthiest persons. Tesla's interpretation of the automobile had always been characterized by contradictions. The company defined its brand in opposition to traditional automaking while subscribing to important establishment assumptions. Nominally green, Tesla cars were electric versions of the large and powerful vehicles that Detroit long insisted were the basis of the American way of life. Where Sebring-Vanguard was iconoclastic, Tesla upheld the tradition of high performance at any price. Sustaining that tradition at scale in turn required resources outwith the US and followed the broader trajectory of Western capital in its shift from industry to finance. For better or for worse, Tesla's fortunes were embedded within the global industrial condominium, a network of partnerships too big to fail.

14 THE LIFE ELECTRIC

Once I am in an electric car, I can tell when the car in front of me doesn't have a catalytic converter, or if there is something wrong with the car. You can smell it. It smells like raw gasoline.

—Tim, electric motorist, May 25, 2015

More than twenty years after Wally Rippel and his team won the Great Electric Car Race, contemporary electric motoring remained in a pioneering phase. Adventurous drivers built on the tradition of seeking technological firsts. One of the most celebrated such events was the attempt by the academician and environmentalist Noel Perrin to drive his converted Ford Escort from California to Vermont in 1991, an adventure widely regarded as a milestone in the contemporary revival of the electric vehicle. Perrin recorded the details of his often-harrowing journey in a book that the famed environmental scientist Donella Meadows said would be remembered for being "astonishingly prescient and amusingly quaint" when everyone was driving electric cars thirty years hence.[1]

As a literal prediction, this forecast fell short. By the mid-2010s, gasoline-fueled internal combustion engine (ICE) propulsion still overwhelmingly dominated the US light-duty fleet. Still, Meadows correctly anticipated the general timeline of the commercialization of the electric car, and in a figurative sense, the shift to a new automobile world. By the mid-2010s, electrics of all types had become relatively common in certain parts of the US, especially California. From 2005 to 2015, the national stock of registered plug-in hybrid and all-battery electric passenger cars went from about 1,000 to more than 400,000 vehicles, and this was in addition to the hundreds of thousands of conventional hybrids on US roads.[2]

Increasingly, electric cars were becoming objects of interest to science and technology studies. Much of this scholarly research involved explorations of the sociotechnical dynamics of green automobility, the construction of user identity, and the problematization of technological determinism. For example, Heidi Gjøen and Mikael Hård addressed the agency of Norwegian users of all-battery electric cars through their reinterpretation of the conventional political and engineering assumptions for the technology.[3] Reid R. Heffner, Kenneth S. Kurani, and Thomas S. Turrentine interpreted the diverse and sometimes contradictory symbolism that users invoked around hybrid electric technology, ranging from thrift to technological sophistication to environmental virtue-signaling.[4] In their study of Prius owners, Ritsuko Ozaki, Isabel Shaw, and Mark Dodgson found, unsurprisingly, that the ill-defined quality of sustainability was not inherent to the technology. Instead, it was coproduced in the interaction between the driver and the car's computerized systems.[5]

As reflexive as these studies were, they sometimes verged on essentializing behavior and experience owing to a tendency to abstract at the level of the individual user-machine. As the scholar of environmental history Christopher Wells noted, discussions of automobile culture around the comparative qualities of technology often occlude social and environmental context.[6] In this chapter, I explore the affective embodied experiences (to paraphrase an idea of the sociologist Mimi Sheller) of several motorists, some of whom were affiliated with the East Bay chapter of the Electric Auto Association (EAA) in the San Francisco Bay Area, in the context of California automobility as it underwent electrification in the mid-2010s.[7] I organized these stories around behaviors and practices that I classified as evangelism, pragmatism, idealism, aestheticism, futurism, and skepticism, associating particular users (whose identities I have pseudonymized) with particular classes of behaviors and practices that may in some sense be interpreted as archetypal. However, I also assumed that all of these behaviors and practices could manifest in a single individual, either singly or in combination, as a particular state over time as sociotechnical and envirotechnical conditions changed. A number of users identified as idealists and enthusiasts in one way or another and some possessed technical skills and experience in conversion culture, where identity and belonging were determined by the capacity to build, operate, and maintain one's own vehicle. Others were less directly involved in the material culture of electric vehicles but nevertheless tested the affordances of new technology, especially batteries, in daily real-world conditions. All of these activities yielded crucial experiential knowledge at a time when scientists were just beginning to understand how semi-commercial and commercial battery packs for electric vehicles aged over time.[8]

With the proliferation of commercial electric cars of all kinds and their adoption by users motivated by values in addition to or other than energy and environmental sustainability, other sorts of considerations informed the user-technology interaction, and other types of contextually determined effects were generated that varied according to the type of vehicle and the mode of its application. In newer vehicles, these effects were importantly mediated by sophisticated onboard computers designed to assist the driver in understanding how energy was converted to vehicular motion. In the Prius, whose brand identity valorized fuel thrift and its assumed environment effects, the computer prompted users to limit their energy use and rewarded those who responded with graphical affirmations of "saved" resources.[9] Other users of the Prius rejected this ethos and subjected cars to hard driving without suffering significant penalties in terms of range.[10]

However, users of the smaller all-battery electrics of the mid-2010s could not always rely on computers to provide accurate information on range. In the all-battery electric duty cycle, there is much greater stress on the battery than in the hybrid electric duty cycle, and that stress varies dynamically in envirotechnical context in ways that the computers of the day could not always accurately predict. These dynamics affected all types of all-battery electric cars but the consequences were more severe for users of cars with smaller batteries because they had less energy at their disposal than cars equipped with larger power sources. As a result, users of such vehicles were compelled to interpret computer information more critically than users of hybrid and large all-battery electric cars and monitor and moderate their driving behavior more closely, at least if they wished to avoid running out of stored energy and becoming stranded.[11]

EVANGELISM

I began my interviews in the summer of 2015 with Tom, then-president of East Bay EAA. He greeted me in the driveway of his modest bungalow in a working-class neighborhood of west Alameda, gesturing toward my rented

Chevy Spark and asking if I needed to plug it in. It was a case of mistaken identity. The car was actually conventional, but its electric version was one of the more common compliance cars on California roads at that time.[12]

Tom had a broad practical knowledge of electric cars and machinery more generally. A retired Coast Guard warrant officer and a mechanic by trade, he worked on everything from patrol boats to helicopters and fixed-wing aircraft and had restored a number of electrics, including a 1976 vintage Sebring-Vanguard Citicar. Tom's primary automobile was a conventional Prius, but he also had an Electron, a 1989 Ford Escort converted to all-battery electric drive by Solar Electric Engineering, the company that prepared a similar car for Noel Perrin.[13]

The Electron was a fixer-upper whose history well illustrated the problems issuing from the temporal mismatch of the components of all-battery electric propulsion technology. Tom acquired the car from another enthusiast at no cost in 2011. Its decade-old battery pack was nearly exhausted, with just enough energy to allow Tom to roll the car onto a trailer. On the other hand, the motor and drivetrain seemed in relatively good condition after almost twenty years of use. Tom added a throttle control, necessitated by the car's dangerously touchy reverse mode, and upgraded the instrumentation. The most serious and costly problem was energy storage. Over two decades, the car had gone through three lead-acid battery packs. Tom built a replacement consisting of eighteen 6-volt lead-acid golf car batteries, at a cost of $1,300.

I asked Tom for a test drive, and he was happy to oblige. Worn and weathered, the Electron had a certain do-it-yourself charm. I noted the homemade oval "electric" logo pasted over the Ford emblem on the hood, a relic of an earlier era of electric vehicle culture. Apart from the absence of a tailpipe, electric conversions are not easily distinguished from the original equipment, so enthusiasts often advertised the propulsion system with stenciled lettering and decals. We drove around Tom's neighborhood and then toward the nearby former Naval Air Station (NAS) Alameda, a vast and largely abandoned military base that turned out to be an ideal place for a test drive. We passed the crumbling guardhouse, and Tom turned the wheel over to me. It was my first time at the controls of an all-battery electric car, and my first impression was that at low speeds, the Electron drove much like any other quarter-century-old car without power steering, except that it was largely noiseless and the "gas" pedal was more responsive than its ICE counterpart.[14]

It soon became clear that the Electron did not have the power or endurance to be safely taken out on the freeway. However, the car acquitted itself well in the deserted streets of the old base. Tom used the vehicle mainly as an educational tool to advise and assist initiates in electric automobility and promote awareness of electric vehicle technology. When Tom joined the East Bay EAA in the mid-2000s, its membership consisted mainly of people like him, "mechanic/technician types," usually white males at or near retirement age who built and maintained their own converted electric vehicles. At their meetings, the enthusiasts talked shop, discussed the newest controller and battery technologies, and took passengers like me on test drives, giving them their first taste of personal electric automobility.

By 2015, the club's membership was aging and less active. A key factor in the decline, Tom held, was that the public became less interested in converted electric cars after the first commercial all-battery electric cars began appearing around 2010. Initially, electric vehicle clubs collaborated with manufacturers in promoting the new Volt hybrids and Leaf all-battery electrics, but as public interest grew and the novelty of the technology wore off, automakers preferred to do their own marketing. To Tom, these were positive signs. He invoked the parallel of the personal computing revolution to describe what he saw happening on Bay Area roads. The first users of personal computers were hobbyists. Decades later, the vast majority of people who purchased computers (and electric cars) did not care exactly how these consumer technologies worked, so long as they worked reliably.

Nevertheless, Tom maintained the club's educational outreach programs and informally advised motorists on technology use. Not all new electric vehicle technology was operationally intuitive, especially some plug-in hybrid formats. Tom recalled an episode at a pilot project for plug-ins that he had participated in at the local Coast Guard base, where drivers had been improperly trained to use the charger for the Chevy Volt. Drivers plugged the car in but failed to activate the charger, allowing the battery to drain. These events went unreported precisely because the Volt was designed to operate when the battery was depleted. In that circumstance, the ICE drove a motor-generator that provided electricity to the main electric motor. For this group of users, it proved easier to top up the car with gasoline than recharge it. In effect, the Volt enabled motorists who were encultured to ICE automobility to maintain their old habits. The problem only came to light two years into the program, when Tom's turn to use the vehicle came around and he found that the battery was completely dead.

Tom also spoke of how the proliferation of different types of electric vehicles outpaced charging infrastructure and how that created new problems that regulatory interventions often only exacerbated. Through the early years of the electric car revival, enthusiasts developed the practice of plug-sharing to make the most efficient use of limited charging infrastructure, posting written acknowledgments on their cars indicating that it was acceptable to unplug vehicles from chargers if their batteries were fully charged, a condition not always easily determined because automakers used different graphical symbols to signify states of charge. In 2002, California passed a law restricting parking in some charging spaces to Zero Emission Vehicles (ZEVs) and requiring users of these spaces to display a ZEV decal authorized by the Department of Motor Vehicles (DMV) at a time when the only type of ZEV was the all-battery electric car. But the commercialization of plug-in hybrids stimulated demand for public charging space and led the state to revise the rules to accommodate the technological shift in 2011. This was Assembly Bill 475, a measure backed by GM that did away with the requirement for the ZEV decal as the legal basis for parking in a designated charging space and created a new criterion based on physical connection to the charger for "charging purposes." The measure dropped ZEV nomenclature entirely and instead defined vehicles eligible to park in charging spaces as electric vehicles including plug-hybrid electrics. The law made it illegal to park vehicles in such spaces if they were not plugged into the charger but it was not illegal if a parked plugged-in vehicle had already recharged its battery and was no longer charging. Experienced users like Tom argued that AB 475 outlawed plug-sharing and reduced the efficiency of charging infrastructure. Vehicles that did not have the capability to lock charger cables to their cars and that were unplugged for whatever reason, a situation that occurred frequently with Tom's Electron, risked being towed.[15]

I recalled our test drive. The Electron seemed well suited to the space of the former NAS Alameda, which had the footprint and building density of a small town. In this environment, Tom had effectively demonstrated the principle of the city car, the original mode of electric automobility, suggesting the possibility of low-cost and effective personal automobile transportation in a car-dependent landscape. But we had enjoyed traffic-free motoring due largely to the fact that we were driving in an urban space

that was also a sacrifice zone. One of a number of former military instal-
lations in the Bay Area, the old base was a superfund site heavily polluted
with polychlorinated biphenyl (PCB), asbestos, and radioactive waste. As
we passed row after row of abandoned barracks, hangers, and jet engine
test stands, it seemed likely that some areas of the base would never be fully
converted to civilian use. The tableau recalled Noel Perrin's description of
urban California as a sea of asphalt. Perhaps some of the rolling stock could
be replaced, Perrin suggested, but society had inherited legacy transportation infrastructure that could not easily be renovated or disposed of.[16]

PRAGMATISM

I wanted to speak with a motorist who used the city car in a more conventional suburban setting and found one in Morton, a retired engineer who
worked in the semiconductor industry for twenty years. Like Tom, Morton
had a kind of hands-on charisma. He designed and built his family home,
an attractive, cedar-paneled structure in an otherwise nondescript part of
east Fremont. The house was tastefully appointed in a southwestern style,
with a central hall that served as an impressively large and eclectic library.
In the kitchen, Morton's wife, Betty, prepared roasted red peppers for lunch,
remarking that her ophthalmologist told her that the orange ones are best
for keeping eyes healthy.[17]

The couples' electric car was certainly eye-catching. It was a 2011 Th!nk
City, a red plastic–paneled, all-battery electric two-seater with a vaguely toy-
like quality. The car looked a bit like something Lego might have built if it had
gone into the electric auto business, exuding a pleasing sense of purposeful,
utilitarian design. Morton professed a lifelong interest in electric vehicles dating back to the great race between the Massachusetts Institute of Technology
(MIT) and the California Institute of Technology (Caltech), a contest whose
victor, he noted with pride, was the latter, his alma mater. The story of how
Morton and Betty acquired their electric runabout spoke to the checkered
history of the business of the small all-battery electric car. The Th!nk City was
very similar in concept to the Sebring-Vanguard Citicar and the later Kewet
EL Jet/Buddy, and only around 1,000 were built. Th!nk was born as Pivco
in Baerum, Norway, in 1991 and briefly entered Ford's orbit in the late
1990s as a ready-made supplier of compliance cars. With the ZEV mandate
neutralized, Ford sold Th!nk's Norway-based manufacturing plant in 2003

to Kamkorp, a private holding company that also owned Frazer-Nash, a UK-based research engineering firm specializing in electric powertrain systems.

Then, in 2006, Th!nk was acquired by a group of Norwegian investors led by the engineer-entrepreneur Jan-Olaf Willums and went on to help pioneer the post-EV1 revival of the all-battery electric car. The company's revamped City attracted high-profile interest, including from Google founders Larry Page and Sergey Brin. Launched on the eve of the Great Recession, like so many other electric car programs, the revived Th!nk had nowhere near the resources available to mainstream automakers to weather the crisis, and in 2011 the company again went bankrupt.[18]

Morton and Betty first became aware of the ensuing fire sale of assets through East Bay EAA. In retrospect, said Morton, their decision to purchase came down to economics. The couple acquired a new City for the bargain price of $15,500, less than the cost of the battery pack for the Model S of the time. The City's operating costs were very low, and Morton and Betty spent practically nothing in their first year of ownership. The City had room for Morton's six-foot frame, and the trunk had more space than some larger subcompacts. Utility aside, the car looked fun to drive, which Morton confirmed in a test drive. Like many contemporary electrics, the City accelerated and cornered with alacrity, thanks in part to its low center of gravity. All the heavy materials, including the batteries, were located below the axles.

In using the City, Morton had to relearn some driving behaviors. Like many electric motorists, he discovered that the relationship between rated battery capacity and range was not straightforward. In the US, the Environmental Protection Agency (EPA) made official estimates based on ideal driving conditions but range in real-world conditions was typically less, often considerably so. The City was equipped with a 23 kilowatt-hour EnerDel lithium ion battery pack, comparable to the type that Nissan used in its popular first-generation Leaf, and credited by the EPA with a range of up to 100 miles. Morton reported that the car's actual range was between 70 and 75 miles, and perhaps 80 miles if he did not speed. Users of all-battery electric vehicles with smaller powerpacks like Morton tended to pay closer attention to their driving that most motorists. Many such vehicles were equipped with economy and performance modes, and a common driving tactic was to switch between them according to conditions. The top speed of the City, reported Morton, was around 70 miles an hour, more than enough to cope with freeway movement. With freeway ingress, Morton selected performance mode, and with freeway egress, he switched to "economy mode," activating regenerative braking and using electromotive force to brake and put charge back into the battery.

For users of smaller all-battery electrics, topography mattered in a way that it did not to users of ICE cars, hybrids, and large all-battery electrics. Uphill climbs gobbled stored energy, whereas downhill runs controlled by regenerative braking restored some of it. For Morton and Betty, an appropriate trip was anything within a 35-to 40-mile radius in Oakland and northern San Jose. This was sufficient for 90 percent of their needs, making the City the couple's primary vehicle. The couple also used the vehicle for leisure touring in the wine country in nearby Milpitas and Livermore. Reliability was generally good, but there were occasional systems problems. After a trouble-free first year, Morton and Betty's fleet dealer issued a recall for routine maintenance that involved installing a new heater and computer software into the City. The repairs were duly made, but a month later, the car suddenly and mysteriously became as inert as a brick, or "bricked" in consumer electronics parlance. Stumped, the dealer had to replace both the computer and the battery. Morton reported that the work was covered under warranty by a third party, although any subsequent work would not be.

The conversation drifted to Tesla, whose giant plant was nearby. The Model S, said Morton with a trace of a grin, was a "wild animal," not a rational form of transportation. What Tesla was really selling, he said, was a lifestyle that recalled lost youth. In this respect, Morton reckoned, the company's great achievement was to change perceptions of electric cars. The couple needed no such encouragement to help them appreciate the virtues of relatively humble automobile technology. I thought of the market logic that enabled Morton and Betty to acquire a cheap battery electric city car and considered how the couple had made the technology work for them in the sprawling exurbs of the East Bay. I also considered the systems failure that immobilized their vehicle, as well as problems of the battery electric aftermarket for parts and service in circumstances where the original equipment manufacturer stopped manufacturing replacement equipment or went out of business altogether.

IDEALISM

My next stop was the Palo Alto neighborhood of Barron Park, where I paid a visit to Felicia and Peter, a couple whose experience in many ways was a composite of that of Tom, Morton, and Betty. Felicia and Peter spanned the do-it-yourself and commercial eras of electric automobility. In the early 1990s, the couple purchased a homebrew conversion. Around the turn of the millennium, they built their own electric car, and then in the 2010s, they purchased one of the first commercial electrics. Barron Park is situated in the historic heartland of the information technology revolution, close to Stanford University and the nondescript garage at 367 Addison Avenue where William Hewlett and David Packard founded Hewlett-Packard (HP), a site designated by the state as the birthplace of Silicon Valley. Across the street from Felicia and Peter's photovoltaic-roofed home, a white Model S recharged.

Felicia had degrees in electric engineering and law and advocated for workers' rights and health and safety. Peter was a retired semiconductor designer but revealed that his real passion was environmentalism, an interest he shared with his partner. In the early 1990s, motorists in Barron Park started going electric. The community then boasted six or seven electric cars, most either built by hobbyists or professionally converted. Like so many enthusiasts, Felicia and Peter looked to Solar Electric, acquiring a Ford Escort station wagon equipped with lead-acid batteries that gave a range of 20 to 30 miles.[19]

The car proved adequate for Peter's commute to his job at the nearby HP campus, but changing employment circumstances pushed the vehicle to its limits. In the late 1990s, Peter switched jobs and went to work for LSI Logic in Santa Clara, some 13 miles away from home. Around 2001, he was posted to Milpitas, which added a few more miles to his commute. By then, the car's controller was starting to fail, and Peter had never found the Escort particularly comfortable anyway. He and Felicia decided to build a replacement, drawing on their experience of upgrading the Ford. Conversion was a fairly simple process from an engineering standpoint, held Peter. The chief variable, recalled Felicia, was how battery chemistry related to available chassis and body shells. One of the main criteria in selecting a conversion platform, she noted, was whether batteries could readily be fitted into it. Nickel-metal hydride and lithium ion rechargeables were not then commercially available for electric vehicles, so most enthusiasts had to make do with lead-acid battery packs. For a converted car to have useful range, held Felicia, its chassis had to be sufficiently robust to accommodate a lot of these heavy power sources.

The couple decided that the 1974 BMW 2002 met this requirement. Into this platform Felicia and Peter wedged twenty 6-volt golf cart batteries worth $1,000. Working as a team, and with some welding help from a neighbor friend, the couple completed the conversion over the course of 2002. Peter drove the vehicle directly to the DMV and had it licensed to drive on public roads. The couple used the car for a decade, gaining some local fame in the process. One of their favorite pastimes was attending local conventional car clubs, whose members tended to be socially conservative males: "These guys spent thousands of dollars on their cars, and every year we would take ours down, in part to see how they would react to having one of their own cars butchered and converted." But Felicia and Peter received a warm response, and twice won the prize for conversion car of the year from the local electric auto club.

For all its advantages over the Ford, however, the BMW did not have that much more range. Because Felicia and Peter rejected ICE propulsion entirely, they had to adjust their driving behavior accordingly. In the intervening years, the couple started a family, meaning that they had to abandon spontaneous trips and plan practically all their automobile movements. When the couple's two children left for college, the range calculus became even more complicated. Felicia and Peter went on longer excursions, and planning could not always make up for low battery capacity. One of the most important lessons the couple learned was that deeply discharging the battery pack could damage it.

Another change in employment caused further complications. At one point, Peter started to work for Stanford University, an institution that in his view was not particularly hospitable to electric cars. At HP and LSI, Peter had reserved parking space with free charging, but Stanford wanted to charge him $900 a year for this service. However, the university did have a charging outlet, and a few times Peter "borrowed" electrons, figuring that he would "ask for forgiveness rather than permission."

These lifestyle changes came at a time when the first of the post-EV1 all-battery electrics were beginning to appear on the market. The couple's philosophy was to try to maintain just one automobile, but few electrics met all their needs. Plug-in hybrids offered trade-offs that Felicia and Peter found unpalatable. The plug-in Prius had a short all-electric range, and while the Volt offered better performance in this respect, the couple believed that the American car had inferior fuel efficiency and internal volume. Economic all-battery electric power was the deciding criterion. Tesla cars were too expensive, so the couple selected a Nissan Leaf, whose 80-mile range was sufficient for most of their local commuting needs. Felicia and Peter also decided to retain a conventional 2002 Honda Civic for longer trips.

In the end, this couple chose a kind of hybrid solution after all. It is tempting to read their story as a microcosm of the conundrums confronting middle-class electric car enthusiasts in the mid-2010s. After years of development, all-battery electric propulsion offered capabilities comparable to ICE propulsion, at least in principle. But it still cost a premium to acquire the kind of battery capacity necessary to fully negotiate the space of sprawling autocentric cities built around the energetics of gasoline.

AESTHETICISM

Having engaged motorists motivated by economics and environmentalism, I felt I needed to speak to someone attracted by the sporting potential of battery electric propulsion. I found such a user in Daniel in the city of Menlo Park. An affable and cherubic person, Daniel presented himself in the coffee shop of the local Draeger's Market in a Tesla-branded baseball cap and windbreaker. He seemed an archetypal affluent Californian. A native of Long Beach and a self-described "technology geek," Daniel earned a degree in psychology but learned electronics after he was drafted into the Air Force in the 1960s. He worked for HP for many years and had no special interest in electric vehicles until a chance viewing of *Who Killed the Electric Car?* on Netflix. Like many Californians, Daniel and his wife experienced a "severe emotional response" to the story of the destruction of the EV1, and they resolved to purchase the first all-battery electric that came on the market. That happened to be the Tesla Roadster, a vehicle that the couple intended to supplement their 2006 Chevy Envoy sport utility vehicle (SUV). But the Roadster quickly became the couple's primary automobile. Daniel had no children from his first marriage and his wife's children were already grown, so there was room in their budget for some extravagance. The couple later became a two-Tesla household when they acquired a Model S for their longer trips.

Much of our conversation took place inside Daniel's Roadster, a bright orange car trimmed with a gray racing stripe. The vehicle was much lower and closer to the pavement than I had imagined, and sliding into the passenger seat was rather awkward. Noting my discomfiture, Daniel supportively remarked that he still had trouble getting into the car. As we prepared to leave Draeger's parking lot, passers-by audibly expressed their admiration for the striking automobile. Moments later, we were speeding north on Interstate 280 into the Santa Cruz mountains. I was surprised by the relative lack of motor noise, even at cruising speed and with the canvas roof panels removed.[20] As I admired the scenery, Daniel remarked, apropos of nothing, that every now and then his wife would tell him to punch it: "So I have to punch it!" With that, Daniel suddenly elongated his body and mashed the pedal, pushing me back into the seat and causing me to involuntarily clutch at the door handle and console. Daniel had just demonstrated the phenomenon of torque, the rotational force resulting from the conversion of energy. In an electric car, torque is much more forceful than in a piston-engined vehicle because stored chemical energy is converted to mechanical action nearly instantaneously. In contrast, ICEs have difficulty gulping sufficient air necessary to sustain the combustion reaction that produces force. In an electric car, the effect of "punching it" produces what aficionados refer to as the "EV smile."

The story of how Daniel and his wife came to acquire their Roadster reflected the dynamics of upper-middle-class life in California and the close-knit nature of enthusiast culture in Silicon Valley. The couple moved in the same circles as Martin Eberhard. Daniel recalled how his wife made the purchase decision at a "challenge dinner" hosted by Eberhard in 2009. They paid a premium for a first-year model because it featured a mock stick shift used to switch the various driving modes, in contrast to the push-button system featured in the second-year model. A stick, Daniel felt, was the mark of a real sports car. His wife demurred, but when Eberhard raised his right hand and asked her to "high-five" him, she took the plunge. Tesla's cofounder sealed the transaction by giving Daniel a personalized license plate ("My Tesla"), now a treasured possession. The plate had originally been intended for Caroline, Martin's wife, but when the executive left Tesla, the couple figured they would not buy another electric car for a while. Eberhard's own orange-on-gray Roadster was already a legend in Woodside, a town on the San Francisco Peninsula. The couple discovered the vehicle on a used car lot in Los Angeles with less than 200 miles on the odometer and realized that the car was special. It had been owned by Justine Musk, the Canadian novelist and former wife of Elon Musk, acquired as part of her divorce settlement in 2008, according to Eberhard. Justine had apparently disposed of the car at the first opportunity.[21]

It was no surprise to learn that Daniel's other passion besides Tesla was Disney, an organization that also specialized in marketing enchantment. His experience showed how much more was at stake in the Tesla project than improving air quality. As a first user of means with a sense of adventure, Daniel was a kind of local celebrity by virtue of his ownership of a rare and iconic automobile. Morton's remarks about the nontransportation aspects of automobility had me thinking that I needed some philosophical perspective on the question of the electric car.

FUTURISM

Peter recommended that I speak with Harvey, a friend and fellow aficionado of electric vehicles. Harvey was a doctor of organizational psychology, a profession, he told me, that was necessarily future-oriented. He ran a consultancy specializing in "innovation strategy and best practices, organization design, and leadership development," especially as they pertained to the environmental effects of transportation systems. I met Harvey at his home in a rustic corner of Barron Park. A Swiss immigrant, he spoke about the psychological and sociological aspects of electric cars. For Harvey, electric vehicle technology was "just one of those experiments in life." In the early 1990s, he purchased a Solar Electric Ford Escort conversion, a car that he bequeathed to his teenage sons in the 2000s and that in turn was acquired by Tom, a transaction that underscored the close-knit nature of the Bay Area enthusiast community. Harvey replaced the Electron with a Th!nk City ("a beautiful plastic car") that he rented from Hertz for two years. As empty-nesters, Harvey and his wife shared a Nissan Leaf and a Prius plug-in hybrid.

All told, Harvey experienced three generations of electric vehicle technology over a quarter-century. One of the qualities of his first car that most impressed him was its effect on women. They seemed attracted to "the quiet and the clean, of not having to handle gasoline, the idea of plugging it in at home." The comparison conjured the separate spheres marketing campaigns of long-vanished turn-of-the century marques like Argo, Baker Electric, and Pope Manufacturing.[22]

Harvey also believed that the power of the Tesla business model lay in its ability to eliminate the cognitive dissonance of received notions of cost and capability. People who looked askance at paying a premium for an electric vehicle, he averred, never questioned consumers who paid $50,000 and more for a premium imported ICE vehicle when a $20,000 Chevrolet would have met their needs equally well, in least in terms of automobility narrowly construed as transportation infrastructure. This double standard was undermined, argued Harvey, by the advent of Tesla's Roadster, Model S, and Model X, vehicles that appealed to a demographic that would not normally have cared about electric cars.

My host was also interested in autonomous vehicles, a technology that figured increasingly prominently in sustainable automobile discourse through the 2010s and was sometimes elided with electric automobility itself.[23] Harvey had grown up in a society with excellent public transit and was fascinated by the prospect of mobility services, a form of public transportation that retained the general form of the personal automobile while radically depersonalizing it. In this imaginary, standardized robot pods arrived where and when they were needed, including at one's front door, like a kind of "train that jumped the tracks." One of the salutary effects of such a system, said Harvey, would be to remove "ego identification" with the vehicle. Over time, he believed, automakers would become more like internet-based service providers like Uber and Lyft. They would produce smaller numbers of cars whose styling would be "neutralized" as in commercial aviation, where it has become difficult to distinguish a Boeing from an Airbus.

Harvey was hardly an outlier in these views, which echoed the broader reassessment of automobility underway in the early twenty-first century. Much of Tesla's success traced to the company's ability to produce high-performance cars that gratified egos, seemingly undermining the rationale for robopod services. Yet Tesla also expended a good deal of energy marketing autopilot while offering leases in conjunction with sales. This suggested that Tesla understood that changing economic and envirotechnical states determined automobility as an empowering or alienating experience, warranting technology and marketing for all conditions.

SKEPTICISM

The discussion with Harvey seemed to bring me back to questions of the economics of all-battery electric automobility, so I headed to Oakland, where I met Raoul, a computer programmer employed at Lawrence Berkeley National Laboratory. Raoul seemed like the kind of person that Harold had in mind when he spoke about depersonalized personal transport. For Raoul, the motor vehicle was primarily a means to the end of achieving the good life, which to judge by the well-stocked kitchen of his Oakmore home, revolved around food, friends, and family. Raoul liked to bicycle and was also fond of his motorcycle, although he largely retired it after he and his wife started a family. He regarded cars as a necessary evil, "a box to go from point A to point B." In 1995, Raoul paid cash for a Ford Ranger, and twenty years later, the pickup was still performing stalwart service.[24]

Then in 2016, Raoul received an email from the University of California, the manager of Lawrence Berkeley, advertising a special deal that Nissan was offering to university employees for its Leaf. Intrigued, Raoul and his wife went to the local dealer in downtown Oakland, and the test drive impressed them. The Leaf accelerated smartly and had a number of amenities, including air conditioning, an MP3 player, and Bluetooth. But it was the financing that made the couple pay closer attention. Nissan offered the car on a three-year lease, and there were $12,000 in government incentives, including $2,500 cash back from the state of California. The state also supported a buyback program, so trading in Raoul's Ranger would generate an additional $1,000. Then the couple factored in the money that they would save in gasoline, which worked out to between $50 and $70 a month. When they did the math, the couple were shocked to discover that the car was essentially free over the span of the three-year lease, with an option to purchase it at that time for $10,000. They decided to take the deal, a move that in retrospect Raoul regarded as out of character: the couple had visited the dealership, test-driven the car, and signed the lease all on the same day.

In effect, public policy had enlisted Raoul in the project to reengineer the American system of automobility. The Leaf became the couple's chief commuting vehicle, with duties that were not all that onerous by Bay Area standards. Depending on the weather, Raoul could bicycle to work. That meant that his wife was the primary user of the Leaf, and her commute was only four miles. For longer trips, the family relied on an ICE minivan. When Raoul did drive the Leaf, he was motivated mainly by the novelty of experiencing free technology, as well as competing in good-natured virtue-signaling. He had friends with Priuses who had what he described as the Prius "halo" but with his Leaf, Raoul felt that he had one-upped them in the game of greener-than-thou.

Most of the time, Raoul used the Leaf's eco-mode to preserve range, a set-ting that made for less responsive driving because it increased the electromo-tive resistance of regenerative braking. Sometimes, however, Raoul allowed himself a "guilty pleasure" and used the Sport setting. As with all smaller all-battery electrics, journeys of more than 40 to 50 miles were risky when the round trip was factored in necessitating research on Google Maps to locate nearby charging stations. In routine return commuting back home to hilly Oakmore, Raoul used techniques he learned on his motorcycle to pre-serve as much battery capacity as possible, tailing semitrailers to take advan-tage of the vacuum effect. He also discovered that the Leaf's computer was not always a good judge of range, in part because it had difficulty interpreting the effects of complex geography and itineraries on battery capacity. The distance between Raoul's home and his gym was 20 miles. The trip to the gym was downhill, and if Raoul started with 100 percent battery capacity, interpreted by the computer as 85 to 92 miles of range, he might arrive with an additional 10 miles of range registered on the computer thanks to regenerative braking. On the ride home, the situation was reversed. The computer might indicate he had around 90 miles of range, but he might arrive home with only around 50 miles of registered battery capacity.

Weather was another important determinant of battery capacity. Battery packs of first-generation Leafs were susceptible to heat degradation in places like Arizona and Texas, partly because Nissan omitted a thermal manage-ment system to save money.[25] The automaker tried to solve this problem by altering the Leaf battery chemistry but faced problems at the other end of the temperature spectrum. An Oakland winter might seem balmy to mid-westerners and northeasterners, but Raoul found that temperatures as com-paratively mild as 4 to 5 degrees Celsius could significantly reduce battery capacity. Wind shear had a similar effect, especially in the evenings, when fogs moved east across the San Francisco Bay.

The lesson was that commuting in the Leaf required more attention to terrain, climate, and traffic than an ICE vehicle, hybrid electric, or any kind of Tesla or other large-battery electric car. To Raoul, the main problem was traffic congestion. In the Bay Area, as in other high-density US cities, park-ing a car could be time-consuming, frustrating, and expensive. An attrac-tive option, Raoul felt, was an electric scooter or bicycle, an increasingly popular way for individuals to negotiate the spaces of urban cores.[26] Long-distance family outings, of course, demanded different solutions.

I conducted these interviews in the years before midsize, long-range, all-battery electrics like the Tesla Model 3 and Chevy Bolt reached the market. Such cars incarnated the dream of the affordable zero emission automobile for the middle class. For most Americans in the 2020s, however, the economic logic of owning relatively large, new-build, all-electric cars was clouded by the coronavirus pandemic and global recession, as well as unresolved questions of battery replacement costs. To be sure, leasing allowed increasing numbers of consumers to access such vehicles, and the growing market for used electrics represented an affordable ownership option. But owners of used all-battery electrics equipped with middle-aged batteries will experience the most serious operational and economic consequences of the temporal mismatch of motor and battery. This emerging cohort of electric motorist, as well as the electric automobile aftermarket, will operate on the frontiers of knowledge, a space where analogies and experience derived from used ICE vehicles will have little practical value.

CONCLUSION

I was hoping there would be more rationality in the progress of the electric vehicle industry. It has been disheartening to see that many decisions are quite irrational and it will take decades, through engineering evolution rather than intelligent design, to reach a more optimal solution.

—Alan Cocconi, electric vehicle pioneer, 2020

The electric cars of the early twenty-first century were products of a protracted conflict between the automaking establishment and public policymakers on the question of how to construct the sustainable passenger automobile of the future. From the early 1960s, energy and environmental planners and the major car companies valued different metrics of automobile performance, driving a politics of compromise. Regulators forced automakers to build relatively cleaner and more efficient conventional automobiles, but the improved fleets failed to yield the expected sociotechnical outcomes. Air pollution and energy consumption increased with the growth of the fleet over time, and these network effects compelled a range of actors to consider alternative propulsion technologies, triggering a parallel debate between industry and regulators on the efficacy of relatively simple nearer-term solutions as opposed to relatively complex longer-term ones. Industry resistance to forced technological change biased the technologies of the sustainable automobile toward complexity.

Over a period of more than a half-century, these pressures exposed automakers to diverse influences, including the appropriate technology movement, the aerospace and consumer electronics sectors, and the national developmental state and its regulatory and research and development arms. From the late 1950s, a series of enthusiast-experts including C. Russell Feldmann, Victor Wouk, Robert Aronson, Bob Beaumont, Ronald Gremban, and Wally Rippel began building electric vehicles from existing automotive components, constituting the first wave of a conversion culture that persisted into the twenty-first century. During this period, the national developmental state institutionalized research on advanced electric propulsion systems but offered scant support to small entrepreneurs like Beaumont. From the late 1980s, a more robust set of public policy instruments were catalyzed by California's Low Emission Vehicle (LEV) program, indirectly nurturing another cohort of enthusiast-expert including Alec Brooks, Alan Cocconi, and James and Anita Worden. This new generation was committed to building purpose-built electric vehicles, on the premise that a lightweight chassis and frame equipped with advanced motor controls could make optimal use of the lead-acid rechargeable battery. Ordinary users including Noel Perrin and many others were also influenced by and indirectly benefited from the new public policy environment and contributed to knowledge-making in this period.

The story of the contemporary electric car casts problems of technology transfer in a new light. Automakers calculated their engagements with the diverse groups of outsiders that public policy injected into their affairs in terms of how these interactions could serve to preserve industry control over processes of invention, innovation, and above all, manufacturing, engendering behavior ranging from passive aggression to tacit and formal collaboration. Automakers negotiated reduced Zero Emission Vehicle (ZEV) mandate quotas with regulators and US manufacturers sought to keep electric propulsion technology out of their core operations by outsourcing as much of it as possible. For all automakers, the longer-term strategy was to neutralize the hated all-battery electric format by reinterpreting the meaning of the ZEV via better battery discourse, an approach that invoked the hydrogen fuel cell and electric supercar imaginaries and deepened the engagement of the auto industry with the US national developmental state.

Having played the chief role in stimulating the mandate through its partnership with AeroVironment in creating the Impact, General Motors (GM) tried to turn back the clock. The automaker sought to control and limit use of nickel-metal hydride rechargeable technology in electric cars through a joint venture with Ovonic Battery Company (OBC) that it later divested to the oil industry, which further restricted the ability of automakers to use this power source. Toyota and Honda responded to GM-Ovonic and US fuel efficiency and air quality regulations by commercializing the hybrid electric car, a platform that narrowed, if not eradicated, the temporal mismatch of battery and motor.

The automaking establishment's liquidation of its all-battery electric programs in the early 2000s heralded another phase in the electric vehicle revival. The death of the electric car triggered a popular backlash in the US that empowered certain enthusiast-experts at a time when investors were looking for the next big thing in the wake of the bursting of the dot-com bubble. With the aid of tech capital, enthusiast-experts revived the all-battery electric format around the induction motor and lithium ion rechargeable batteries, a formula pioneered by AC Propulsion (ACP) and the nascent Tesla Motors. Sustained by popular goodwill and venture capital, and later by public stimulus, Tesla sought to market the electric supercars that automakers had invoked as a chimera to indefinitely delay action on mandated technological change. Mainstream automakers felt compelled to follow suit, marketing all-battery electrics equipped with relatively large and powerful battery packs.

Tortuous and protracted though it was, the electric vehicle revival may be interpreted as a triumph of science-based industrial innovation and the principle of public-private partnership. From the 1950s, practitioners boosted the energy density of rechargeable battery chemistries by a factor of ten, and the capacity of electric vehicle battery packs from tens to hundreds of thousands of watt-hours.[1] This progress dramatically increased the range of the all-battery electric car from several dozen to several hundred miles, such that the most capable electrics were comparable to average gasoline-fueled internal combustion engine (ICE) vehicles in this respect. The investment of hundreds of billions of dollars in research, development, and industrial manufacturing yielded electric cars that were stylish, comfortable, and capable, with operating qualities widely admired by users. Such vehicles at least matched and often surpassed the handling characteristics of comparable ICE vehicles, especially in terms of torque.[2]

This emerging new fleet also problematized the longstanding assumption that electric propulsion was technologically simpler, more efficient, and cheaper to operate than ICE propulsion, thanks largely to the influence of the California context on design and manufacturing. Automakers convinced California air quality regulators that it was not enough for automobiles produced for this environment to emit little or no pollution. California cars also had to have a range commensurate with standards of convenience determined by the capacities of gasoline-fueled ICE propulsion in the state's built spaces. It was for this reason that regulators accepted the notional capabilities of fuel cell electric propulsion as the performance standard for the ZEV. To be sure, first users demonstrated that conversions and smaller electrics in the city car and subcompact classes could be usefully applied in California cities, so long as users changed their learned driving behaviors and planned their movements, moderated their speed, and observed changing envirotechnical conditions.

But most manufacturers eschewed the small electric car. Better battery discourse and competitive pressure informed an industry-regulatory bias for large, high-performance electrics that were relatively expensive in resources, complex to manufacture, and, if battery replacement was taken into account, potentially costly to maintain. Paradoxically, the decision of planners, engineers, and financiers to develop electric cars that matched or bettered the road performance and manners of ICE cars reproduced the very features of American automobility that critics long argued were undesirable. From the 1990s, manufacturers of electrics followed the general trend in automobile design toward larger, heavier, and more powerful vehicles.[3]

The electric supercar confounded other expectations for efficiency around infrastructure. For decades, converted all-battery electrics used the standard 120-volt socket, and Victor Wouk and others believed that one of the main advantages of the conventional hybrid electric car was its capacity to use the existing electric and gasoline infrastructures.[4] From the late 2000s, however, automakers devised several charging standards reflecting the national preferences of US, German, and Japanese industry regarding the new generation of plug-in electrics. Moreover, the trend in larger batteries led car companies to supplement low-voltage, slow-charging systems with costlier high-voltage, fast-charging systems.[5] By the early 2010s, California policymakers worried that infrastructure trailed the deployment of plug-in electrics and sought to accommodate the multiple fast-charging standards in public spaces.[6] The state invested heavily in public charge points, predicating planning on the 100-mile-range all-battery electric that then represented the state of the art. However, the increasing popularity of large-battery electrics with a range of over 200 miles threatened to make the new infrastructure redundant. Most users of such vehicles, and, indeed, of plug-in electrics of all types, tended to rely on overnight home charging.[7]

These difficulties spoke to the broader complexities of integrating automobility and electricity as energy conversion infrastructures. In 2001, California reregulated electricity, resolving by conventional political means the problem for which vehicle-to-grid had been conjured as a silver bullet.[8] The proliferation of large-battery plug-in electrics later in the decade triggered research, development, and start-up enterprises devoted to bidirectional electric vehicle power in the US and Europe.[9] These initiatives revealed that suturing distributed generation and storage technologies into legacy grid systems faced a host of challenges.[10] In principle, many of these problems could be solved with new technology, but devising appropriate market models remained a challenge. With the resolution of California's electricity crisis, recalled Alec Brooks, the inflated prices that had made ancillary services an attractive market disappeared.[11] Some emerging suppliers of the equipment and services of bidirectional electric vehicle power experimented with and ruled out the frequency regulation market in the mid-2010s on the grounds that it was too small and, crucially, unscalable.[12]

Interest in bidirectional electric vehicle power had the potential to increase with the growth of the electric fleet, but in the early 2020s the prospect of average users becoming power entrepreneurs appeared more distant. The notional everyday participant in vehicle-to-grid faced the complex task of arbitraging the difference between wholesale and retail prices while still paying the retail rate.[13] Advocates argued that advances in software and charging technology, in concert with market models bolstered by public subsidies, could reconcile this conundrum and enable owners of such vehicles to pay for their batteries.[14] But market models tended not to engage the technoscience of battery entropy. Owners of electric vehicles would be no further ahead if they substantially degraded their batteries in the process of paying them off. Understanding how batteries age in electric vehicles would seem to be a prerequisite for meaningfully engaging the sociotechnical problems of repurposing electric vehicle rechargeables for stationary applications.

Historically, however, the political economy of advanced power source research and development disincentivized innovation in the field of battery durability. One of the signal ironies of the electric vehicle revolution is that few of the basic advanced rechargeable battery chemistries originated in research expressly devoted to electric vehicles, let alone stationary generation. Research in nickel-metal hydride and lithium ion rechargeables traced to the energy crisis of the 1970s, but these chemistries were first applied in consumer electronics, and in this context, technologists privileged energy and power, not durability and cost. Other technologists adapted these chemistries for use in electric vehicles, and policymakers subsidized their manufacture to help scale production to cost parity with gasoline-fueled ICE propulsion. In 2006, M. Stanley Whittingham, one of the pioneers of the lithium rechargeable, suggested that the focus in electric vehicle battery research was likely to shift to durability.[15]

Nevertheless, the dominant battery formulas in the electric vehicles of the 2010s were informed by the bias for energy and power.[16] Moreover, industry's treatment of the price of battery power as a trade secret mystified cost and its relationship with durability and safety. The experiences of users on the front lines of the cost-durability dilemma constituted an important resource in the emerging study of battery longevity.[17]

The ZEV mandate and the electric vehicle revival had important unintended consequences for industrial manufacturing in the context of deepening globalization. Air quality regulators and national developmental planners had substantially different goals for the electric car, with the former interested mainly in the technology's potential to realize favorable environmental outcomes and the latter interested in stimulating US industry in achieving these outcomes. These imperatives did not inherently conflict but the US auto industry's rejection of both had the effect of accentuating the asymmetries of the global consumer electronics industry, the locus of advanced battery manufacturing, which was situated mainly in East Asia. If the prospect of the commercial all-battery electric car forced automakers to reconsider the parts and service aspects of their business model, the possibility of an aftermarket in replacement batteries presented one avenue for profit, at least in theory.

To US automakers in the 1990s, 2000s, and 2010s, all options in this regard were unpalatable, so they chose to resist public policy. Industry's experience in the 2010s seemed to vindicate their position. For most of this period, automakers struggled to make money from all-battery electrics, and of the major car companies, only Nissan attempted to produce its own battery cells, an enterprise that it eventually sold to Chinese interests. As we have seen, Toyota and Matsushita/Panasonic attempted to reconcile the interests of automaking and batterymaking in their partnership around the Prius platform. From the late 2010s and early 2020s, other automakers (including Volkswagen) sought to vertically integrate cell, module, and battery pack manufacturing, with BYD being the first to achieve this goal.[18]

Over time, environmental imperatives eclipsed the national industrial imperatives of US public policy. American car companies had little choice but to deal with makers of battery cells in the global consumer electronics complex, deepening industrial dependency. Stimulus helped LG become a major supplier of cells for electric vehicle batteries and stoked the Seoul-based corporation's ambitions to supply finished battery packs.[19] Public policy helped shape an even more intimate and important relationship between Panasonic and Tesla. Tesla's function in socializing the air quality and climate obligations of the established automakers depended on public subsidies and Panasonic core industrial content. Thanks to its monopoly of cell supply for the Prius, Panasonic became the world's leading manufacturer of cells for electric cars, and its alliance with Tesla enabled the company to become an important producer of lithium ion cells for plug-in electrics as well. Japanese industrial technology was an indispensable ingredient in Tesla's rise during the reformation of US automaking that followed the Great Recession of 2007–2009.[20]

The electric vehicle revival also highlighted the linkages and tensions between the objectives of clean air and plentiful clean energy, between state and federal interests, between air quality policy and energy policy, and between the Zero Emission Vehicle and the efficient ICE vehicle as the primary technological instruments of these policy objectives. As the green car wars unfolded through the 1990s, 2000s, and 2010s, regulators and automobile engineers quietly made major strides in cleaning up the gasoline-fueled ICE fleet and reducing smog.[21] Where a 1970-era ICE vehicle produced nearly 2,000 pounds of smog-forming pollutants over its service life, a 2010-era ICE vehicle produced around 10 pounds. So successful were these efforts that it became possible to compare the emissions of gasoline-fueled ICE cars with the upstream effluent produced by fossil-fueled energy conversion systems supplying electricity to electric cars.[22]

Increasingly, the policy rationale for deploying electric cars shifted to the mitigation of climate-changing carbon dioxide, a substance that the US Environmental Protection Agency (EPA) had long refused to treat as an air pollutant on the grounds that it did not directly threaten human health, unlike the substances that caused smog. That changed in the 2000s when lawmakers at the state and federal levels took a series of measures to control automobile emissions of greenhouse gases, efforts expressed not in terms of air quality and the technologies of the ZEV but in terms of energy efficiency and the technologies of the efficient gasoline-fueled ICE vehicle. Toyota made a crucial contribution to this policy template. Through its use of the Third Conference of the Parties to the United Nations Framework Convention on Climate Change (UNFCCC) to market the Prius, the automaker did more than any single entity to promote the association between high fuel efficiency, greenhouse gas reduction, and climate change mitigation. From 2002, the California assembly member Frances Pavley introduced a series of measures to cut per-mile greenhouse gas emissions which, in effect, boosted fuel efficiency standards. To be sure, the relationship between an ICE automobile's fuel efficiency and the amount of carbon dioxide it emitted was not clear.[23] Automobile energy conversion efficiency and emissions production vary according to driving conditions in envirotechnical context.

Nevertheless, the California Air Resources Board (CARB) adopted the Pavley regulations in 2005. The regulations implied fuel efficiency standards higher than those of the federal government and created a condition of dual power in energy policy that precipitated a protracted jurisdictional battle. In 2007, the US Supreme Court ruled that greenhouse gases were pollutants that could be regulated under the Clean Air Act (CAA). In 2008, however, the George W. Bush administration's EPA denied California's request for a waiver that would have given the state the right to control mobile sources of such emissions. In 2009, the Obama administration granted a waiver for the Pavley regulations and brokered a compromise by harmonizing federal and California fuel efficiency standards, reinforcing these actions by supporting the development of plug-in hybrids through the American Recovery and Reinvestment Act. In 2013, the Obama administration's EPA granted another waiver covering California's greenhouse gas and ZEV standards. In 2019, the Trump administration revoked this waiver, nullifying California's ZEV standards (but not its LEV program), promoting instead a uniform fuel economy standard that was lower than the Obama-era harmonized standard.[24] In the ensuing legal confrontation, automakers sided with the Trump administration but abandoned their support with the advent of the Biden administration, which signaled that it would restore the Obama-era agreement and California's special powers to regulate emissions.[25]

California's growing economic and political potential gave the state a significant say in the formulation of federal environmental, energy, and industrial policies, as well as in automaking on a global scale. Influenced by envirotechnical factors and years of resistance from the car companies and their political allies, California regulators ultimately compelled the auto industry to produce a mixed fleet of clean ICE vehicles and alternative propulsion vehicles, mostly hybrid electrics of various types. All these vehicles had environmental implications that went far beyond local and global air quality. Where plug-in electrics of all types were concerned, it was long axiomatic in environmental discourse that such vehicles were only as green as the primary energy conversion systems that supplied their electricity, and preliminary lifecycle analyses emerging in the mid-2010s painted an even more complicated picture.[26] Environmental assessments both for and against plug-in electrics tended to assume that such vehicles would consume only one battery during their ten- to fifteen-year service lives, a pivotal conjecture because even analyses in favor of electric cars acknowledged that manufacturing electric cars produced more pollution than ICE cars owing to the higher energy costs of fabricating materials and components.[27]

The question of battery replacement in the fleet of large-battery plug-in electrics as it aged, especially in the used car market, was thus central in lifecycle and environmental assessment. It was also vital in establishing the technology and economics of electric vehicle battery recycling, an emerging enterprise with two potential models, both of which required the solving of a host of major sociotechnical problems. One model was to recondition used electric vehicle batteries for the automobile aftermarket or repurpose them for stationary applications. Another was to recover materials from cell scrap and spent battery packs.[28] By the early 2020s, most efforts in electric vehicle battery recycling focused on materials recovery, an enterprise that had to compete with dedicated suppliers of commodity materials.[29] Industrial materials recycling required subsidies and moreover caused fresh environmental complications. The process of recovering materials from consumer electronics was then based on hydrometallurgy, a technology of powerful and highly toxic solvents.[30] Unless green hydrometallurgies could be developed, scaled materials production from waste electric vehicle batteries had the potential to seriously damage soil and groundwater quality.

The resource requirements and environmental constraints of electric cars of all types were no less serious than their ICE counterparts. Alan Cocconi observed that the pressure of competition led manufacturers of electrics to supplant the induction motor with the permanent magnet motor, a design more amenable to miniaturization and power control.[31] Induction motors could be built from cheap common metals like iron, copper, and aluminum, and while permanent magnet motors could also be constructed from cheap materials, automakers preferred magnets fabricated of costly rare earth elements like neodymium and dysprosium because these substances enabled the smallest and most powerful magnets. Some analysts forecast that scaled production of electrics would result in shortages of these and other strategic elements, including lithium and cobalt, and trigger intensive exploitation of these substances, causing further environmental and social harm, especially to indigenous peoples.[32] Some observers believed that hybrid electrics were especially resource-hungry and identified the Prius program as the single largest consumer of rare earths, as well as a major consumer of palladium for use in the catalytic converter.[33] China's possession of the world's largest proven reserves of rare earth elements constituted an additional geopolitical concern for some Western analysts, who believed that Chinese planners sought to manipulate this market in favor of China's domestic and foreign policy priorities.

The paradoxes of the electric vehicle revival traced ultimately to a public policy worldview that perceived consumer convenience, corporate profit, and environmental and economic sustainability as complementary, not mutually exclusive. All-battery electric cars generally lost money in the 2010s because of high production costs and uneven demand, and yet public policy support for the technology boosted the stock of enterprises like Tesla as part of the speculative bubble inflated by the federal government's policy of quantitative easing in the aftermath of the Great Recession. The regulatory, monetary, and fiscal framework and much of the basic technology of the contemporary electric car all could be traced to arms of the US national developmental state, and yet much (if not most) of the core industrial content was created by enterprises headquartered in Asian nation-states. Much of the added value of new technology and employment was offshored, along with the most damaging environmental effects of the industries of the electric automobile. Another way of expressing this relationship is that improving the air quality of US and especially Californian society came at the cost of exacerbating the air, water, and soil pollution of industrializing Asian societies.

Reasonably enough, US policies to ameliorate automobile pollution first focused on emissions that degraded local air quality. Policies designed to cut the automobile pollution responsible for climate change necessarily lagged, mainly because the improved and constantly growing ICE fleet itself became the largest single source of greenhouse gases even as it produced less smog, but partly because some of that effluent was bound up in the emissions of the global industrial infrastructure that manufactured automobiles. On its own, the project to electrify significant portions of the light-duty fleet could not avert climate change because that project depended heavily on fossil fuels. Coal, petroleum, and natural gas became the building blocks of contemporary industrial civilization because they are the most concentrated forms of stored primary energy and occur in the form of mineral resources, properties that make these nonrenewable forms of energy industrially versatile in ways that renewable forms of energy are not. Besides driving almost all automobile propulsion and most stationary electricity generation, fossil fuels are a primary industrial feedstock in the production of chemicals and metals, especially steel.[34]

Policymakers willing to consider the larger conundrum of how to absorb the nonrenewable energy costs of renewable energy have long looked to hydrogen as a universal energy carrier, fuel, and feedstock. Most of the technologies of a zero-carbon hydrogen economy are technologically feasible. It is another question entirely whether the social reorganization implied by any of the net-zero energy imaginaries, including the hydrogen economy, can be accommodated by current sociopolitical and cultural norms and value systems. Global financial and political elites followed the global science community in recognizing the risks of climate change, but they were unable to agree on appropriate remedies at the multilateral level, largely because of the conflict of interest between representatives of transnational capital and representatives of the nation-states transformed by capital into regions increasingly optimized for resource extraction, manufacturing, and/ or services.

Such considerations were largely excluded from the program to reform the automobile system. The electric car imaginary derives partly from the parallax view that scaled personalized automobility imposes on its users, a system that limits perspective even as it shrinks local time and space. The system socializes individual users to see the world from behind the steering wheel in idealized conditions and to take for granted the dependencies that the real system inculcates. Electric cars of all types were cleaner than ICE cars in terms of effluent at the point of use, but they were as subject to the network effects of the system of automobility as any automobile. Congestion, accidents, air quality alerts, gasoline shortages, electricity outages, and myriad other circumstances continually reminded users of all kinds of vehicles of the fragile interconnectivity of energy and transportation infrastructure. In California, the perception that these limits had serious consequences for public health first developed in relation to smog, a dramatic envirotechnical phenomenon that provoked a protracted reappraisal of the technology of the personal passenger automobile, but not the system of automobility as a dynamic organic machine.

The limited reappraisal of automobility will surely continue with the scaling of the electric car and the deepening of the environmental crisis. As the age of auto electric unfolds amid other dramatic registers of social and environmental distress besides local air quality, Americans may come to believe that making their own immediate environments cleaner and more livable is somehow connected to ensuring that other environments, far out of view in other parts of the country (not to mention the world) are equally clean and pleasant. One day, they may reconsider personal ownership of new passenger automobiles as an economic investment, and perhaps even the personal car as the primary means of traveling through local space. In the nearer future, for policymakers and entrepreneurs and for increasing numbers of ordinary motorists who depend on automobility but also recognize its social and environmental harms, the electric car will remain the cathartic car, expiating the guilt of gasoline and restoring faith in technological ingenuity as the key to the sustainable good life.

NOTES

CHAPTER 1

1. This statistic includes both all-battery and plug-in hybrid electric cars; International Energy Agency (IEA), *Global EV Outlook 2021: Accelerating Ambitions Despite The Pandemic*, 17, accessed February 21, 2022, https://www.iea.org/reports/global-ev-outlook-2021/trends-and-developments-in-electric-vehicle-markets.

2. See Seth Fletcher, *Bottled Lightning: Superbatteries, Electric Cars, and the New Lithium Economy* (New York: Hill and Wang, 2011); International Battery Materials Association (IBA), "Special Symposium to Honor Michael Thackeray," last modified March 10, 2013, http://congresses.icmab.es/iba2013/images/stories/PDF/mt.pdf; University of Texas, "UT Austin's John B. Goodenough Wins Engineering's Highest Honor for Pioneering Lithium-Ion Battery," last modified January 6, 2014, https://news.utexas.edu/2014/01/06/ut-austins-john-b-goodenough-wins-engineerings-highest-honor-for-pioneering-lithium-ion-battery/.

3. Royal Swedish Academy of Sciences (RSAC), "The Nobel Prize in Chemistry 2019: The Royal Swedish Academy of Sciences Has Decided to Award the Nobel Prize in Chemistry 2019 to John B. Goodenough, M. Stanley Whittingham, and Akira Yoshino for the Development of Lithium-Ion Batteries: They Created a Rechargeable World," last modified October 9, 2019, https://www.nobelprize.org/uploads/2019/10/press-chemistry-2019-2.pdf.

4. See the documentary filmmaker Chris Paine's *Who Killed the Electric Car?* (Sony Pictures Classics, 2006) and *Revenge of the Electric Car* (WestMidWest Productions/Area 23a Films, 2011).

5. Most such work has been done by journalists; see, for example, John J. Fialka, *Car Wars: The Rise, the Fall, and the Resurgence of the Electric Car* (New York: Thomas Dunne Books, 2015); Ashlee Vance, *Elon Musk: How the Billionaire CEO of SpaceX and Tesla Is*

Shaping Our Future (HarperCollins, 2015); Fletcher, *Bottled Lightning*; Sherry Boschert, *Plug-in Hybrids: The Cars That Will Recharge America* (Gabriola Island, Canada: New Society Publishers, 2006); and Jack Doyle, *Taken for a Ride: Detroit's Big Three and the Politics of Pollution* (New York and London: Four Walls Eight Windows, 2000).

6. Gijs P. A. Mom, *The Electric Vehicle: Technology and Expectations in the Automobile Age* (Baltimore: Johns Hopkins University Press, 2004); David A. Kirsch, *The Electric Vehicle and the Burden of History* (New Brunswick, NJ: Rutgers University Press, 2000).

7. David A. Kirsch and Gijs P. A. Mom, "Visions of Transportation: The EVC and the Transition from Service- to Product-Based Mobility," *Business History Review* 76, no. 1 (2002): 75–110; for an analysis of the early use of electric vehicles as means of storing off-peak electricity, see Mom, *The Electric Vehicle*, 206–210, 253–254.

8. Kirsch and Mom, "Visions of Transportation," 109.

9. Some scholars nominally committed to constructivism also have sought to understand how, if at all, the qualities or affordances of technologies make possible or necessary social outcomes; see, for example, Edmund Russell, James Allison, Thomas Finger, et al., "The Nature of Power: Synthesizing the History of Technology and Environmental History," *Technology and Culture* 52, no. 2 (April 2011): 246–259; Christophe Lécuyer and David C. Brock, "The Materiality of Microelectronics," *History and Technology* 22, no. 3 (September 2006): 301–325; Frank N. Laird, "Constructing the Future: Advocating Energy Technologies in the Cold War," *Technology and Culture* 44, no. 1 (2003): 27–49; Langdon Winner, "Do Artifacts Have Politics?" *Daedalus* 109, no. 1 (Winter 1980): 121–136.

10. Cyrus C. M. Mody, *The Long Arm of Moore's Law: Microelectronics and American Science* (Cambridge, MA: MIT Press, 2016), 9–10.

11. Richard H. Schallenberg, *Bottled Energy: Electrical Engineering and the Evolution of Chemical Energy Storage* (Philadelphia: American Philosophical Society, 1982), 391–392.

12. Sheila Jasanoff, "Future Imperfect: Science, Technology, and Imaginations of Modernity," in *Dreamscapes of Modernity: Sociotechnical Imaginaries and the Fabrication of Power*, eds. Sheila Jasanoff and Sang-Hyun Kim (Chicago: University of Chicago Press, 2015), 19.

13. On the origins of technofuturism as a worldview, see David F. Noble, *The Religion of Technology: The Divinity of Man and the Spirit of Invention* (Penguin, 1999), 6, 71.

14. David E. Nye, "Technological Prediction: A Promethean Problem," in *Technological Visions: The Hopes and Fears That Shape New Technologies*, eds. Marita Sturken, Douglas Thomas, and Sandra J. Ball-Rokeach (Philadelphia: Temple University Press, 2004), 159–161. The literature on the history of technological futurism and utopianism has been complemented by an emerging sociology of expectation devoted to exploring speculative enterprises of emerging technology; see Neil Pollock and Robin Williams, "The Business of Expectations: How Promissory Organizations Shape Technology and Innovation," *Social Studies of Science* 40, no. 4 (2010):

525–548; Michael Fortun, *Promising Genomics: Iceland and DeCODE Genetics in a World of Speculation* (Berkeley: University of California Press, 2008); and "Mediated Speculations in the Genomics Future Markets," *New Genetics and Society* 20, no. 2 (2001): 139–156; Nik Brown and Mike Michael, "A Sociology of Expectations: Retrospecting Prospects and Prospecting Retrospects," *Technology Analysis and Strategic Management* 15, no. 1 (2003): 3–18; Nik Brown, "Hope Against Hype: Accountability in Biopasts, Presents and Futures," *Science Studies* 16, no. 2 (2003): 3–21.

15. See Howard P. Segal, *Technological Utopianism in American Culture* (Chicago: University of Chicago Press, 1985); and Michael D. Gordin, Helen Tilley, and Gyan Prakash, "Utopia and Dystopia Beyond Space and Time," in *Utopia/Dystopia: Conditions of Historical Possibility*, eds. Michael D. Gordin, Helen Tilley, and Gyan Prakash (Princeton, NJ: Princeton University Press, 2010), 1–17.

16. Jon Gertner, *The Idea Factory: Bell Labs and the Great Age of American Innovation* (New York: Penguin Press, 2012); Bernadette Bensaude-Vincent, "The Construction of a Discipline: Materials Science in the United States," *Historical Studies in the Physical and Biological Sciences* 31, no. 2 (2001): 223–248.

17. Robert F. Heizer, "The Background of Thomsen's Three-Age System," *Technology and Culture* 3, no. 3 (1962): 259–266.

18. William O. Baker, "The National Role of Materials Research and Development" in *Properties of Crystalline Solids: ASTM Special Technical Publication No. 283* (Baltimore, MD: American Society for Testing Materials, 1961), 1–7. Progression in materials thinking in science policymaking can be traced through a series of major policy reports; see National Academy of Sciences, *Materials and Man's Needs, Vol. 1: The History, Scope, and Nature of Materials Science and Engineering* (Washington, DC: National Academy of Sciences, 1975); National Research Council, *Materials Science and Engineering for the 1990s: Maintaining Competitiveness in the Age of Materials* (Washington, DC: National Academies Press, 1989); National Research Council, *Condensed-Matter and Materials Physics: The Science of the World Around Us* (Washington, DC: National Academies Press, 2007); National Science and Technology Council, *Materials Genome Initiative for Global Competitiveness* (Washington, DC: Executive Office of the President of the United States, 2011).

19. On the origins of linear ideology, see Benoît Godin, "The Linear Model of Innovation: The Historical Construction of an Analytical Framework," *Science, Technology, & Human Values* 31, no. 6 (2006): 639–667; Karl Grandin, Nina Wormbs, and Sven Widmalm, eds., *The Science-Industry Nexus: History, Policy, Implications* (Sagamore Beach, MA: Science History Publications, 2005); Michael Aaron Dennis, "'Our First Line of Defense:' Two University Laboratories in the Postwar American State," *Isis* 85, no. 3 (1994): 427–455.

20. On problems of intrafirm technology transfer, see Benjamin Gross, *The TVs of Tomorrow: How RCA's Flat-Screen Dreams Led to the First LCDs* (Chicago: University of Chicago Press, 2018); Gertner, *The Idea Factory*; Hyungsub Choi, "The Boundaries

of Industrial Research: Making Transistors at RCA, 1948–1960," *Technology and Culture* 48, no. 4 (2007): 758–782.

21. On the political crisis of state-sponsored, undirected basic science, see Chalmers W. Sherwin and Raymond S. Isenson, "Project Hindsight," *Science* 156, no. 3782 (1967): 1571–1577; Daniel S. Greenberg, *The Politics of Pure Science* (Chicago: University of Chicago Press, 1967); and Daniel S. Greenberg, *Science, Money, and Politics: Political Triumph and Ethical Erosion* (Chicago: University of Chicago Press, 2001).

22. Joseph D. Martin, *Solid State Insurrection: How the Science of Substance Made American Physics Matter* (Pittsburgh: University of Pittsburgh, 2018), 1–10. On the development of federally funded academic materials research, see Cyrus Mody and Hyungsub Choi, "From Materials Science to Nanotechnology: Interdisciplinary Center Programs at Cornell University, 1960–2000," *Historical Studies in the Natural Sciences* 43, no. 2 (2013): 121–161; Hyungsub Choi and Brit Shields, "A Place for Materials Science: Laboratory Buildings and Interdisciplinary Research at the University of Pennsylvania," *Minerva* 53, no. 1 (2015): 21–42; Stuart W. Leslie, *The Cold War and American Science: The Military-Industrial-Academic Complex at MIT and Stanford* (New York: Columbia University Press, 1993).

23. On the military origins of the integrated circuit and digital computers, see Kent C. Redmond and Thomas M. Smith, *From Whirlwind to MITRE: The R&D Story of the SAGE Air Defense Computer* (Cambridge, MA: MIT Press, 2000); Paul E. Ceruzzi, *A History of Modern Computing* (Cambridge, MA: MIT Press, 2003); Leslie Berlin, *The Man Behind the Microchip: Robert Noyce and the Invention of Silicon Valley* (New York: Oxford University Press, 2005).

24. On the concept of material-as-device, see Lécuyer and Brock, "The Materiality of Microelectronics," 304.

25. The battery researcher Johan Coetzer noted the tendency of power-source communities to privilege energy and power over durability, cost, and safety; see Johan Coetzer, "A New High Energy Density Battery System," *Journal of Power Sources* 18, no. 4 (1986): 377–380, on 377. See also Matthew N. Eisler, *Overpotential: Fuel Cells, Futurism, and the Making of a Power Panacea* (New Brunswick, NJ: Rutgers University Press, 2012).

26. This expression refers to attempts to make observation accord with theory. It is often attributed to Plato as an injunction to reconcile the theory of uniform circular motion in the heavens with observed nonuniform circular motion, a task that the physicist and historian and philosopher of science Pierre Duhem claimed was the prime motivation of ancient and medieval astronomy; see Bernard R. Goldstein, "Saving the Phenomena: The Background to Ptolemy's Planetary Theory," *Journal for the History of Astronomy* 28 (1997): 1–12.

27. Fred Block, "Swimming Against the Current: The Rise of a Hidden Developmental State in the United States," *Politics and Society* 36, no. 2 (2008): 169–206;

Meredith Woo-Cumings, ed., *The Developmental State* (Ithaca, NY: Cornell University Press, 1999); Ziya Öniş, "The Logic of the Developmental State," *Comparative Politics* 24, no. 1 (1991): 109–126.

28. Brian Balogh, *Chain Reaction: Expert Debate and Public Participation in American Commercial Nuclear Power, 1945–1975* (Cambridge: Cambridge University Press, 1991), 12–13.

29. I first used the expression "quasi-planning" in Matthew N. Eisler, "Energy Innovation at Nanoscale: Case Study of an Emergent Industry," *Science Progress*, May 23, 2011, http://scienceprogress.org/2011/05/innovation-case-study-nanotechnology-and-clean -energy/.

30. An important consequence of the emergence of the national developmental state was that it accelerated the materials turn in federal science. By the late 1980s, federal physics laboratories were increasingly hosting research devoted to characterizing nonliving and biological materials; see Matthew N. Eisler, "'The Ennobling Unity of Science and Technology:' Materials Sciences and Engineering, the Department of Energy, and the Nanotechnology Enigma," *Minerva* 51, no. 2 (2013): 225–251; Catherine Westfall, "Retooling for the Future: Launching the Advanced Light Source at Lawrence's Laboratory, 1980-1986," *Historical Studies in the Natural Sciences* 38, no. 4 (2008): 569–609; Park Doing, *Velvet Revolution at the Synchrotron: Biology, Physics, and Change in Science* (Cambridge, MA: MIT Press, 2009); Peter J. Westwick, *The National Labs: Science in an American System, 1947–1974* (Cambridge, MA: Harvard University Press, 2003).

31. There is a large body of literature on these reforms and their unintended consequences; for example, see Philip Mirowski, "The Future(s) of Open Science," *Social Studies of Science* 48, no. 2 (2018): 171–203; Rebecca Lave, Philip Mirowski, and Samuel Randalls, "Introduction: STS and Neoliberal Science," *Social Studies of Science* 40, no. 5 (2010): 659–675; Philip Mirowski and Esther-Mirjam Sent, "The Commercialization of Science and the Response of STS," in *Handbook of Science and Technology Studies*, 3rd ed., eds. Edward J. Hackett, Olga Amsterdamska, Michael Lynch, and Judy Wajcman (Cambridge, MA: MIT Press, 2008), 635–689; Ann Johnson, "The End of Pure Science: Science Policy from Bayh-Dole to the NNI," in *Discovering the Nanoscale*, eds. Davis Baird and Alfred Nordmann (Amsterdam: IOS Press, 2004), 217–230; David C. Mowery, "Collaborative R&D: How Effective Is It?" *Issues in Science and Technology* 15, no. 1 (1998): 37–44; David C. Mowery, Richard R. Nelson, Bhaven N. Sampat, and Arvids A. Ziedonis, "The Growth of Patenting and Licensing by US Universities: An Assessment of the Effects of the Bayh-Dole Act of 1980," *Research Policy* 30, no. 1 (2001): 99–119; Daniel Sperling, *Future Drive: Electric Vehicles and Sustainable Transportation* (Washington, DC: Island Press, 1995).

32. Dirk Breitschwerdt, Andreas Cornet, Sebastian Kempf, Lukas Michor, and Martin Schmidt, *The Changing Aftermarket Game and How Automotive Suppliers Can Benefit from Arising Opportunities* (McKinsey and Company, 2017), 9; Morris A. Cohen,

Narendra Agrawal, and Vipul Agrawal, "Winning in the Aftermarket," *Harvard Business Review*, May 2006, https://hbr.org/2006/05/winning-in-the-aftermarket.

33. One industry-linked study held that the process of bringing a new power source to market could take almost twenty years; see Ralph J. Brodd, *Factors Affecting US Production Decisions: Why Are There No Volume Lithium-Ion Battery Manufacturers in the United States?* (Gaithersburg, MD: National Institute of Standards and Technology, 2005), 18.

34. For important analyses of high-technology industry, see Christophe Lécuyer and David C. Brock, "High Tech Manufacturing," *History and Technology*, 25, no. 3 (2009): 165–171; Christophe Lécuyer and David C. Brock, "From Nuclear Physics to Semiconductor Manufacturing: The Making of Ion Implantation," *History and Technology* 25, no. 3 (September 2009): 193–217; Lécuyer and Brock, "The Materiality of Microelectronics."

35. Richard Chase Dunn and Ann Johnson, "Chasing Molecules: Chemistry and Technology for Automotive Emissions Control," in *Toxic Airs: Body, Place, Planet in Historical Perspective*, eds. James Rodger Fleming and Ann Johnson (Pittsburgh: University of Pittsburgh Press, 2014), 109–126.

36. On the enduring influence of progressive-era conservation ideology in US culture and politics, see William Cronon, "Foreword: Revisiting Origins: Questions That Won't Go Away," in *Conservation in the Progressive Era: Classic Texts*, ed. David Stradling (Seattle: University of Washington Press, 2004), vii–ix.

37. For exemplary envirotechnical studies, see Etienne S. Benson, "Infrastructural Invisibility: Insulation, Interconnection, and Avian Excrement in the Southern California Power Grid," *Environmental Humanities* 6, no. 1 (2015): 103–130; Dolly Jørgensen, "Mixing Oil and Water: Naturalizing Offshore Oil Platforms in Gulf Coast Aquariums," *Journal of American Studies* 46, no. 2 (2012): 461–480; Ashley Carse, "Nature as Infrastructure: Making and Managing the Panama Canal Watershed," *Social Studies of Science* 42, no. 4 (2012): 539–563; and *Beyond the Big Ditch: Politics, Ecology, and Infrastructure at the Panama Canal* (Cambridge, MA: MIT Press, 2014); Sara B. Pritchard, "An Envirotechnical Disaster: Nature, Technology, and Politics at Fukushima," *Environmental History* 17, no. 2 (April 2012): 219–243; Sara B. Pritchard, *Confluence: The Nature of Technology and the Remaking of the Rhône* (Cambridge, MA: Harvard University Press, 2011); David Blackbourn, *The Conquest of Nature: Water, Landscape, and the Making of Modern Germany* (New York: Norton, 2006); Mark Cioc, *The Rhine: An Eco-Biography, 1815–2000* (Seattle: University of Washington Press, 2002).

38. Richard White, *The Organic Machine: The Remaking of the Columbia River* (New York: Hill and Wang, 1995).

39. The term "automobility" seems to have been coined, or at least first explicitly theorized, by the sociologist John Urry in "The 'System' of Automobility," *Theory, Culture and Society* 21, nos. 4–5 (2004): 25–39. For exemplary analyses of the

automobile as a social system, see Deborah Clarke, *Driving Women: Fiction and Automobile Culture in Twentieth-Century America* (Baltimore: Johns Hopkins University Press, 2007); Sally H. Clarke, *Trust and Power: Consumers, the Modern Corporation, and the Making of the United States Automobile Market* (Cambridge: Cambridge University Press, 2007); Tom McCarthy, *Auto Mania: Cars, Consumers, and the Environment* (New Haven, CT: Yale University Press, 2007); Matthew Paterson, *Automobile Politics: Ecology and Cultural Political Economy* (Cambridge: Cambridge University Press 2007); Mimi Sheller and John Urry, "The City and the Car," *International Journal of Urban and Regional Research* 24, no. 4 (2000): 737–757; David Gartman, *Auto Opium: A Social History of American Automobile Design* (London: Routledge, 1994); and Virginia Scharff, *Taking the Wheel: Women and the Coming of the Motor Age* (Albuquerque: University of New Mexico Press, 1991).

40. Virtually all energy conversion/carrier systems, including electricity, are integrated into industrial structures and include industrial/metallurgical coal, industrial gas, utility natural gas, petrochemicals, transportation fuel, and domestic heating oil. In his pioneering studies of the development of early electricity systems, Thomas P. Hughes acknowledged the geophysical qualities of the primary energy resources of coal, oil, and hydro as sociotechnical factors in electricity generation, but he did not root the social relations of primary energy in the context of energy conversion/carrier regimes; see Thomas P. Hughes, *Networks of Power: Electrification in Western Society, 1880–1930* (Baltimore: Johns Hopkins University Press, 1983), 262, 367, 406, 418.

41. One estimate held that by the 1990s, the US light duty fleet had around sixteen times more capacity than US stationary plants (12,000 gigawatts versus 750 gigawatts); see Willett Kempton and Steven E. Letendre, "Electric Vehicles as a New Power Source for Electric Utilities," *Transportation Research Part D: Transport and Environment* 2, no. 3 (1997): 157–175, on 159.

42. The environmental historian Martin V. Melosi held that 86 million tons of the 146 million tons of pollutants emitted in the US in 1966 originated in the motor vehicle fleet; see Melosi, "The Automobile and the Environment in American History," accessed February 2022, http://www.autolife.umd.umich.edu/Environment /E_Overview/E_Overview3.htm#:~:text=The%20Automobile%20and%20the%20 Environment%20in%20American%20History.,production%20of%20motor%20 vehicles%20with%20internal%20combustion%20engines.

43. For a representative view, see David R. Keith, Samantha Houston, and Sergey Naumov, "Vehicle Fleet Turnover and the Future of Fuel Economy," *Environmental Research Letters* 14 (2019): 021001.

44. US Department of Transportation (DOT), "Summary of Fuel Economy Performance," December 15, 2014, https://www.nhtsa.gov/sites/nhtsa.gov/files/performance -summary-report-12152014-v2.pdf.

45. Sudhir Chella Rajan, *The Enigma of Automobility: Democratic Politics and Pollution Control* (Pittsburgh: University of Pittsburgh Press, 1996), 25–28.

46. For a discussion on the efficiency trade-offs of exhaust gas recirculation, see Haiqiao Wei, Tianyu Zhu, Gequn Shu, Linlin Tan, and Yuesen Wang, "Gasoline Engine Exhaust Gas Recirculation: A Review," *Applied Energy* 99 (2012): 534–544; see also Dunn and Johnson, "Chasing Molecules," 116–118.

47. US Environmental Protection Agency (EPA), Air Pollutant Emissions Trends Data, "Criteria Pollutants National Tier 1 for 1970–2020," https://www.epa.gov /air-emissions-inventories/air-pollutant-emissions-trends-data. The energy analyst Vaclav Smil observed that the idea that secular improvements in energy conversion efficiency equated with an aggregate decline in energy consumption was first debunked by the English economist William Stanley Jevons in 1865; see Vaclav Smil, *Energy in Nature and Society: General Energetics of Complex Systems* (Cambridge, MA: MIT Press, 2008), 271–272.

48. Daniel Sperling and Deborah Gordon, *Two Billion Cars: Driving Towards Sustainability* (New York: Oxford University Press, 2009), 20–21.

49. Poliana Rodrigues de Almeida, Akira Luiz Nakamura, and José Ricardo Sodré, "Evaluation of Catalytic Converter Aging for Vehicle Operation with Ethanol," *Applied Thermal Engineering* 71 (2014): 335–341.

50. Between 1970 and 1990, the US light duty fleet almost doubled from around 111 million to 193 million registered vehicles; see US Department of Transportation, Bureau of Transportation Statistics (2021), "Number of US Aircraft, Vehicles, Vessels, and Other Conveyances." Between 1970 and 1990, the US transportation sector increased its consumption of petroleum from 15,311 trillion British thermal units to 21,626 trillion British thermal units; see US Energy Information Agency, *August 2020; Monthly Energy Review*, 44–45. For the trend in rising greenhouse gas emissions, see US Environmental Protection Agency, *Inventory of US Greenhouse Gas Emissions and Sinks*, 1990–2019, EPA 430-R-21-005 (April 14, 2021), 2–30. Daniel Sperling and Deborah Gordon recognized that increased vehicle fuel efficiency historically did not translate into increased fleet fuel economy, but they assumed that increased fleet fuel economy could in principle translate into reduced aggregate pollution, especially greenhouse emissions, despite explicitly recognizing (as indicated in the title of their 2009 book *Two Billion Cars: Driving Towards Sustainability*) that it was the continual growth of the automobile fleet that undid both efficiency and emissions controls.

51. Riki Therivel and Bill Ross argued that cumulative effects assessment was a required but underdeveloped aspect of national environmental impact assessment policies from the 1990s; see Therivel and Ross, "Cumulative Effects Assessment: Does Scale Matter?" *Environmental Impact Assessment Review* 27 (2007): 365–385.

52. US EPA, *Inventory of US Greenhouse Gas Emissions and Sinks*, 2–1, 2–2, 2–30.

53. By 2019, renewable energy was used in about 17 percent of electricity generation, with solar and especially wind accounting for most of the growth in renewable

capacity during the 2010s; see US Energy Information Agency (EIA), *August 2020 Monthly Energy Review*, 129.

54. Hydroelectric power was favored for this role because dammed water represents a large store of energy that can rapidly be converted to electricity. However, the prolonged drought that began in the US Southwest around 2001 complicated this vital systems-balancing function in the broader energy conversion complex; see Dominique M. Bain and Thomas L. Acker, "Hydropower Impacts on Electrical System Production Costs in the Southwest United States," *Energies* 11, no. 2 (2018): 368.

55. On the history of vehicle-to-grid, see Matthew N. Eisler, "Vehicle-to-Grid and the Energy Conversion Imaginary," in *Rethinking Electric History: From Esoteric Knowledge to Invisible Infrastructure to Fragile Networks*, eds. W. Bernard Carlson and Erik M. Conway (forthcoming, University of Virginia Press, 2023).

56. In 2019, renewable energy (defined by the EIA as biomass) accounted for 5 percent of primary energy converted for transportation (1410 of 28,206 trillion British thermal units) in the US; see US EIA, *August 2020 Monthly Energy Review*, 44–45.

57. Glen R. Asner, "The Linear Model, the US Department of Defense, and the Golden Age of Industrial Research," in *The Science-Industry Nexus: History, Policy, Implications*, eds. Karl Grandin, Nina Wormbs, and Sven Widmalm (Sagamore Beach, MA: Science History Publications/USA, 2004), 3–30; Benoît Godin, "The Linear Model of Innovation: The Historical Construction of an Analytical Framework," *Science, Technology & Human Values* 31, no. 6 (2006): 639–667; Miles MacLeod, "What Makes Interdisciplinarity Difficult? Some Consequences of Domain Specificity in Interdisciplinary Practice," *Synthese* 195 (2018): 697–720.

58. W. Patrick McCray, *The Visioneers: How a Group of Elite Scientists Pursued Space Colonies, Nanotechnologies, and a Limitless Future* (Princeton, NJ: Princeton University Press, 2013), 1–10.

59. In STS parlance, the ZEV is an exemplary boundary object of air-quality discourse; on boundary objects, see Susan Leigh Star, "The Structures of Ill-Structured Solutions: Boundary Objects and Distributed Heterogeneous Problem Solving," in *Readings in Distributed Artificial Intelligence*, eds. M. Huhns and L. Glasser (San Mateo, CA: Morgan Kauffmann), 37–54; and "This Is Not a Boundary Object: Reflections on the Origin of a Concept," *Science, Technology, & Human Values* 35, no. 5 (2010): 602.

60. For classic studies of dramaturgy in science and engineering communities, see Steven Shapin and Simon Schaffer, *Leviathan and the Air-Pump: Hobbes, Boyle, and the Experimental Life* (Princeton, NJ: Princeton University Press, 1985); Stephen Hilgartner, *Science on Stage: Expert Advice as Public Drama* (Stanford, CA: Stanford University Press, 2000); Michael Fortun, "Mediated Speculations in the Genomics Futures Markets," *New Genetics and Society* 20, no. 2 (2001): 139–156. For an analysis of dramaturgical parallels in science and art, see Megan K. Halpern, "Negotiations and Love Songs: Integration, Fairness, and Balance in an Art–Science Collaboration,"

in *Routledge Handbook of Art, Science, and Technology Studies*, eds. Hannah Star Rogers, Megan K. Halpern, Dehlia Hannah, and Kathryn de Ridder-Vignone (Abingdon, UK: Routledge, 2021), 319-334. For an analysis of the uses of rhetoric in relation to material displays designed to generate and promote ideas, see Hannah Star Rogers, *Art, Science, and the Politics of Knowledge* (Cambridge, MA: MIT Press, 2022).

61. See Eisler, *Overpotential*.

62. On this subject, see, for example, William A. Pizer, "A Tale of Two Policies: Clear Skies and Climate Change," in *Painting the White House Green: Rationalizing Environmental Policy Inside the Executive Office of the President*, eds. Randall Lutter and Jason F. Shogren (Washington, DC: Resources for the Future, 2004), 10–45.

63. On the commercialization of the lithium-cobalt oxide rechargeable battery in consumer electronics, see Yoshio Nishi, "Lithium Ion Secondary Batteries: Past 10 Years and the Future," *Journal of Power Sources* 100, nos. 1–2 (2001): 101–106; Yoshio Nishi, "The Development of Lithium Ion Secondary Batteries," *The Chemical Record* 1 (2001): 406–413; Yoshio Nishi, "Foreword: My Way to Lithium-Ion Batteries," in *Lithium-Ion Batteries: Science and Technologies*, eds. Masaki Yoshio, Ralph J. Brodd, and Akiya Kozawa (New York: Springer, 2009), v–vii; see also Matthew N. Eisler, "Exploding the Black Box: Personal Computing, the Notebook Battery Crisis, and Postindustrial Systems Thinking," *Technology and Culture* 58, no. 2 (2017): 368–391.

64. Trevor Pinch, "'Testing—One, Two, Three . . . Testing!' Toward a Sociology of Testing," *Science, Technology & Human Values* 18, no. 1 (1993): 27–31.

65. On this phenomenon, see, for example, Mody, *The Long Arm of Moore's Law*; Choi, "The Boundaries of Industrial Research;" Lécuyer and Brock, "The Materiality of Microelectronics;" Stuart W. Leslie, "Blue Collar Science: Bringing the Transistor to Life in the Lehigh Valley," *Historical Studies in the Physical and Biological Sciences* 32, no. 1 (2001): 71–113.

66. I define embodied practice both as the mutual performativity between user and technology as understood by Ritsuko Ozaki, Isabel Shaw, and Mark Dodgson in their study of Prius users, and the active shaping of the materiality of technology; see Ozaki, Shaw, and Dodgson, "The Coproduction of 'Sustainability': Negotiated Practices and the Prius," *Science, Technology, & Human Values* 38, no. 4 (2013): 518–541.

67. McCarthy, *Auto Mania*, 7, 12.

68. Ian Bogost, "The Tesla Model 3 Is Still a Rich Person's Car," *The Atlantic*, April 7, 2016, https://www.theatlantic.com/technology/archive/2016/04/tesla-model-3-/477243/m; "Motorists Don't Make Socialists, They Say; Not Pictures of Arrogant Wealth, as Dr. Wilson Charged," *New York Times*, March 4, 1906, 12, https://timesmachine.nytimes.com/timesmachine/1906/03/04/issue.html.

69. Clarke, *Driving Women*, 18–19. For an analysis of separate spheres marketing, see Scharff, *Taking the Wheel*, 35–50.

70. See Paterson, *Automobile Politics*, 225.

71. Peter Freund and George Martin, *The Ecology of the Automobile* (Black Rose Books: Montréal, 1993), 3.

CHAPTER 2

1. In 1962, Wouk founded the Electronic Energy Conversion Corporation to develop miniaturized solid-state rectifiers, devices that converted alternating current to direct current, selling the technology to companies (including IBM) for use in experimental applications; see Victor Wouk, interview by Judith R. Goodstein, New York, New York, May 24, 2004, Oral History Project, California Institute of Technology Archives, 49–50, https://oralhistories.library.caltech.edu/92/; Barbara E. Taylor, *The Lost Cord: The Storyteller's History of the Electric Car* (Columbus, OH: Greyden Press, 1995), 113.

2. Dunn and Johnson, "Chasing Molecules," 115–116.

3. Chip Jacobs and William J. Kelly, *Smogtown: The Lung-Burning History of Pollution in Los Angeles* (New York: Overlook, 2008); Mike Davis, *Ecology of Fear: Los Angeles and the Imagination of Disaster* (New York: Vintage Books, 1998), 95.

4. Dunn and Johnson, "Chasing Molecules," 115–116.

5. See Jacobs and Kelly, *Smogtown*; Zus Haagen-Smit, interview by Shirley K. Cohen, Pasadena, California, March 16 and 20, 2000, Oral History Project, California Institute of Technology Archives, https://oralhistories.library.caltech.edu/42/.

6. Sarah S. Elkind, *How Local Politics Shape Federal Policy: Business, Power, and the Environment in Twentieth-Century Los Angeles* (Chapel Hill: University of North Carolina Press, 2011), 67–70.

7. Arie Jan Haagen-Smit, "Chemistry and Physiology of Los Angeles Smog," *Industrial and Engineering Chemistry* 44, no. 6 (1952): 1342–1346; James Bonner, "Arie Jan Haagen-Smit, 1900–1977: A Biographical Memoir," in *Biographical Memoirs* 58 (Washington, DC: National Academies Press, 1989), 189–216.

8. Dunn and Johnson, "Chasing Molecules," 113.

9. Diana Clarkson and John T. Middleton, "The California Control Program for Motor Vehicle Created Air Pollution," *Journal of the Air Pollution Control Association* 12, no. 1 (1962): 22–28.

10. *Air Quality Act of 1967*, Public Law 90–148, *US Statutes at Large* 81 (1967): 501.

11. California Air Resources Board, "History," accessed April 7, 2021, https://ww2.arb.ca.gov/about/history.

12. Dunn and Johnson, "Chasing Molecules," 114.

13. In his message to Congress in support of the Air Quality Act of 1967, President Lyndon Johnson cited an especially serious smog incident in New York City associated

with the deaths of eighty people; Lyndon B. Johnson, "Special Message to the Congress: Protecting Our Natural Heritage," (speech, Washington, DC, January 30, 1967), The American Presidency Project, University of California at Santa Barbara, https://www.presidency.ucsb.edu/documents/special-message-the-congress-protecting-our-natural-heritage.

14. Horace Heyman, US Congress, Senate, Committee on Commerce and the Subcommittee on Air and Water Pollution, *Electric Vehicles and Other Alternatives to the Internal Combustion Engine: Joint Hearings Before the Committee on Commerce and the Subcommittee on Air and Water Pollution of the Committee on Public Works*, 90th Cong., 1st sess., March 14-17 and April 10, 1967, 164–165. Heyman added that there were around 60,000 battery electric forklift, utility, and industrial work trucks in the UK, and British industry produced around 12,000 such units per year.

15. Harry F. Barr, *Electric Vehicles*, 250–255.

16. Alan S. Boyd, *Electric Vehicles*, 77–78.

17. Federal Power Commission, *Electric Vehicles*, 5–8.

18. Andy Leparulo, *Electric Vehicles*, 193.

19. J. A. McIlnay, *Electric Vehicles*, 154.

20. Warren G. Magnuson, *Electric Vehicles*, 1.

21. Some studies of redlining hint at the envirotechnical effects of the systematic denial of urban services, linking air pollution with higher rates of asthma in redlined parts of Californian cities; see Anthony Nardone et al., "Associations Between Historical Residential Redlining and Current Age-Adjusted Rates of Emergency Department Visits Due to Asthma Across Eight Cities in California: An Ecological Study," *Lancet Planet Health* 4 (2020): 24–31.

22. Boyd, US Congress, Senate, Committee, *Electric Vehicles*, 70–71.

23. For an exemplary history of materials science, see Bensaude-Vincent, "The Construction of a Discipline." For accounts of the development of solid-state electronics, see Michael Riordan and Lillian Hoddeson, *Crystal Fire: The Invention of the Transistor and the Birth of the Information Age* (Norton, 1997); and Gertner, *The Idea Factory*.

24. Advanced batteries are often considered to have energy densities greater than 30–40 watt-hours per kilogram, the limits of classical battery chemistries such as lead-acid and nickel-cadmium around the last third of the twentieth century; see Chen-Xi Zu and Hong Li, "Thermodynamic Analysis on Energy Densities of Batteries," *Energy and Environmental Science* 4 (2011): 2614–2624, on 2615.

25. Among US science and technology agencies, the National Aeronautics and Space Administration (NASA) was a leading promoter of the idea of the power source spin-off; see John E. Condon, "Practical Values of Space Exploration," October 10, 1962, Record Number 18530 IX: Technology Utilization, Addresses, Speeches, 3–6, NASA Headquarters Archive, Washington, DC (hereafter cited as NASA Technology

Utilization); and James T. Dennison, "Contributions of Aerospace Research to the Business Economy," September 26, 1963, NASA Technology Utilization.

26. Baker, "The National Role of Materials Research and Development." Years later, William O. Baker further expounded on these ideas in "Advances in Materials Research and Development," in *Advancing Materials Research*, eds. Peter A. Psaras and H. Dale Langford (Washington, DC: National Academies Press, 1987), 3–22. According to James R. Killian, Jr., the tenth president of the Massachusetts Institute of Technology (MIT) and a key organizer of the postwar science advisory apparatus, Baker exerted considerable authority in science and defense policy circles in an unusually long, parallel career; see James R. Killian, Jr., *Sputnik, Scientists, and Eisenhower: A Memoir of the First Special Assistant to the President for Science and Technology* (Cambridge, MA: MIT Press, 1977); see also Gertner, *The Idea Factory*.

27. Richard H. Schallenberg expressed this point in his pioneering study of battery technology; see Schallenberg, *Bottled Energy*.

28. Judgments of similarity extrapolated from simple controlled environments to more complicated ones are an important characteristic of reductive laboratory testing; see Pinch, "'Testing—One, Two, Three . . . Testing!'"

29. On the idea of the material as the device, see Lécuyer and Brock, "The Materiality of Microelectronics."

30. See Eisler, *Overpotential*.

31. Paul R. Hayes, "Auto Facts: 200 Motorists to Give Chrysler Turbine 'Ride-and-Drive' Test for Year," *Philadelphia Inquirer*, May 20, 1963, p. 24; General Motors, "Firebird III," accessed February 21, 2022, https://www.gmheritagecenter.com/docs/gm -heritage-archive/historical-brochures/1958-firebird-III/1958_Firebird_III_Brochure .pdf.

32. See Barr, *Electric Vehicles*, 252–253.

33. It is noteworthy that at that time, the Corvair was becoming the subject of a great deal of scrutiny over what critics claimed was a potentially fatal problem with handling stability caused by the design's rear-engine format and independent rear suspension; see Ralph Nader, *Unsafe at Any Speed: The Designed-in Dangers of the American Automobile* (New York: Grossman Publishers, 1965).

34. Jalal T. Salihi, Paul D. Agarwal, and George J. Spix, "Induction Motor Scheme for Battery-Powered Electric Car (GM Electrovair-I)," *IEEE Transactions on Industry and General Applications* 3, no. 5 (September/October 1967): 463–469; E. A. Rishavy, W. D. Bond, and T. A. Zechin, "Electrovair: A Battery Electric Car," *SAE Transactions* 76 (1968): 981–991, 1023–1028.

35. For a brief history of Yardney's silver-zinc battery, see A. P. Karpinski et al., "Silver-Zinc: Status of Technology and Applications," *Journal of Power Sources* 80, nos. 1–2 (1999): 53–60.

36. Salihi et al., "Induction Motor Scheme"; Rishavy et al., "Electrovair."

37. Grove referred to the device as a "gaseous voltaic battery;" see William Robert Grove, "On a Gaseous Voltaic Battery," *Philosophical Magazine and Journal of Science* 21, S.3 (December 1842): 417.

38. For an excellent synoptic history of fuel cells, see Harold D. Wallace, Jr., "Fuel Cells: A Challenging History," *Substantia* 3, no. 2 (2019): 83–97.

39. Matthew N. Eisler, "'A Modern Philosopher's Stone': Techno-Analogy and the Bacon Cell," *Technology and Culture* 50, no. 2 (2009): 345–365.

40. At the outset of this project, some Army officials evinced skepticism about hydrocarbon fuel cells, cautioning that the "present enthusiasm" had only a "sketchy" basis in science; see ARPA, "Summary of Proposal of Research on Energy Conversion," February 6, 1961, box 4, Project Lorraine, Energy Conversion, 1958–1966 Official Correspondence Files, Materials Sciences Office, ARPA, accession number 68-A-2658, Record Group 330, National Archives and Records Administration, College Park, Maryland (hereafter cited as OCF-MSO, ARPA).

41. Charles F. Yost to R. L. Sproull, "Memorandum for Dr. Sproull: Subject: Project Lorraine," September 12, 1963, box 4, OCF-MSO, ARPA; Memorandum by Charles F. Yost, July 13, 1962, box 2, AO 247-Esso Research and Engineering, OCF-MSO, ARPA.

42. A 1967 Army Mobility Command analysis ranked reforming as the most difficult of all the various fuel cell configurations; see James R. Huff and John C. Orth, "The USAMECOM-MERDC Fuel Cell Electric Power Generation Program," in *Fuel Cell Systems II: 5th Biennial Fuel Cell Symposium Sponsored by the Division of Fuel Chemistry at the 154th Meeting of the American Chemical Society, Chicago, Illinois, September 12–14, 1967* (Washington, DC: American Chemical Society, 1969), 318.

43. Craig Marks, Edward A. Rishavy, and Floyd A. Wyczalek, "Electrovan: A Fuel Cell Powered Vehicle," *SAE Transactions* 76 (1968): 992–1002, 1023–1028.

44. Barr, *Electric Vehicles*, 255.

45. Leparulo, *Electric Vehicles*, 196.

46. As a response to the Enfield 8000, Ford's autonomous British division in 1967 built two prototypes of the Comuta, a two-seat electric city car equipped with lead-acid batteries; see Michael H. Westbrook, *The Electric Car: Development and Future of Battery, Hybrid and Fuel-Cell Cars* (London: Institution of Electrical Engineers, 2001), 22–23, 67, 79–80.

47. Wouk, interview by Goodstein, 51.

48. Wouk, interview by Goodstein, 54.

49. Joseph T. Kummer and Neill Weber, "A Sodium-Sulfur Secondary Battery," *SAE Transactions* 76 (1968): 1003-1007.

50. John B. Goodenough, interview by author, July 11, 2013.

51. Hervé Arribart and Bernadette Bensaude-Vincent, "Beta-Alumina," Caltech Library, February 16, 2001, http://authors.library.caltech.edu/5456/1/hrst.mit.edu /hrs/materials/public/Beta-alumina.htm.

52. Rebecca Slayton also suggested the problem of employment redundancy in her study of Lincoln Laboratory; see "From a 'Dead Albatross' to Lincoln Labs: Applied Research and the Making of a Normal Cold War University," *Historical Studies in the Natural Sciences* 42, no. 4 (2012): 255–282.

53. Goodenough, interview.

CHAPTER 3

1. Stan Luger, *Corporate Power, American Democracy, and the Automobile Industry* (Cambridge: Cambridge University Press, 2000), 76–96.

2. Ralph Nader, "The Management of Environmental Violence: Regulation or Reluctance," in *Environment in Peril*, ed. Anthony N. Wolbarst (Washington, DC: Smithsonian Institution Press, 1991), 2–25; and Robert Gottlieb, *Forcing the Spring: The Transformation of the American Environmental Movement* (Washington, DC: Island Press, 2005), 179.

3. James C. Williams, *Energy and the Making of Modern California* (Akron, OH: University of Akron Press, 1997), 321–322; E. F. Schumacher, *Small Is Beautiful: A Study of Economics as if People Mattered* (London: Blond and Briggs, 1973).

4. Andrew G. Kirk, *Counterculture Green: The Whole Earth Catalog and American Environmentalism* (Lawrence: University Press of Kansas, 2007), 9, 182.

5. See Amory B. Lovins, "Energy Strategy: The Road Not Taken?" *Foreign Affairs* (October 1976): 65–96; Paul Hawken, Amory B. Lovins, and L. Hunter Lovins, *Natural Capitalism: Creating the Next Industrial Revolution* (New York: Little, Brown, and Company, 1999). For accounts of the appropriate technology movement, see Carroll Pursell, "The Rise and Fall of the Appropriate Technology Movement in the United States, 1965–1985," *Technology and Culture* 34, no. 3 (1993): 629–637; Williams, *Energy and the Making of Modern California*, 320–322; Samuel P. Hays, *Beauty, Health, and Permanence: Environmental Politics in the United States, 1955–1985* (Cambridge: Cambridge University Press, 1987), 261–262.

6. Between 1970 and 1980, annual emissions of nitrogen oxides from highway vehicles fell from around 12.6 million tons per year to around 11.5 million tons per year; see US Environmental Protection Agency (EPA), "Criteria Pollutants National Tier 1 for 1970–2020."

7. Michael Lamm, "PM Owners Report: Electric Cars," *Popular Mechanics*, March 1977, 90–93, 137.

8. Wally Rippel, interview by author, May 21, 2019.

9. Rippel, interview.

10. Leon S. Loeb, "Across the USA with MIT's Electric Car," *Popular Mechanics*, November 1968, 52J, 52K, 184,186, 188; Fernanda Ferreira, "The Great Big Headache of 1968," *MIT Technology Review*, February 26, 2020, https://www.technologyreview.com/2020/02/26/905991/the-great-big-headache-of-1968/.

11. "Cambridge or Bust, Pasadena or Bust: Both Teams in the Great Electric Car Race Made It—and Busted Too," *Engineering and Science* 32, no. 1 (October 1968): 10–17.

12. According to Rippel, Caltech president Lee DuBridge once informed him that the Union Oil Company of California had given the university a large endowment on the condition that it did not engage in activities harmful to the oil industry; Rippel, interview.

13. Massachusetts Institute of Technology, "MIT Electric Vehicle Team History: The Clean Air Car Race," accessed February 16, 2022, http://web.mit.edu/evt/Clean AirCarRace.html.

14. Taylor, *The Lost Cord*, 187–195.

15. Taylor, *The Lost Cord*, 190–193.

16. Taylor, *The Lost Cord*, 264; Johs Jensen, Jorgen Lundsgaard, and Carol M. Perram, *Electric Vehicles for Urban Transport: A Preliminary Investigation into the Possibilities for Introduction of Electric Buses and Other Electric Vehicles in Odense, Denmark* (Odense, Denmark: Odense University Press, 1980), 50; Westbrook, *The Electric Car*, 24.

17. Mike Knepper, "Citicar: Have You Heard the One About the Voltswagen?" *Motor Trend*, November 1976, 60–63.

18. Taylor, *The Lost Cord*, 331–334.

19. Robert W. Irvin, "The Revival of Electric Vehicles: Passim [sic] Fancy or Car of the Future?" *New York Times*, April 7, 1974.

20. US Department of Transportation (DOT), *Summary of Fuel Economy Performance*, December 15, 2014.

21. The act provided for the federal government to purchase or lease 2,500 electric or hybrid electric vehicles and another 5,000 "advanced" electric vehicles; see *Electric and Hybrid Vehicle Research, Development, and Demonstration Act of 1976*, Public Law 94–413, *US Statutes at Large* 90 (1976): 1260–1272, on 1264.

22. Public Law 94–413, 1260, 1264; see also Taylor, *The Lost Cord*, 430–432.

23. Pandit G. Patil, "Prospects for Electric Vehicles," *IEEE AES Systems Magazine* (December 1990): 15–16.

24. This bias was reflected in a 1973 study on alternative propulsion systems commissioned by Ford on the suggestion of company president Lee Iacocca and conducted by scientists and engineers from NASA's Jet Propulsion Laboratory and the Caltech Environmental Quality Laboratory. Unsurprisingly, the study reaffirmed

previous research concluding that a breakthrough in battery performance was necessary before electric cars could be competitive with ICE cars. It observed that the energy economy of lead-acid battery electric propulsion was "competitive or slightly superior to" ICE vehicles at ranges of around thirty to fifty miles but rapidly dropped as both range and battery weight increased. The report viewed the hybrid battery electric platform as the worst option, holding that the technology yielded only modest improvements in fuel efficiency and had high materials and maintenance costs; see R. Rhoads Stephenson et al., *Should We Have a New Engine? An Automobile Power Systems Evaluation Volume II, Technical Reports* (Pasadena: Jet Propulsion Laboratory, California Institute of Technology, 1975), 8–19, 8–20, 9–3, 9–18. I thank Alec Brooks for providing me with a copy of this document.

25. See the address delivered by Richard Nixon, November 7, 1973; US President, Proclamation, "The Energy Emergency," *Federal Register* 9, no. 45 (November 12, 1973): 1312–1318.

26. These issues were raised in a major DOE review of its programs. Among other criticisms, the review noted that the DOE did not usually consider the market or assess consumer needs before starting new technology programs; see Robert W. Fri et al., *Energy Research at DOE: Was It Worth It? Energy Efficiency and Fossil Energy Research, 1978 to 2000* (Washington, DC: National Academy Press, 2001), 36.

27. Pietro S. Nivola, *The Politics of Energy Conservation* (Washington, DC: Brookings Institution, 1986), 86–87.

28. Esso Research and Engineering Company, "Proposal for the Continuation of Government Contract Research on Fuel Cells; Program Period–Calendar Year 1965," July 24, 1964, Esso Research and Engineering Company, box 2, AO 247-Esso Research and Engineering Company, OCF-MSO, ARPA.

29. Invented by Thomas Edison, nickel-zinc is an inexpensive rechargeable battery chemistry with higher energy density and much higher power than the lead-acid rechargeable but with a shorter lifetime. GM cancelled the Electrovette, but the program provided an education for its manager, Kenneth Baker. In the 1990s, Baker would become manager of the program to turn GM's Impact concept car into the production car known as the EV1, a vehicle that would play a central role in California air quality politics; see also GM Heritage Center, "GM Vehicle Technologies," accessed https://gmheritagecenter.com/featured/Alt-Fuel.html.

30. Goodenough, interview.

31. M. S. Whittingham, "Electrical Energy Storage and Intercalation Chemistry," *Science* (New Series) 192, no. 4244 (1976): 1126–1127.

32. John B. Goodenough, "Rechargeable Batteries: Challenges Old and New," *Journal of Solid-State Electrochemistry* 16, no. 6 (2012): 2019–2029, on 2022; John B. Goodenough and Youngsik Kim, "Challenges for Rechargeable Li Batteries," *Chemistry of Materials Review* 22, no. 3 (2010): 587–603, on 592.

33. According to Goodenough, his new employers assumed he was a chemist because he had headed a ceramics laboratory engaged in solid-state chemistry. Goodenough claimed that his hiring alienated British inorganic chemists who had coveted the post. See Goodenough, interview.

34. See Goodenough, "Rechargeable Batteries," and interview with the author. See also K. Mizushima, P. C. Jones, P. J. Wiseman, and J. B. Goodenough, "Li$_x$CoO$_2$: A New Cathode Material for Batteries of High Energy Density," *Materials Research Bulletin* 15, no. 6 (1980): 783–789; J. B. Goodenough, K. Mizushima, and T. Takeda, "Solid-Solution Oxides for Storage-Battery Electrodes," *Japanese Journal of Applied Physics* 19 (1980): 305–313.

35. Ralph J. Brodd, "Chapter 1: Synopsis of the Lithium-Ion Battery Markets," in *Lithium-Ion Batteries: Science and Technologies*, eds. Masaki Yoshio, Ralph J. Brodd, and Akiya Kozawa (New York: Springer, 2009), 1–7, on 1.

36. Kazunori Ozawa, "Lithium-Ion Rechargeable Batteries with LiCoO$_2$ and Carbon Electrodes: The LiCoO$_2$/C System," *Solid-State Ionics* 69, nos. 3–4 (1994): 212–221, on 212. See also Yoshio Nishi, "My Way to Lithium-Ion Batteries," in *Lithium-Ion Batteries: Science and Technologies*, eds. Masaki Yoshio, Ralph J. Brodd, and Akiya Kozawa (New York: Springer, 2009), v–vii; and Masaki Yoshio, Akiya Kozawa, and Ralph J. Brodd, "Introduction: Development of Lithium-Ion Batteries," in *Lithium-Ion Batteries: Science and Technologies*, eds. Masaki Yoshio, Ralph J. Brodd, and Akiya Kozawa (New York: Springer, 2009), v–vii, xvii–xxvi.

37. Andrea Wong, "The Untold Story Behind Saudi Arabia's 41-Year Debt Secret," *Bloomberg News*, May 30, 2016, http://www.bloomberg.com/news/features/2016-05-30/the-untold-story-behind-saudi-arabia-s-41-year-u-s-debt-secret.

38. GE and NASA's Jet Propulsion Laboratory contributed the HTV-1, an advanced plug-in hybrid electric utilizing a lead-acid battery built by Johnson Controls that gave a range of thirty miles in electric-only mode; see Patil, "Prospects for Electric Vehicles," 16. On the founding of EPRI, see Brent Barker, "Electric Power Research Institute: Born in a Blackout," *EPRI Journal* (Summer 2012): 14–17.

39. The idea of hydrogen as fuel was first popularized by Jules Verne (*The Mysterious Island*, Hetzel, 1874) and Max Pemberton (*The Iron Pirate: A Plain Tale of Strange Happenings on the Sea*, London: Cassell and Company, 1893). For representative examples of hydrogen futurist discourse in the 1970s, see Eduard Justi, *Leitungsmechanismus und Energieumwandlung in Festkörpern* (Göttingen, Germany: Vandenhoeck and Ruprecht, 1965); John Bockris and A. J. Appleby, "The Hydrogen Economy: An Ultimate Economy? A Practical Answer to the Problem of Energy Supply and Pollution," in *Environment This Month: The International Journal of Environmental Science* 1, no. 1 (July 1972): 29–35; Lawrence W. Jones, "Hydrogen: A Fuel to Run Our Engines in Clean Air," *Saturday Evening Post*, Spring 1972, 34; D.P. Gregory et al., *A Hydrogen-Energy System* (Institute of Gas Technology/American Gas Association, 1973); Cesare Marchetti, "From the Primeval Soup to World Government: An Essay on Comparative Evolution," *International Journal of Hydrogen Energy* 2, no. 1 (1977): 1–5.

40. See E. Eugene Ecklund, "Federal Hydrogen Energy Activities in the United States of America," in *Hydrogen Energy Progress IV: Proceedings of the Fourth World Hydrogen Energy Conference, Pasadena, California, USA., June 13–17, 1982*, vol. 4, eds. T. N. Veziroglu, W. D. Van Vorst, and J. H. Kelley (Oxford, UK: Pergamon Press, 1982), 1431–1434.

41. See J. Byron McCormick and James R. Huff, "The Case for Fuel-Cell–Powered Vehicles," *Technology Review* (August/September 1980): 54–65; D. A. Freiwald and W. J. Barattino, "Technical Note: Alternative Transportation Vehicles for Military-Base Operations," *International Journal of Hydrogen Energy* 6, no. 6 (1981): 631–636; Ecklund, "Federal Hydrogen Energy Activities," 1431–1434.

42. D. G. Kingwill, *The CSIR: The First 40 Years* (Pretoria, South Africa: Scientia Printers, CSIR, 1990), 6–9, 32.

43. See Coetzer, "A New High Energy Density Battery System;" and Michael Thackeray, "20 Golden Years of Battery R&D at CSIR, 1974–1994," *South African Journal of Chemistry* 64 (2011): 61–66.

44. Thackeray, "20 Golden Years of Battery R&D," 64.

45. J. L. Sudworth, "Zebra Batteries," *Journal of Power Sources* 51, nos. 1–2 (1994): 105–114, on 114.

46. W. D. Van Vorst, J. H. Kelley, and T. N. Veziroglu, "WHEC-IV," *International Journal of Hydrogen Energy* 8, nos. 11–12 (1983): 858–859; Helmut Buchner and R. Povel, "The Daimler-Benz Hydride Vehicle Project," *International Journal of Hydrogen Energy* 7, no. 3 (1982): 259–266.

47. Lillian Hoddeson and Peter Garrett, *The Man Who Saw Tomorrow: The Life and Inventions of Stanford R. Ovshinsky* (Cambridge, MA: MIT Press, 2018), 123–124.

48. For an analysis of the innovation dynamics of Edison's laboratories, see Paul. B. Israel, "Inventing Industrial Research: Thomas Edison and the Menlo Park Laboratory," *Endeavour* 26, no. 2 (2002): 48–54, on 51; see also Paul B. Israel, *Edison: A Life of Invention* (New York: John Wiley, 1998).

49. Hoddeson and Garrett, *The Man Who Saw Tomorrow*, 123–146.

50. Hoddeson and Garrett, *The Man Who Saw Tomorrow*, 187.

51. Hoddeson and Garrett, *The Man Who Saw Tomorrow*, 187–192.

52. Michael A. Fetcenko et al., "Recent Advances in NiMH Battery Technology," *Journal of Power Sources* 165, no. 2 (2007): 545–546; see also S. R. Ovshinsky, M. A. Fetcenko, and J. Ross, "A Nickel-Metal Hydride Battery for Electric Vehicles," *Science* 260, no. 5105 (April 9, 1993): 176–181.

53. M. L. Perry and T. F. Fuller, "A Historical Perspective of Fuel Cell Technology in the 20th Century," *Journal of the Electrochemical Society* 149, no. 7 (2002): S59–S67, on S60.

54. Ernst M. Cohn, "The Growth of Fuel Cell Systems," August 1965, Record No. 13761: Propulsion, Auxiliary Power: Fuel Cells, 1961–1999, 5–7, NASA Headquarters

Archive; Barton C. Hacker and James M. Grimwood, *On the Shoulders of Titans: A History of Project Gemini* (Washington, DC: National Aeronautics and Space Administration, 1977), 149.

55. Perry and Fuller, "A Historical Perspective," S60, S64.

56. Tom Koppel, *Powering the Future: The Ballard Fuel Cell and the Race to Change the World* (Toronto: John Wiley and Sons Canada, 1999), 1–36.

57. Keith B. Prater, "The Renaissance of the Solid Polymer Fuel Cell," *Journal of Power Sources* 29, nos. 1–2 (1990): 243.

58. Koppel, *Powering the Future*, 66, 93–94.

59. Koppel, *Powering the Future*, 127–132.

CHAPTER 4

1. See US Environmental Protection Agency (EPA), "Criteria Pollutants National Tier 1 for 1970–2020."

2. US Department of Transportation (US DOT), Bureau of Transportation Statistics, "Number of US Aircraft, Vehicles, Vessels, and Other Conveyances."

3. Robert W. Fri et al., *Energy Research at DOE: Was It Worth It? Energy Efficiency and Fossil Energy Research 1978 to 2000* (Washington, DC: National Academies Press, 2001), 9.

4. Jananne Sharpless, interview by author, September 5, 2019.

5. The LEV set rolling quotas for automakers with annual sales of more than 35,000 light-duty vehicles in California to produce progressively larger numbers of transitional low emission, low emission, ultralow emission, and zero emission vehicles (TLEVs, LEVs, ULEVs, and ZEVs, respectively); see California Air Resources Board (CARB), "Proposed Regulations for Low-Emissions Vehicles and Clean Fuels: Technical Support Document," August 13, 1990, I-4–I-16; see also Gustavo Collantes and Daniel Sperling, "The Origin of California's Zero Emission Vehicle Mandate," *Transportation Research Part A: Policy and Practice* 42, no. 10 (2008): 1302–1313, on 1305.

6. Sharpless, interview.

7. George Harrar, "Technology: The 'Concept Car' Pushes Change," *New York Times*, July 1, 1990, Section 3, p. 5.

8. Paul B. MacCready, "Sunraycer Odyssey: Winning the Solar-Powered Car Race Across Australia," *Engineering and Science* (Winter 1988): 3–13. Brooks held that he had independently learned about the Tholstrup race and was planning his own entry around the time MacCready was contacted by Ellion; Alec Brooks, interview by author, December 6, 2019.

9. Stempel had been part of the team that designed the 1966 Oldsmobile Toronado, Detroit's first front-wheel-drive vehicle in the post–World War II era. As part of the

engineering department of Chevrolet, Stempel worked on the front-wheel-drive/ transverse engine configuration that industry was increasingly favoring as the most efficient powertrain configuration for ICE propulsion. Stempel also promoted the use of computer technology to control fuel and ignition systems for cleaner and more efficient energy conversion; see Betsy Ancker-Johnson and Bruce MacDonald, "Robert C. Stempel, 1933-2011," in *Memorial Tributes: National Academy of Engineering, Volume 16* (Washington, DC: National Academies Press, 2012), 310–311.

10. Allan Abbott and Alec Brooks, "Flying Fish, The First Human-Powered Hydrofoil to Sustain Flight," accessed January 8, 2020, https://flyingfishhydrofoil.com/.

11. Brooks, interview.

12. Alan Cocconi, interview by author, November 11, 2020.

13. Brooks, interview.

14. Cocconi, interview.

15. Brooks, interview.

16. Paul Dean, "It's a Bird . . . It's a Plane; It's Weird—But It Can Fly," *Los Angeles Times*, February 16, 1986, https://www.latimes.com/archives/la-xpm-1986-02-16-vw -8846-story.html.

17. Hughes selected specialized power sources then used largely for military applications, using some of its own gallium arsenide photovoltaic cells and silver-zinc batteries made by Eagle-Picher; see Bill Tuckey, *Sunraycer* (Hornsby, Australia: Chevron Publishing Group, 1989), 13–20, 30, 36, 43–49; Michael Shnayerson, *The Car That Could: The Inside Story of GM's Revolutionary Electric Vehicle* (New York: Random House, 1996), 14–15.

18. MacCready, "Sunraycer Odyssey," 4.

19. Tuckey, *Sunraycer*, 45.

20. Roger B. Smith, "A Message from the Chairman," introduction to *Sunraycer* by Tuckey, 9.

21. Brooks, interview; Shnayerson, *The Car That Could*, 21.

22. According to Rippel, Cocconi played a decisive role in this debate that was indicative of the esteem in which the engineer was held at AeroVironment. Working as a consultant to AeroVironment in 1985, Rippel had proposed developing a high-performance electric car, only to be rebuffed by MacCready. According to Rippel, two years later, as the AeroVironment team deliberated on the format of Sunraycer's successor, Cocconi rejected MacCready's plan for a delivery truck, arguing that a high-performance electric was the best way to push the technological frontier. Cocconi's intervention, averred Rippel, "changed everything;" see Rippel, interview.

23. AeroVironment and Howard G. Wilson, *Final Study Report and Proposal: The Electric Vehicle—Time for a New Look* (Monrovia, California: AeroVironment, Inc., 1988). Alec Brooks is the uncredited author of this document. I thank him for providing me with a copy.

24. Brooks, interview; and email communication with the author, March 3, 2020.

25. Shnayerson, *The Car That Could*, 77–81.

26. Doron P. Levin, "GM to Begin Production of Battery-Powered Car," *New York Times*, April 19, 1990, D5. Of the two major New York–based national newspapers, the *Wall Street Journal* was the more skeptical of Smith's claims; see Rick Wartzman, "GM Unveils Electric Car with Lots of Zip But Also a Battery of Unsolved Problems," *Wall Street Journal*, January 4, 1990, A1; Joseph P. White, "GM Says It Plans an Electric Car, But Details Are Spotty," *Wall Street Journal*, April 19, 1990, B1.

27. Tom Cackette, interview by author, September 27, 2019; see also Collantes and Sperling, "The Origin of California's Zero Emission Vehicle Mandate," 1306.

28. Cackette, interview.

29. Sharpless, interview.

30. Sharpless, interview. See also Collantes and Sperling, "The Origin of California's Zero Emission Vehicle Mandate," 1308.

31. In his closing remarks at the twelfth Electric Vehicle Symposium in December 1994, Stempel said that he preferred incentives that provided "market pull;" see Robert C. Stempel, "Challenge for Tomorrow: Forging the Road Ahead," EVS-12 Symposium, December 7, 1994, 8, box 49, Business Admin/Miscellaneous (2) 1993–1995; Stanford R. Ovshinsky Papers, Bentley Historical Library, University of Michigan at Ann Arbor (hereafter cited as SROP).

32. Brooks, interview.

33. General Motors, "Impact's Aluminum Frame Provides Lightweight Support," undated photo, GM Electric Vehicles, received October 14, 1993, R.C. Stempel, box 31, Electric Vehicles Miscellaneous, Robert C. Stempel Papers, Bentley Historical Library, University of Michigan at Ann Arbor (hereafter cited as RCSP).

34. Rippel, interview; Shnayerson, *The Car That Could*, 47.

35. Shnayerson, *The Car That Could*, 61, 63, 64, 70.

36. David Lawder, "GM Names Smith Chairman as Smale Steps Down," *Buffalo News*, December 4, 1995.

37. Shnayerson, *The Car That Could*, 99, 102, 121, 166–167.

38. Matthew L. Wald, "Expecting a Fizzle, GM Puts Electric Car to Test," *New York Times*, January 28, 1994, D4.

39. William B. Wylam to Subhash Dhar, July 12, 1990; Dennis A. Corrigan to Michael A. Fetcenko, June 15, 1990, box 54, ECD Subsidiaries/Ovonic Battery Co./Misc., SROP; see also Shnayerson, *The Car That Could*, 39–41.

40. Ovshinsky's vision of nickel-metal hydride power did win over one GM researcher. Dennis Corrigan, an electrochemist, had worked in the Physical Chemistry Department of GM's Research Laboratories and was involved in the 1990 talks.

Greatly impressed by Ovshinsky, Corrigan left GM and joined OBC in 1992; see Corrigan to Fetcenko, June 15, 1990, SROP.

41. Srinivasan Venkatesan, Subhash Dhar, Stanford Ovshinsky, and Michael Fetcenko, "Ovonic Nickel-Metal Hydride Batteries for Industrial and Electric Vehicle Applications," *Proceedings of the Sixth Annual Battery Conference on Applications and Advances* (1991): 59–73, on 61.

42. The arrangement yielded a concept electric car shown at the Tokyo Automotive Show in October 1993; ECD, "News Release: Energy Conversion Devices and Its Subsidiary, Ovonic Battery Company, Announce Battery Agreement with a Major Japanese Automobile Manufacturer," September 19, 1991; ECD, "News Release: ECD/OBC Announces Agreement with Honda," January 4, 1994, box 50, ECD/Press Kits/Press Releases, SROP.

43. Ovshinsky et al., "A Nickel-Metal Hydride Battery."

44. For analyses that address the influence of government policy in the semiconductor industry, see Clair Brown and Greg Linden, *Chips and Change: How Crisis Reshapes the Semiconductor Industry* (Cambridge, MA: MIT Press, 2009); and Berlin, *The Man Behind the Microchip*; see also Block, "Swimming Against the Current," 181–188.

45. William J. Clinton and Albert Gore, Jr., *Technology for America's Economic Growth, A New Direction to Build Economic Strength*, February 22, 1993, White House: Office of the Press Secretary, 9.

46. Hoddeson and Garrett, *The Man Who Saw Tomorrow*, 194–195; and Shnayerson, *The Car That Could*, 175–179.

47. S. K. Dhar, S. R. Ovshinsky, P. R. Gifford, D. A. Corrigan, M. A. Fetcenko, and S. Venkatesan, "Nickel/Metal Hydride Technology for Consumer and Electric Vehicle Batteries: A Review and Update," *Journal of Power Sources* 65, nos. 1–2 (1997): 1–7, on 5; see also Shnayerson, *The Car That Could*, 171–181.

48. By late 1991, the collective national mandate quotas for automakers stood at around 70,000 ZEVs by 1998 and around 500,000 by 2003; see Matthew L. Wald, "A Tough Sell for Electric Cars: Technology Lagging as Markets Emerge," *New York Times*, November 26, 1991, D1.

49. North Dakota Office of the Governor, "Harry J. Pearce," August 11, 2004, https://www.governor.nd.gov/theodore-roosevelt-rough-rider-award/harry-j-pearce.

50. Doron P. Levin, "Mr. Pearce's Growing Domain," *New York Times*, November 15, 1992; Donald W. Nauss, "GM's Man Who Bested NBC Helps Rouse Sleeping Giant," *Los Angeles Times*, February 17, 1993.

51. Levin, "Mr. Pearce's Growing Domain."

52. Shnayerson, *The Car That Could*, 187–190.

53. Shnayerson, *The Car That Could*, 198–200.

54. Leon J. Krain to Stanford R. Ovshinsky, March 4, 1994, box 47, ECD/Corporate Partners/Joint Ventures, 1971–2004, SROP.

55. On OBC patents, see Shiuan Chang, Kwo-hsiung Young, Jean Nei, and Cristian Fierro, "Reviews on the US Patents Regarding Nickel/Metal Hydride Batteries," *Batteries* 2, no. 10 (2016): 2–3.

CHAPTER 5

1. Noel Perrin, *Solo: Life with an Electric Car* (New York: W. W. Norton and Company, 1992), 13–14, 19, 30.

2. Matthew L. Wald, "Company News: Electric Car Venture Set with Itochu," *New York Times*, June 10, 1994, D0000.3.

3. "Solectria Unveils United States' First Mass Producible All Composite Ground-Up Electric Vehicle," *Business Wire*, December 2, 1994.

4. Victor Wouk, "Hybrids: Then and Now," *IEEE Spectrum* 32, no. 7 (July 1995): 16–21.

5. Mom, *The Electric Vehicle*, 124–126, 193–195.

6. Wouk believed that the project was terminated on the personal orders of Eric Stork, then the EPA deputy assistant administrator for mobile source air pollution control; Wouk, interview by Goodstein, 65.

7. Stephenson, *Should We Have a New Engine?* 9–14, 9–16. General Motors (GM) experimented with a vehicle utilizing a 500-pound pack of lead-acid auxiliary batteries linked to a Stirling engine, a low-power device that operated on the principle of the cyclic compression and expansion of gas at a temperature differential.

8. A 1982 collaboration between General Electric (GE) and the DOE resulted in a prototype of what was claimed as the first modern hybrid electric car employing computer controls. Andrew Burke, one of the principals in the project, dubbed the HTV-1, recalled that the hybrid electric system did little to overcome the poor quality of the batteries selected for it. Burke recalled that the car's lead-acid rechargeable had such a short life span that the device was virtually spent by the time GE finished testing the car and turned it over to the DOE for further testing. Burke reported that the DOE substituted the spent pack with one made of nickel-cadmium cells, then used mainly in consumer electronics; see Andrew Burke, "The First Modern Hybrid Car, HTV-1, 1978–1982," accessed December 9, 2016, https://www.youtube.com/watch?v=p_pqT21eLdI.

9. William J. Clinton and Albert Gore Jr., "High Technology Policy Initiatives," filmed February 22, 1993, San Jose California, *C-SPAN*, https://www.c-span.org/video/?38171-1/high-technology-policy-initiatives.

10. United States Council for Automotive Research (USCAR), "USCAR as Umbrella for Big Three Research," undated, box 31, Electric Vehicles Miscellaneous, 1–4 on 1,

RCSP; United States Council for Automotive Research (USCAR), "Who We Are," accessed June 23, 2013, https://www.uscar.org/guest/history.php.

11. Clinton and Gore, "Technology for America's Economic Growth," 33.

12. Brent D. Yacobucci, "The Partnership for a New Generation of Vehicles: Status and Issues," *Congressional Research Service, Report RS20852*, 1–6, on 1, last modified January 22, 2003, https://wikileaks.org/wiki/CRS:_The_Partnership_for_a_New_Generation_of_Vehicles:_Status_and_Issues,_January_22,_2003.

13. National Research Council, *Review of the Research Program of the Partnership for a New Generation of Vehicles: Third Report* (Washington, DC: National Academies Press, 1997), 71–73; Robert M. Chapman, *The Machine That Could: PNGV, A Government-Industry Partnership* (Santa Monica, CA: RAND, 1998), 33–34.

14. Matthew L. Wald, "Government Dream Car: Washington and Detroit Pool Resources to Devise a New Approach to Technology," *New York Times*, September 30, 1993, A1.

15. Robert W. Crandall, "The Effects of US Trade Protection for Autos and Steel," *Brookings Papers on Economic Activity* 1 (1987): 271–288, on 274–275.

16. US Environmental Protection Agency (EPA), *Light-Duty Automotive Technology, Carbon Dioxide Emissions, and Fuel Economy Trends: 1975 Through 2016* (EPA-420-R-16–010, November 2016), 51.

17. David L. Levy and Sandra Rothenberg, "Heterogeneity and Change in Environmental Strategy: Technological and Political Responses to Climate Change in the Global Automobile Industry," in *Organizations, Policy, and the Natural Environment: Institutional and Strategic Perspectives*, eds. Andrew Hoffman and Marc Ventresca (Stanford, CA: Stanford University Press, 2002), 179–180; Paterson, *Automobile Politics*, 207–208.

18. National Highway Traffic Safety Administration (NHTSA), "Summary of CAFE Fines Collected," last modified July 24, 2014, file:///Users/mne/Downloads/cafe_fines-07-2014.pdf.

19. Hideshi Itazaki, *The Prius That Shook the World: How Toyota Developed the World's First Mass-Production Hybrid Vehicle* (Tokyo: Nikkan Kogyo Shimbun, 1999), 107, 153–154.

20. Itazaki, *The Prius That Shook the World*, 19–21.

21. Former CARB member Daniel Sperling suggested that the PNGV inspired the creation of the Toyota's advanced propulsion program; see Daniel Sperling, "Public-Private Technology R&D Partnerships: Lessons from US Partnership for a New Generation of Vehicles," *Transport Policy* 8, no. 4 (2001): 247–256, on 251.

22. Itazaki, *The Prius That Shook the World*, 71–73. So effective was Toyota at concealing the Prius that even Wouk was unaware of it. In an October 1997 article, Wouk held that no hybrid electric cars were near volume production; see Victor Wouk, "Hybrid Electric Vehicles," *Scientific American* 277, no. 4 (1997): 70.

23. Itazaki, *The Prius That Shook the World*, 115; Dave Lesher, "Midwest Governors Give Wilson's Campaign a Jolt over Electric Cars," *Los Angeles Times*, May 20, 1995.

24. Floyd A. Wyczalek, "Market Mature 1998 Hybrid Electric Vehicles," *IEEE AES Systems Magazine* 14, no. 3 (March 1999): 41–44, on 43.

25. Jerry Patchell, "Creating the Japanese Electric Vehicle Industry: The Challenges of Uncertainty and Cooperation," *Environment and Planning A: Economy and Space* 31, no. 6 (1999): 997–1016, on 998; see also Banri Asanuma, "Manufacturer-Supplier Relationships in Japan and the Concept of the Relation-Specific Skill," *Journal of the Japanese and International Economies* 3, no. 1 (1989): 1–30.

26. Itazaki, *The Prius That Shook the World*, 75–6, 83, 94, 102, 210–212.

27. Akihiro Taniguchi, Noriyuki Fujioka, Munehisa Ikoma, and Akira Ohta, "Development of Nickel/Metal-Hydride Batteries for EVs and HEVs," *Journal of Power Sources* 100, nos. 1–2 (2001): 117–124; Panasonic, "Panasonic Battery History," accessed June 29, 2020, https://www.panasonic.com/global/consumer/battery/about_us/hist; Panasonic, "PEV: Battery for Pure Electric Vehicles," accessed January 28, 2018, http://www.evnut.com/rav_battery_data_sheet.html.

28. Itazaki, *The Prius That Shook the World*, 263–265; Taniguchi et al., "Development of Nickel/Metal-Hydride Batteries," 119–121.

29. Shnayerson, *The Car That Could*, 138.

30. Thomas N. Young, "Civil Action 96–70919: Memorandum of Law in Support of Ovonic Battery Company Inc's Motion for Preliminary Injunction," March 3, 1996, box 32, untitled dustcover (Matsushita), 5-6, RCSP. Young represented the Troy, Michigan-based law firm Young & Basile.

31. Kent A. Jordan, "Civil Action No. 96–101: Matsushita Battery Industrial Co., Ltd., versus Energy Conversion Devices, Inc., and Ovonic Battery Company, Inc.," February 28, 1996, box 32, untitled dustcover (Matsushita), 8, RCSP. Jordan represented the Wilmington, Delaware-based law firm Morris, James, Hitchens & Williams.

32. In April 1992, Kawauchi wrote a short letter to Ovshinsky that included the following lines: "As you know, according to Buddhist philosophy, we all have a unique occasion to meet the right person at the right time. I regard our meeting in Japan as one of those occasions." Over the years, Ovshinsky constantly asked his secretary to show him this letter; Shosuke Kawauchi to Stanford Ovshinsky, June 12, 1992, box 8, Matsushita, 2007, SROP.

33. Itazaki, *The Prius That Shook the World*, 247, 264–265.

34. Satoshi Ogiso, "The Story Behind the Birth of the Prius, Part 2," last modified December 13, 2017, https://newsroom.toyota.co.jp/en/prius20th/challenge/birth/02/.

35. Toyota Motor Corporation, "Toyota RAV4 Electric Vehicle," last modified August 1999, https://media.toyota.co.uk/wp-content/uploads/sites/5/1324550080rav4_ev_whole.pdf.

36. Itazaki, *The Prius That Shook the World*, 115–116.

37. A battery's state of charge is defined as the level of charge relative to battery capacity, ranging from empty to full; see Nasser H. Kutkut, Herman L. N. Wiegman, Deepak M. Divan, and Donald W. Novotny, "Design Considerations for Charge Equalization of an Electric Vehicle Battery System," *IEEE Transactions on Industry Applications* 35, no. 1 (1999): 28–35.

38. Nickel-metal hydride cells are moderately tolerant of overdischarge because the reaction yields hydrogen that, if not produced too rapidly, may be reabsorbed by the anode. OBC promoted this quality of its metal hydride materials as a way of dispensing with costly pack management technology; see Robert C. Stempel, Stanford R. Ovshinsky, Paul R. Gifford, and Dennis A. Corrigan, "Nickel-Metal Hydride: Ready to Serve," *IEEE Spectrum* 35, no. 11 (1998): 29–34.

39. I thank the battery expert Jack Johnson, cofounder of Volta Power Systems, for these insights; see Jack Johnson, interview by author, January 26, 2017.

40. Itazaki, *The Prius That Shook the World*, 274–278.

41. A further complicating factor in battery pack management is that even cells from the same final assembly batch can develop different states of charge depending on their placement in the pack; see Victor Tikhonov, "Simple Analog BMS for the Tinkerer: Part 1," *Current Events* 44, no. 12 (2012): 1, 34.

42. Clinton and Gore, "High Technology Policy Initiatives."

43. Michael Parrish, "Electric Vehicle Firm Struggles to Go On," *Los Angeles Times*, March 21, 1995.

CHAPTER 6

1. Alex Taylor III, "The Star-Crossed Career of a Fallen GM CEO," *Fortune*, May 11, 2011, https://archive.fortune.com/2011/05/10/autos/gm_ceo_robert_stempel.fortune /index.htm; Clarke, *Trust and Power*, 127.

2. Shnayerson, *The Car That Could*, 233.

3. Robert C. Stempel, presentation to EVS-12 Symposium, "Challenge for Tomorrow: Forging the Road Ahead," December 7, 1994, 1–15, box 49, Business Admin/ Miscellaneous (2) 1993–1995, SROP; Shnayerson, *The Car That Could*, 250.

4. Robert C. Stempel to Stanford R. Ovshinsky, November 26, 1994, box 49, Business Admin/Miscellaneous (2) 1993–1995, SROP.

5. Donald W. Nauss, "Autos: GM Group Forms Unit to Make, Sell Parts for Electric Vehicles," *Los Angeles Times*, September 22, 1994, https://www.latimes.com/archives /la-xpm-1994-09-22-fi-41631-story.html.

6. Shnayerson, *The Car That Could*, 250; Frederick C. Ingram, "Delphi Automotive Systems Corporation," in *International Directory of Company Histories: Encyclopedia*

.com, January 24, 2022, https://www.encyclopedia.com/books/politics-and-business -magazines/delphi-automotive-systems-corporation.

7. These divergent assumptions are illustrated in a comparison of a draft and the final version of the press release announcing the formation of the GM-Ovonic management team. The draft, likely written by OBC personnel, quoted Adams as stating that GM-Ovonic's evaluation of the OBC battery was the "next logical step in the commercialization process . . . as you move from the laboratory to the marketplace." The draft also quoted Ovshinsky as stating that "opening the manufacturing facility" was necessary to achieve the company's cost and production goals. The official GM press release omitted any mention of Ovshinsky and included only a more moderate claim from Adams that the automaker was encouraged by its technical evaluation and GM-Ovonic was committed to moving the technology "from the laboratory to the marketplace." Stempel objected to the inclusion of the word "laboratory" and unsuccessfully lobbied GM to have it omitted from the press release; see, respectively, Ovonic Battery Company (OBC), "GM-Ovonic Forms Management Team, Readies Manufacturing Facility," August 30, 1994, and General Motors (GM), "GM-Ovonic Forms Management Team, Names Board of Managers," September 9, 1994, box 47, ECD/Corporate Partners/Joint Ventures, 1971–2004, SROP; Robert C. Stempel to Stanford R. Ovshinsky, September 9, 1994, box 47, ECD/Corporate Partners/Joint Ventures, 1971–2004, SROP.

8. General Motors (GM), "Manufacturing EV1 and S-10 Electric NiMH Batteries," December 3, 1998, box 31, Electric Vehicle Miscellaneous, 1–3, RCSP.

9. GM-Ovonic, "Slide 4 [GM-Ovonic organizational tree]," April 26, 1995, box 49, ECD Misc, 1963–2002, SROP.

10. Leon J. Krain to Stanford R. Ovshinsky, March 4, 1994, 1–3; OBC_PLN3.DOC, June 13, 1994, box 32, GM-Ovonic, USABC John Adams, 4, RCSP; Shnayerson, *The Car That Could*, 101, 204.

11. The initial business plan stipulated that GM would not fund OBC directly but would instead facilitate resources from "organizations such as USABC"; see OBC_ PLN3.DOC, June 13, 1994, RCSP; Shnayerson, *The Car That Could*, 233.

12. Shnayerson, *The Car That Could*, 234–235.

13. Shnayerson, *The Car That Could*, 233–241.

14. David A. Kirsch noted this "odd twist" in the negotiating stance of the automakers; see Kirsch, *The Electric Vehicle*, 207.

15. Shnayerson, *The Car That Could*, 249.

16. To get the cost in the range of around $300 per kilowatt-hour, GM-Ovonic projected that it needed to produce at least 2,200 cells per day and 800,000 annually, the equivalent of around 2,800 packs; GM-Ovonic, "Production Status" and "Cost Projections," April 26, 1995, 5, 7, 10, box 54, Ovshinsky/Career/ECD/Subsidiaries/ Ovonic Battery Company/GM-Ovonic, SROP.

17. California Air Resources Board (CARB), "1998 Zero-Emission Vehicle Biennial Program Review," July 6, 1998, i–iii, 4, 23.

18. Lawrence M. Fisher, "GM, in a First, Will Sell a Car Designed for Electric Power This Fall," *New York Times*, January 5, 1996, A10.

19. Young, "Civil Action 96–70919," 7, RCSP.

20. Donald W. Nauss, "GM Rolls Dice with Roll-Out of Electric Car," *Los Angeles Times*, December 5, 1996, https://www.latimes.com/archives/la-xpm-1996-12-05-mn -6000-story.html.

21. General Motors (GM), Form 8-K, "Harry J. Pearce," February 1, 2000, 10–11.

22. Robert C. Stempel, handwritten notes on meeting with John W. Adams, February 9, 1996, box 32, GM-Ovonic, USABC John Adams, RCSP.

23. Srinivasan Venkatesan, Michael Fetcenko, Benny Reichman, and Kuochih C. Hong, "Development of Ovonic Rechargeable Metal Hydride Batteries," *Proceedings of the 24th Intersociety Energy Conversion Engineering Conference* 3 (1989): 1659–1664; Shiuan Chang, Kwo-hsiung Young, Jean Nei, and Cristian Fierro, "Reviews on the US Patents Regarding Nickel/Metal Hydride Batteries," *Batteries* 2, no. 10 (2016): 1–29, on 3–4.

24. Robert C. Stempel, handwritten on meeting with John Adams, February 9, 1996, 1–2, box 32, GM-Ovonic, USABC John Adams, RCSP; Chang et al., "Reviews on the US Patents," 2–3.

25. Thomas N. Young, "Complaint and Jury Demand," February 29, 1996, 6–7, and Young, "Civil Action No. 96–70919," 6, box 32, untitled dustcover (Matsushita), RCSP.

26. Morton Amster to Chester T. Kamin, February 28, 1996, box 32, untitled dust-cover (Matsushita), RCSP. Amster represented the New York–based law firm Amster, Rothstein, and Ebenstein.

27. Stanford Ovshinsky to Shosuke Kawauchi, February 5, 1996, box 32, untitled dustcover (Matsushita) RCSP.

28. Chester Kamin to Morton Amster, February 26, 1996, box 32, untitled dustcover (Matsushita) RCSP.

29. Thomas N. Young and Carl H. von Ende, "Civil Action No. 96–70919: Toyota Motor Sales USA, Inc.'s Memorandum in Support of Motion to Stay This Proceeding Pending Decision of Delaware Court on Motion to Stay, Dismiss, or Transfer," March 12, 1996, 2, box 32, RCSP. Carl H. von Ende represented the Detroit-based law firm Miller, Canfield, Paddock, and Stone.

30. Kamin to Amster, February 26, 1996; Jordan, "Civil Action No. 96–101," 11–12.

31. Young, "Civil Action No. 96–70919," 11–12.

32. Jordan, "Civil Action No. 96–101," 5–12.

33. Young, "Civil Action No. 96–70919," 3–7.

34. A number of large manufacturing enterprises, mainly Japanese, had extensively patented nickel-metal hydride cell construction methods and components; see Chang et al., "Reviews on the US Patents," 7–9.

35. Subhash K. Dhar, Stanford R. Ovshinsky, Paul R. Gifford, Dennis A. Corrigan, Michael A. Fetcenko, and Srinivasan Venkatesan, "Nickel/Metal Hydride Technology for Consumer and Electric Vehicle Batteries: A Review and Update," *Journal of Power Sources* 65, nos. 1–2 (1997): 3.

36. Stanford Ovshinsky to Andrew Ng, December 20, 1996, 1–3, box 54, S. Ovshinsky/Career/ECD/Subsidiaries/Ovonic Battery Co./Correspondence-Unhappy Corporate Partners, December 1996, SROP.

37. Jordan, "Civil Action 96–101," 15; see also Young and von Ende, "Civil Action No. 96–70919: Toyota Motor Sales USA," 9.

38. Robert C. Stempel to Rich Piellisch, March 6, 1996, box 32, *Fleets and Fuels* inquiry, March 6, 1996, RCSP.

39. Because OBC's claim of infringement was based on tests of Matsushita consumer cells, not its large electric vehicle battery, Toyota argued that it had the right to take discovery and request documents, admissions, and depositions; see Young and von Ende, "Civil Action No. 96–70919: Toyota Motor Sales USA," 10.

40. Young, "Civil Action 96–70919," 15.

41. Stempel attempted to interest Ford chair and CEO Alex J. Trotman in OBC technology. Doubtless aware of Ford's troubles with its sodium-sulfur battery, Stempel reminded Trotman that as a member of the USABC, Ford had helped build up OBC and should consider reaping the benefits of its investment: "If you decide to enter the EV race," wrote Stempel, "we would like to be considered as your EV battery supplier;" see Robert C. Stempel to Alex J. Trotman, February 21, 1995, box 49, Business admin/miscellaneous (1) 1993–1995, 1–2, SROP.

42. Honda, "Honda EV Plus: The Dream of an Electric Vehicle," https://global .honda/heritage/episodes/1988evplus.html, accessed July 4, 2020.

43. According to OBC lawyer Thomas Young, Honda agreed to delay use of Matsushita batteries pending resolution of Matsushita's pre-suit discussions with OBC; see Young, "Civil Action No. 96–70919," 16.

44. Robert C. Stempel, handwritten notes, "H. J. Pierce [sic] Office, 9:30 AM 13 Aug. 1997," box 32, unmarked folder 2, RCSP.

45. Srinivasan Venkatesan, Matt van Kirk, Lynn Taylor, and Jim Strebe to Stanford R. Ovshinsky, Subhash K. Dhar, Michael A. Fetcenko, Dennis A. Corrigan, and Paul R. Gifford, "Subject: Pilot Cells for Honda Delivery in June 1996 (4 modules)," May 6, 1996, box 32, unorganized/misc 1, dustcover 1, RCSP.

46. David Sedgwick, "Battery Maker Faces High Cost, Low Demand," *Automotive News Europe*, June 23, 1997, http://europe.autonews.com/article/19970623/ANE

/706230848?template=printartANE; The journalist Marc Geller noted that the Honda EV Plus used the same Panasonic EV-95 modules as the Toyota RAV4 EV; see Geller, "Wired Blogger Takes on Nissan Leaf," *Plug and Cars,* January 25, 2010, https://plugsandcars.blogspot.com/2010_01_25_archive.html.

CHAPTER 7

1. Clinton and Gore, "Technology for America's Economic Growth," 34.

2. Henry Kelly and Robert H. Williams, "Fuel Cells and the Future of the US Automobile" (unpublished manuscript, December 7, 1992). I thank Dr. Kelly for providing me with a copy of this unpublished paper. A nearly identical paper was published by Robert H. Williams as "Fuel Cells, Their Fuels, and the US Automobile," in *Proceedings: First Annual World Car 2001 Conference, June 21–24, 1993* (California Institute of Technology, 1993), 73–75.

3. E-mail communication with the author, January 31, 2011. The transportation analyst Daniel Sperling believed, at least in 1995, that fuel cells had been the "impetus" for the creation of the PNGV; see Sperling, *Future Drive,* 84.

4. John Templeman, "Daimler's New Driver Won't Be Making Sharp Turns," *Bloomberg,* July 4, 1994, https://www.bloomberg.com/news/articles/1994-07-03/daimlers -new-driver-wont-be-making-sharp-turns. Edzard Reuter, appointed the chair of the Daimler-Benz board in 1987, continued the corporate policy of acquiring aerospace assets. Under the previous chair, Werner Breitschwerdt, Daimler-Benz purchased Dornier as well as the engineering firm MAN's 50 percent stake in MTU, their aero- engine joint venture, in 1985. In 1989, Daimler-Benz acquired the aircraft developer MBB and merged its aviation assets, along with the electrical equipment maker AEG, into Deutsche Aerospace AG.

5. Matthew L. Wald, "A Tough Sell for Electric Cars: Technology Lagging as Markets Emerge," *New York Times,* November 26, 1991, D1.

6. Sudworth, "Zebra Batteries," 109.

7. Wald, "A Tough Sell;" Daimler AG, "Electric Motors as an Alternative to Combustion Engines," last modified November 9, 2007, http://media.daimler.com/mars MediaSite/en/instance/ko/Electric-motors-as-an-alternative-to-combustion-engines .xhtml?oid=9274529.

8. A. J. Appleby, "Issues in Fuel Cell Commercialization," *Journal of Power Sources* 58, no. 2 (1996): 172.

9. John M. DeCicco, *Fuel Cell Vehicles: Technology, Market and Policy Issues* (Warrendale, PA: Society of Automotive Engineers, 2001), 72.

10. Oscar Suris, "Daimler-Benz Unveils Electric Vehicle, Claiming a Breakthrough on Fuel Cells," *Wall Street Journal,* April 14, 1994, B2.

11. Steven G. Chalk, Pandit G. Patil, and S. R. Venkateswaran, "The New Generation of Vehicles: Market Opportunities for Fuel Cells," *Journal of Power Sources* 61, nos. 1–2 (1996): 10; "Ford, Chrysler Win Auto Fuel-Cell Work," *Wall Street Journal*, July 13, 1994, B2.

12. Lawrence M. Fisher, "California Is Backing off Mandate for Electric Car: Board Finds Shortcomings in Technology," *New York Times*, December 26, 1995, A14.

13. Charles Stone and Anne E. Morrison, "From Curiosity to 'Power to Change the World®,'" *Solid-State Ionics* 152–153 (2002): 8; National Research Council, *Review of the Research Program of the Partnership for a New Generation of Vehicles: Second Report* (Washington, DC: National Academies Press, 1996), 53–54.

14. Nick Nuttall, "Breathtaking: The Vehicle Powered by Air," *The Times*, May 15, 1996, Home News 7.

15. Jason Mark, "Cleaning up Cars," *Washington Post*, August 14, 1996, HO2.

16. *Chicago Tribune*, "Ford Unplugs Electric Vans After Two Fires," June 6, 1994, https://www.chicagotribune.com/news/ct-xpm-1994-06-06-9406060018-story.html.

17. Matthew L. Wald, "Three Guesses: The Fuel of the Future Will be Gas, Gas, or Gas," *New York Times*, October 16, 1997, G16; Valerie Reitman, "Toyota to Sell Hybrid Gas-Electric Car: Auto Maker Cites High Efficiency, Low Emissions," *Wall Street Journal*, March 26, 1997, A 12:1.

18. Brandon Mitchener and Tamsin Carlisle, "Daimler, Ballard Team to Develop Fuel-Cell Engine," *Wall Street Journal*, April 15, 1997, B, 8:4.

19. Matthew L. Wald, "Ford Plans Zero-Emission Fuel Cell Car," *New York Times*, April 22, 1997, D2.

20. John H. Cushman Jr., "Intense Lobbying Against Global Warming Treaty," *New York Times*, December 7, 1997, Section 1, 28.

21. Alex J. Trotman, speech, National Press Club, Washington DC, October 27, 1997, https://www.c-span.org/video/?94065-1/business-environment.

22. Anthony Depalma, "Ford Joins in a Global Alliance to Develop Fuel-Cell Auto Engines," *New York Times*, December 16, 1997, D1; Valerie Reitman, "Ford Is Investing in Daimler-Ballard Fuel-Cell Venture," *Wall Street Journal*, December 16, 1997, B8.

23. Donald W. Nauss, "Ford Investing $420 Million for Fuel-Cell-Powered Auto," *Los Angeles Times*, December 16, 1997, https://www.latimes.com/archives/la-xpm-1997-dec-16-mn-64565-story.html.

24. When the *Fortune* journalist Stuart F. Brown visited Daimler's test facility on the outskirts of Nabern in March 1998, he was given a test-drive in the hydrogen-powered Necar II and implicitly acknowledged the potential for false impressions to arise from the company's involvement in dissimilar fuel systems; see Stuart F. Brown, "The Automakers' Big-Time Bet on Fuel Cells," *Fortune*, March 30, 1998, 122[D].

25. For example, *Times* technology correspondent Nick Nuttall suggested that fuel cells operated equally well on any hydrogen-rich fuel including liquid hydrogen, methanol, ethanol, and gasoline; see Nuttall, "Breathtaking," 7.

26. Christopher E. Borroni-Bird, "Fuel Cell Commercialization Issues for Light-Duty Vehicle Applications," *Journal of Power Sources* 61, nos. 1–2 (1996): 33–48, on 42.

27. National Research Council, *Third Report*, 65; Steven G. Chalk, James F. Miller, and Fred W. Wagner, "Challenges for Fuel Cells in Transport Applications," *Journal of Power Sources* 86, nos. 1–2 (2000): 44; Matthew L. Wald, "In a Step Toward a Better Electric Car, Company Uses Fuel Cell to Get Energy from Gasoline," *New York Times*, October 21, 1997, A14.

28. Jason Mark, "Clean Car's Wrong Turn," *New York Times*, October 26, 1997, WK14.

29. Wald, "Three Guesses," G16.

30. *Spark M. Matsunaga Hydrogen Research, Development, and Demonstration Program Act of 1990*, Public Law 101–566, *US Statutes at Large* 104 (1990): 2797–2801.

31. Warren E. Leary, "Use of Hydrogen as Fuel is Moving Closer to Reality," *New York Times*, April 16, 1995, S1, 15.

32. *Hydrogen Future Act of 1996*, Public Law 104–271, *US Statutes at Large* 110 (1996): 3306–3307.

CHAPTER 8

1. Valerie Reitman, "Toyota to Sell Hybrid Gas-Electric Car: Auto Maker Cites High Efficiency, Low Emissions," *Wall Street Journal*, March 26, 1997, A 12:1.

2. Energy Conversion Devices (ECD), "Potential Settlement Plan," January 28, 1997, box 50, Business Admin ECD/Notes (includes organizational chart) 2005–2006, 1–3, SROP.

3. OBC claimed that its second-generation battery pack (the GMO-2) had an energy density of 80 watt-hours per kilogram and hoped to put it into production by the end of 1997; Energy Conversion Devices (ECD), "Annual Meeting 1997, Slide 48: Family Chart," 58, box 96, ECD Annual Meetings, 1996–1998, SROP.

4. Robert C. Stempel to Robert C. Purcell, April 14, 1997, box 32, GMR Meeting 7/23 Baker, RCSP.

5. Robert C. Stempel to Harry J. Pearce, April 14, 1997, box 32, GMR Meeting 7/23 Baker, RCSP.

6. Michael Saft, Guy Chagnon, Thierry Faugeras, Guy Sarre, and Pierre Morhet, "Saft Lithium-Ion Energy and Power Storage Technology," *Journal of Power Sources* 80, nos. 1–2 (1999): 185.

7. National Research Council, *Review of the Research Program of the Partnership for a New Generation of Vehicles: Third Report* (Washington, DC: National Academies Press, 1997), 72.

8. Subhash K. Dhar to Stanford Ovshinsky, "Subject: Hybrid Electric Vehicle," March 21, 1997, 1–3, box 32, HEV Batteries Ovonic, RCSP; ECD, "Slide 48: Family Chart."

9. Dennis Corrigan to Stanford Ovshinsky and Robert Stempel, "Subject: HEV Presentations to GM," March 21, 1997, 1–3, box 32, HEV Batteries Ovonic, RCSP; Paul Gifford to Stanford Ovshinsky, "Subject: Meeting with GM Hybrid Vehicle Team," March 21, 1997, 1–2, on 2, box 32, HEV Batteries Ovonic, RCSP.

10. Dennis Corrigan to Stanford Ovshinsky, Robert Stempel, and Subhash Dhar, "Subject: Response from USABC Regarding HEV Funding," May 7, 1997, box 32, unorganized/miscellaneous 1, dustcover 3, USABC, RCSP; National Research Council, *Third Report*, 126–127; US Department of Energy (DOE), *PNGV Battery Test Manual: DOE/ID-10597, Rev. 3* (Idaho National Engineering and Environmental Laboratory, 2001), D-1.

11. Phil Gow to Stanford Ovshinsky, Robert Stempel, and Subhash Dhar, "Subject: AeroVironment Testing of OBC Prototype Hybrid Battery Module," April 4, 1997, box 32, unorganized/miscellaneous 1, dustcover 3, RCSP.

12. Larry Oswald to Paul Gifford and Dennis Corrigan, "Re: High Power-to-Energy Ratio Ovonic Batteries," April 17, 1997, box 32, unorganized/miscellaneous 1, dustcover 3, RCSP.

13. Harold Haskins to Dennis Corrigan, May 6, 1997, 1–2, box 32, unorganized/miscellaneous 1, dustcover 3, USABC, RCSP.

14. Corrigan to Ovshinsky, Stempel, and Dhar, "Subject: Response from USABC Regarding HEV Funding."

15. National Research Council, *Third Report*, 69–75.

16. Sedgwick, "Battery Maker Faces High Cost."

17. Sedgwick, "Battery Maker Faces High Cost."

18. Robert C. Stempel, handwritten notes, "H. J. Pierce [sic] Office, 9:30 AM 13 Aug. 1997," "GM Ovonic Battery Manufacturing Operations Current Status," "Future Plant Site," "Ovonic Battery Company as an Investment Opportunity," box 32, unmarked folder 2, RCSP.

19. Subhash Dhar to Stanford Ovshinsky, "EV1 Test Results with Ovonic Batteries," April 17, 1997, box 32, unorganized/miscellaneous 1, dustcover 2, RCSP.

20. Energy Conversion Devices (ECD), "Annual Meeting 1997, Slide 51: Pike's Peak Truck," 61.

21. Robert C. Stempel to Kenneth R. Baker, April 23, 1997, 1–5, box 32, GMR Meeting 7/23 Baker, RCSP.

22. Stanford Ovshinsky to Robert C. Stempel, "Subject: ECD as Resource to General Motors," May 15, 1997, box 32, unorganized/miscellaneous 1, dustcover 3, RCSP.

23. Robert C. Stempel, "Trio Report: Summit of 8 Meetings, Denver, Colorado, June 20, 1997," 2–3, June 30, 1997, box 32, dustcover 3, unorganized/miscellaneous 1, RCSP; Stempel, handwritten notes, "H. J. Pierce [sic] Office, 9:30 AM 13 Aug. 1997," "Global Warming/Climate Change Issue," box 32, unmarked folder 2, RCSP.

24. Stempel, handwritten notes, "H. J. Pierce [sic] Office, 9:30 AM 13 Aug. 1997," "Global Warming/Climate Change Issue."

25. Robert C. Stempel to Kenneth R. Baker, September 15, 1997, 1–2, box 32, GM-Ovonic Board Meeting 09.23.97, Plant 7, RCSP.

26. Michael J. Riezenman, "EV Candidate for Mass Production Does Boston-to–New York Run on One Charge," *IEEE Spectrum* (December 1997): 68–70; Energy Conversion Devices (ECD), "Annual Meeting 1997, Slide 52: Work Truck; Slide 55: Boston-New York-Sunrise," 62, 65.

27. For example, the actor and electric car enthusiast Alan Alda referred to Ovshinsky as a "brain and a moral force to be cherished;" box 50, Business Administration; ECD/Notes (includes organizational chart) 2005–2006, SROP.

28. Ogiso, "The Story Behind the Birth of the Prius," Itazaki, *The Prius That Shook the World*, 354, 361–369.

29. In the preface to his classic 1982 study of Japanese industrial policy, Chalmers A. Johnson wrote that Western analysts tended either to condemn the Japanese state as "overweening" or dismiss it as "merely supportive;" see Chalmers Johnson, *MITI and the Japanese Miracle: The Growth of Industrial Policy, 1925–1975* (Stanford, CA: Stanford University Press, 1982), vii; on US interpretations of and reactions to MITI, see Andrew Pollack, "America's Answer to MITI," *New York Times*, March 5, 1989, Section 3,1, 8.

30. James P. Womack, Daniel T. Jones, and Daniel Roos, *The Machine That Changed the World: How Lean Production Revolutionized the Global Car Wars* (London: Simon and Schuster, 2007), 50; Max Åhman, "Government Policy and the Development of Electric Vehicles in Japan," *Energy Policy* 34, no. 4 (2006): 440–442.

31. Itazaki, *The Prius That Shook the World*, 381.

32. Ogiso, "The Story Behind the Birth of the Prius."

33. Itazaki, *The Prius That Shook the World*, 115, 270; United Nations Framework Convention on Climate Change (UNFCCC), "Report of the Conference of the Parties on Its Second Session, Held at Geneva from 8 to 19 July 1996," FCCC/CP/1996/15/Add.1, October 29, 1996, 2018, https://unfccc.int/documents?search2=&search3=%22Report+of+the+Conference+of+the+Parties+on+its+Second+Session%22, accessed November 15, 2018.

34. Energy Conversion Devices (ECD), "News Release: MBI Litigation Concluded in Favor of ECD," January 5, 1998, box 32, Black Binder (ECD/Ovonic/NiMH Update), RCSP; Energy Conversion Devices (ECD), "Annual Meeting 1997, Slide 61: MBI," 71.

CHAPTER 9

1. Jim Motavalli, *Forward Drive: The Race to Build "Clean" Cars for the Future* (New York: Earthscan, 2001), 50–53; Itazaki, *The Prius That Shook the World*, 370–372.

2. Ralph Kisiel, "Chrysler Designs a Mild Hybrid: Small Battery Only Boosts the Diesel, Costs Just $15,000 More," *Automotive News Europe*, February 2, 1998.

3. Jack Morton Company, "Creative Summary," in HKO Media, "Highlights of GM Advanced Technology Vehicles Press Conference, North American International Auto Show," January 4, 1998, box 96, Interviews (SRO and Stempel), GM Press Conference, and Texaco-Ovonic Systems PR, 1998–2002, SROP.

4. Jack Morton Company, "Creative Summary," 12–13.

5. Boschert, *Plug-In Hybrids*, 67–75; "World News This Week," *AutoWeek* 348, no. 2 (1998): 2–3.

6. Jack Morton Company, "Creative Summary," 19–20.

7. Sharon Beder, *Global Spin: The Corporate Assault on Environmentalism* (Devon, UK: Green Books, 2002), 238.

8. The expression "merchants of doubt" is derived from the title of the 2010 book of the same name by Naomi Oreskes and Erik M. Conway. Oreskes and Conway are among a group of scholars who observed the tendency of corporations to deconstruct scientific knowledge considered problematic to their interests. The shifting views of car companies on climate change in turn problematizes the assumption that industry necessarily viewed environmental science as a social construction; see Naomi Oreskes and Erik M. Conway, *Merchants of Doubt: How a Handful of Scientists Obscured the Truth on Issues from Tobacco Smoke to Global Warming* (New York: Bloomsbury, 2010); and Naomi Oreskes, Erik M. Conway, and Matthew Shindell, "From Chicken Little to Dr. Pangloss: William Nierenberg, Global Warming, and the Social Deconstruction of Scientific Knowledge," *Historical Studies in the Natural Sciences* 38, no. 1 (2008): 109–152.

9. California Air Resources Board (CARB), *2000 Zero Emission Vehicle Program Biennial Review: Executive Summary to the Staff Report*, August 7, 2000, 5; Deborah Salon, Daniel Sperling, and David Friedman, *California's Partial ZEV Credits and LEV II Program: UCTC No. 470* (Berkeley: University of California Transportation Center, 2001), 4; California Air Resources Board (CARB), "1998: LEV II and ZEV," in "California's Zero Emission Vehicle Program," June 2009.

10. Jack Morton Company, "Creative Summary," 11–12.

11. "Detroit Turns a Corner," *New York Times*, January 11, 1998, Section 4, 18.

12. Energy Conversion Devices (ECD), "GM Announcement of HEV and Fuel Cell Vehicles, January 4, 1998," in "ECD/Ovonic NiMH Battery Update," box 32, Black Binder, RCSP.

13. Energy Conversion Devices (ECD), "GM-Ovonic Manufacturing: Next Steps," in "ECD/Ovonic NiMH Battery Update."

14. General Motors (GM) Advanced Technology Vehicles, "Manufacturing EV1 and S-10 Electric NiMH Batteries."

15. Stanford R. Ovshinsky to Robert C. Stempel, "Subject: GMO 3 Status," July 24, 1998, 1–4, box 50, Business Admin ECD/notes (includes org chart) 2005–2006, SROP.

16. Patchell, "Creating the Japanese Electric Vehicle Industry," 1004.

17. Honda Motor Company, "EV Plus: The Dream of an Electric Vehicle/1988," https://global.honda/heritage/episodes/1988evplus.html, accessed July 4, 2020; Mark Rechtin, "Honda Pulls the Plug on EV Plus," *Automotive News*, April 26, 1999, https://www.autonews.com/article/19990426/ANA/904260758/honda-pulls-the-plug-on-ev-plus.

18. Honda Motor Company, "Honda Electric Vehicle Program Enters Next Phase," April 26, 1999, 1–2, box 7, Hitachi Maxell 1997/98, SROP.

19. Paul Gifford, John Adams, Dennis Corrigan, and Srinivasan Venkatesan, "Development of Advanced Nickel/Metal Hydride Batteries for Electric and Hybrid Vehicles," *Journal of Power Sources* 80, nos. 1–2 (1999): 162.

20. Methanol burns with a low-visibility blue flame, so for safety reasons, a luminosity agent had to be added before the fuel could be sold. This placed additional stress on the electrocatalytic reformer of the fuel cell system, shortening its lifetime; see Richard K. Stobart, "Fuel Cell Power for Passenger Cars: What Barriers Remain?" in *Fuel Cell Technology for Vehicles*, ed. Richard Stobart (Warrendale, PA: Society of Automotive Engineers, 2001), 14; see also Doyle, *Taken for a Ride*, 427.

21. Sean Casten, Peter Teagan, and Richard Stobart, "Fuels for Fuel Cell-Powered Vehicles," in *Fuel Cell Technology for Vehicles*, ed. Richard Stobart (Warrendale, PA: Society of Automotive Engineers, 2001), 61–62; National Research Council, *Review of the Research Program of the Partnership for a New Generation of Vehicles, Sixth Report* (Washington, DC: National Academies Press, 2000), 85–87.

22. Brown, "The Automakers' Big-Time Bet on Fuel Cells."

23. Jeffrey Ball, "DaimlerChrysler Unveils Prototype Car Using Fuel Cell, Seeks Sales in 5 Years," *Wall Street Journal*, March 18, 1999, B2.

24. Doyle, *Taken for a Ride*, 426.

25. Andrew Pollack, "Cars and the Environment: Where to Put the Golf Clubs? Right Next to the Hydrogen!" *New York Times*, May 19, 1999, G20.

26. National Academy of Sciences, *Review of the Research Program, Sixth Report*, 65–66.

27. Gregory L. White, "GM Stops Making Electric Car, Holds Talks with Toyota," *Wall Street Journal*, January 12, 2000, A14.

28. National Academy of Sciences, *Review of the Research Program, Sixth Report*, 2–3, 31.

29. "Science and Technology: Hybrid Vigour"? *The Economist* 354, no. 8155 (January 29, 2000): 94–95.

30. Stanford R. Ovshinsky and Robert C. Stempel to *The Economist*, undated, box 50, Business Administration ECD/notes (includes org chart), 2005–2006, SROP.

31. "Ovonic NiMH Batteries Featured in GM Advanced Technology Vehicle Introduced at North American Auto Show," *PR Newswire*, January 13, 2000, 1.

32. Hoddeson and Garrett, *The Man Who Saw Tomorrow*, 199–200; Texaco, "Texaco Response to Proposed US National Energy Policy," May 17, 2001, and Energy Conversion Devices (ECD), Ovonic Battery Company (OBC), and Texaco, "Waiver (Re: Chevron Merger)," July 17, 2001, box 25, Chevron Dinner Meeting, San Francisco, Tuesday, May 29, 2001, RCSP.

33. "Texaco to Acquire General Motors' Share of GM-Ovonic Battery Joint Venture," *Business Wire*, October 10, 2000, 1.

34. Andrew Ross Sorkin and Neela Banerjee, "Chevron Agrees to Buy Texaco for Stock Valued at $36 Billion: Deal Creates World's 4th-Largest Oil Company," *New York Times*, October 16, 2000, A1.

35. California Air Resources Board (CARB), *2000 Zero Emission Vehicle Program*, 4; Brad Heavner, *Pollution Politics 2000: California Political Expenditures of the Automobile and Oil Industries, 1997–2000* (Santa Barbara, CA: California Public Interest Research Group Charitable Trust, 2000), 7–8.

36. John O'Dell, "Car Companies Team up to Fight State's ZEV Rule," *Los Angeles Times*, January 23, 2002, https://www.latimes.com/archives/la-xpm-2002-jan-23-hy-green23-story.html.

37. Gary Polakovic and John O'Dell, "Injunction Holds Up ZEV Program," *Los Angeles Times*, June 15, 2002, https://www.latimes.com/archives/la-xpm-2002-jun-15-me-emisions15-story.html.

38. Jeffrey Ball, "Fuel-Cell Makers Get a Big Boost from Bush's Auto Subsidy Plan," *Wall Street Journal*, January 10, 2002, B2.

39. Jann S. Wenner and Will Dana, "Al Gore: The Rolling Stone Interview," *Rolling Stone*, November 9, 2000, https://www.rollingstone.com/feature/al-gore-the-rolling-stone-interview-62074/. Gore also claimed that the PNGV had helped stimulate the current "massive cutthroat competition" in fuel cell automobility among the major manufacturers.

40. *New York Times*, "Spencer Abraham's Dream Car," January 14, 2002, A14.

41. Jeffrey Ball, "Evasive Maneuvers: Detroit Again Tries to Dodge Pressures for a 'Greener' Fleet," *Wall Street Journal*, January 28, 2002, A1.

42. Peter Pae, "GM Seen to Drop Suits on Zero-Emission-Vehicle Mandate," *Los Angeles Times*, August 12, 2003, https://www.latimes.com/archives/la-xpm-2003-aug -12-fi-zero12-story.html; Danny Hakim, "Automakers Drop Suits on Air Rules," *New York Times*, August 12, 2003, A1.

43. EVAmerica and the US Department of Energy (DOE), "1998 Ford Ranger EV" and "1999 Ford Ranger EV."

44. "Pivco Bankruptcy Takes Car Off Fast Track," *Plastics News*, November 9, 1998, https://www.plasticsnews.com/article/19981109/NEWS/311099998/pivco-bankruptcy -takes-car-off-fast-track; Ford Motor Company, "Th!nk City Electric Vehicle Demonstration Program: Final Project Report, June 2005," June 18, 2004, Award DE-FG26-O1ID14048, 2.

45. Peter Horton, "Peter Buys an Electric Car," *Los Angeles Times*, June 8, 2003, https://www.latimes.com/archives/la-xpm-2003-jun-08-tm-ev123-story.html; Owen Edwards, "The Death of the EV1: Fans of a Battery-Powered Emissions Free Sedan Mourn Its Passing," *Smithsonian Magazine*, June 2006, https://www.smithsonianmag .com/science-nature/the-death-of-the-ev-1-118595941/.

46. Greenpeace, "Th!nk Again: Ford Does a U-Turn," September 17, 2004, https:// web.archive.org/web/20060609043839/http://www.greenpeace.org/international /news/th-nk-again-ford-does-a-u-tur#; Ford, "Th!nk City Electric Vehicle;" Matthew Phenix, "A Recharged Th!nk Contemplates Its Comeback," *Wired*, September 8, 2007, https://www.wired.com/2007/09/post-ford-a-rec/; Doug Demuro, "The RAV4 EV Has Had Two Obscure Generations," *Autotrader*, May 17, 2019, https://www .autotrader.com/car-news/toyota-RAV4EV-has-had-two-obscure-generations-2814 74979930553.

47. According to Hoddeson and Garrett, Ovshinsky held that in the 1970s, Japanese business supported ECD at a time when his science claims were under attack in the US: see Hoddeson and Garrett, *The Man Who Saw Tomorrow*, 162.

48. Energy Conversion Devices (ECD), "News Release: ECD Announces Ovonic Battery Is Filing a Lawsuit Against Matsushita Battery," March 6, 2001, 1–2, box 50, Business Administration, ECD/Press Releases (includes announcement of Stan leaving ECD) 1993–2007, SROP.

49. Energy Conversion Devices (ECD), Form 8-K, July 7, 2004.

50. Taniguchi et al., "Development of Nickel/Metal-Hydride Batteries," 121.

51. Chester T. Kamin to Robert C. Stempel and Stanford R. Ovshinsky, "Re: ChevronTexaco," October 20, 2004, box 47, Various Important Letters, 1992–2007 (includes Board of Directors' decision to remove Stan), SROP.

52. Dave Strand to Stanford R. Ovshinsky, "Subject: Discussion with Takeo Ohta," April 22, 2001, box 47, ECD/Corporate Partners/Relations with Japan, SROP.

CHAPTER 10

1. Bob Pool, "Drivers Find Outlet for Grief Over EV1s," *Los Angeles Times*, July 25, 2003, https://www.latimes.com/archives/la-xpm-2003-jul-25-me-funeral25-story.html.

2. Paris Productions,"EV1 Funeral: Hollywood Forever Cemetery, Thursday, July 24, 2003," accessed December 29, 2021, https://www.youtube.com/watch?v=HZHka8KUj 74andt=302s and https://www.youtube.com/watch?v=zFsOxZPR-eQ.

3. Boschert, *Plug-in Hybrids*.

4. At one board meeting, Eberhard became annoyed when regulators repeatedly rejected the appeals of a BMW executive to be allowed to use hydrogen in an ICE. The automaker, they insisted, had to use hydrogen fuel cells. When Eberhard's turn came to speak, he recalled that he said something to the effect of: "'We all know the hydrogen fuel cell is not the future. But I am delighted that you are pushing the auto manufacturers to keep making fuel cells and I want you to do this as long as possible. And I want you to do this because it keeps people who would otherwise be my competitors off my back. They are all wasting their time doing stupid technology and I can own the electric car market;'" see Martin Eberhard, interview by author, July 25, 2016.

5. Eberhard, interview.

6. For classic accounts of the historical development of Silicon Valley as a manufacturing district, see AnnaLee Saxenian, *Regional Advantage: Culture and Competition in Silicon Valley and Route 128* (Cambridge, MA: Harvard University Press, 1994); Ross Knox Bassett, *To the Digital Age: Research Labs, Start-up Companies, and the Rise of MOS Technology* (Baltimore: Johns Hopkins University Press, 2002); and Christophe Lécuyer, *Making Silicon Valley: Innovation and the Growth of High Tech, 1930–1970* (Cambridge, MA: MIT Press, 2005). For analyses of Silicon Valley that address vertical disintegration and deindustrialization, see Clair Brown, Greg Linden, and Jeffrey T. Macher, "Offshoring in the Semiconductor Industry: A Historical Perspective," in *Brookings Trade Forum: Offshoring White-Collar Work* (Washington, DC: Brookings Institution Press, 2005), 279–333; Brown and Linden, *Chips and Change*.

7. Joan Magretta, "The Power of Virtual Integration: An Interview with Dell Computer's Michael Dell," *Harvard Business Review* (March-April 1998): 72–84.

8. Eberhard, interview; see also Vance, *Elon Musk*, 152–153.

9. Gordon E. Moore, "Cramming More Components onto Integrated Circuits," *Electronics* 38, no. 8 (1965): 114–117. For literature that conceives Moore's Law as a metonym for information technology innovation, see William Aspray, ed., *Chasing Moore's Law: Information Technology Policy in the United States* (Raleigh, NC: SciTech Publishing, 2004); Dale W. Jorgenson and Charles W. Wessner, "Preface," in *Productivity and Cyclicality in Semiconductors: Trends, Implications, and Questions; Report of a Symposium*, eds. Dale W. Jorgenson and Charles W. Wessner (Washington, DC: National Academies Press, 2004), xiii–xviii. Cyrus Mody noted that there are several

definitions of Moore's Law but that the expression generally refers to the doubling of transistors on a single silicon wafer in a period of time that in 1965 averaged around 12 months and by 1975 was 24 months; see Mody, *The Long Arm of Moore's Law*, 7–9.

10. See Mody, *The Long Arm of Moore's Law*, 80–81; David C. Brock and Christophe Lécuyer, "Digital Foundations: The Making of Silicon-Gate Manufacturing Technology," *Technology and Culture* 53, no. 3 (2012): 564; Gordon E. Moore, "Progress in Digital Integrated Electronics," *Technical Digest*, IEEE International Electron Devices Meeting 21 (1975): 11–13.

11. Moore, "Cramming More Components."

12. See, for example, Gordon E. Moore, "The Cost Structure of the Semiconductor Industry and Its Implications for Consumer Electronics," *IEEE Transactions on Consumer Electronics* CE-23, no. 1 (1977): xvi; Gordon E. Moore, "Are We Really Ready for VLSI^2?" *Digest of Technical Papers, IEEE International Solid-State Circuits Conference* (1979): 54–55; and Gordon E. Moore, "VLSI: Some Fundamental Challenges," *IEEE Spectrum* 16, no. 4 (1979): 30.

13. See, for example, Chris Gaither and Dawn C. Chmielewski, "Fears of Dot-Com Crash, Version 2.0," *Los Angeles Times*, July 16, 2006, https://www.latimes.com/archives/la-xpm-2006-jul-16-fi-overheat16-story.html.

14. There was a precedent for capital's shift into transportation in the crisis of textile manufacturing near the end of the first phase of the British industrial revolution in the 1820s. Eric Hobsbawm argued that by the early nineteenth century, it no longer make sense for textile merchants to invest in the textile sector because it was built to capacity, and simple water-powered spinning machine technology could not absorb much capital in any case. Railway systems had all the requirements for large-scale capital investment, and the resulting railway boom from the 1830s underpinned a second phase of industrialization based on coal and steel; Eric Hobsbawm, *Industry and Empire: From 1750 to the Present Day* (London: Pelican, 1968).

15. One problem was how to scale the production of cathode material. The existing process yielded fine and highly reactive lithium-cobalt oxide particles that were easily ignited in the event of a short circuit or external damage to the cell, so Sony had to invent a process to coarsen and enlarge the granules; see Yoshio Nishi, "The Development of Lithium Ion Secondary Batteries," *The Chemical Record* 1 (2001): 409.

16. See Eisler, "Exploding the Black Box;" Brodd, *Factors Affecting US Production Decisions*.

17. Eberhard, interview.

18. Donald MacArthur, George Blomgren, and Robert A. Powers, *Lithium and Lithium Ion Batteries, 2000: A Review and Analysis of Technical, Market, and Commercial Developments* (Westlake, OH: Robert A. Powers Associates, 2000), 17–18.

19. Michael G. Pecht, "Editorial: Re-Thinking Reliability," *IEEE Transactions on Components and Packaging Technologies* 29, no. 4 (2006): 893–894; Robert X. Cringely,

"Safety Last," *New York Times*, September 1, 2006, https://www.nytimes.com/2006/09/01/opinion/01cringely.html?p.

20. Nishi, "The Development of Lithium Ion Secondary Batteries," 411–412.

21. I thank the chemists Mark DeMeuse and Michael Jaffe for these insights. DeMeuse and Jaffe were involved in polymer membrane development at Celanese in the 1980s, and in the 2000s they pursued this work at Celgard, a major producer of battery separators, and at the New Jersey Institute of Technology, respectively; Mark DeMeuse, interview by author, October 10, 2016; Michael Jaffe, interview by author, January 24, 2017.

22. Michael Kanellos, "Can Anything Tame the Battery Flames?" *C/NET*, August 16, 2006, http://news.cnet.com/Can-anything-tame-the-battery-flames/2100-11398_3-6105924.html.

23. Goodenough, interview.

24. Cocconi, interview.

25. Rippel, interview.

26. Cocconi, interview.

27. Cocconi, interview; AC Propulsion, "About Us," accessed November 8, 2018, https://www.acpropulsion.com/index.php/about-us/management.

28. Brooks, interview.

29. Cocconi, interview.

30. Chris Dixon, "Lots of Zoom, with Batteries," *New York Times*, September 19, 2003, F1.

31. Brooks, interview; Eberhard, interview.

32. Contemporary analyses often trace these efforts to the Public Utility Regulatory Policies Act (PURPA). Passed by Congress in 1978, PURPA was designed to foster competition and energy diversity by requiring utilities to purchase power from independent entities if they could produce it more cheaply than the utilities. The legislation encouraged the use of conservation technology and renewable energy resources, as well as the decentralization of energy conversion; see, for example, Paul L. Joskow, "Markets for Power in the United States: An Interim Assessment," *Energy Journal* 27, no. 1 (2006): 1; Ghazal Razeghi, Brendan Shaffer, and Scott Samuelsen, "Impact of Electricity Deregulation in the State of California," *Energy Policy* 103 (2017): 106; Yohanna M. L. Gultom, "Governance Structures and Efficiency in the US Electricity Sector after the Market Restructuring and Deregulation," *Energy Policy* 129 (2019): 1008–1019, on 1009. See *Public Utility Regulatory Policies Act of 1978*, Public Law 95–617, *US Statutes at Large 92 (1978): 3117*–3173.

33. Alfred E. Kahn, "Surprises of Airline Deregulation," *American Economic Review* 78, no. 2 (1988): 316–322.

34. In the list of twenty-one self- and jointly authored publications that Kahn cited in the bibliography of *Whom the Gods Would Destroy, or How Not to Deregulate* (Washington, DC: AEI Press, 2001), only two focused expressly on utility deregulation.

35. David W. Wise, "The Tides of Deregulation," *Public Utilities Fortnightly*, 124, no. 5 (1989): 39–40.

36. William Sweet and Elizabeth A. Bretz, "How to Make Deregulation Work: Alfred E. Kahn, the Father of Airline Deregulation, Firmly Defends It in an Interview with IEEE Spectrum but Is Less Sanguine About the Effect on Electricity and Communications," *IEEE Spectrum* 39, no. 1 (2002): 51–56, on 51.

37. Benjamin Ross, "California's Regulation Debacle," *Dissent* (Spring 2001): 45–47. Policy analyses of electricity do not always address the technological implications of unbundling. For example, University of Arizona researchers Elizabeth Baldwin, Valerie Rountree, and Janet Jock acknowledged that electricity must balance supply and demand to function properly, but they also suggested that the three core electricity services of generation, transmission, and distribution were unproblematically separable. This perspective informed their larger thesis that the sociotechnical problems of distributed generation could be solved with distributed governance, which ostensibly empowered consumers through demand-side management; see Baldwin, Rountree, and Jock, "Distributed Resources and Distributed Governance: Stakeholder Participation in Demand Side Management Governance," *Energy Research and Social Science* 39 (2018): 39.

38. Ross, "California's Regulation Debacle."

39. I credit this insight into the origins of the expression "ancillary services" to David Hawkins, an engineer who served as a chief aide to Kellan L. Fluckiger, vice president of operations for CAISO in the 1990s and early 2000s; Hawkins, interview by author, November 10, 2020.

40. Richard Farmer, Dennis Zimmerman, and Gail Cohen, *Causes and Lessons of the California Electricity Crisis* (Congressional Budget Office, September 2001); Timothy Egan, "Tapes Show Enron Arranged Plant Shutdown," *New York Times*, February 4, 2005, A12.

41. See Kempton and Letendre, "Electric Vehicles as a New Power Source," 159–160; Steven E. Letendre and Willett Kempton, "The V2G Concept: A New Model for Power?" *Public Utilities Fortnightly* 140, no. 4 (2002): 16–26; Steven E. Letendre, Paul Denholm, and Peter Lilienthal, "New Load, or New Resource?" *Public Utilities Fortnightly* 144, no. 12 (2006): 28–33.

42. Cocconi, interview. According to Cocconi, the integrated charging system he developed for Impact also had bidirectional capability, although it was not fully implemented. Kempton claimed Cocconi developed the bidirectional feature for ACP to enable the quick discharge of an electric vehicle battery into the grid

rather than depleting the battery through automobile use, a process that could take many hours; Kempton, email communication with the author, October 13, 2020.

43. Kempton, email communication.

44. Willett Kempton, Jasna Tomić, Steven Letendre, Alec Brooks, and Timothy Lipman, "Vehicle-to-Grid Power: Battery, Hybrid, and Fuel Cell Vehicles as Resources for Distributed Electric Power in California," UC Davis Institute for Transportation Studies, ECD-ITS-RR-01–03, June 2001, xiv.

45. Kempton, email communication.

46. Kempton et al., "Vehicle-to-Grid," 56.

47. Alec N. Brooks, *Final Report: Vehicle-to-Grid Demonstration Project: Grid Regulation Ancillary Service with a Battery Electric Vehicle: Prepared for the California Air Resources Board and the California Environmental Protection Agency*, Contract number 01–313, December 10, 2002 (San Dimas, CA: AC Propulsion), 11–12.

48. Brooks, "Final Report," 1–3, 8–10.

49. Kempton et al., "Vehicle-to-Grid," v.

50. Brooks, "Final Report," 49.

51. Eberhard, interview.

52. Eberhard, interview.

53. Eberhard, interview.

54. Dennis A. Corrigan to Jim Metzger, "Subject: OBC Licensee Royalty Opportunities," December 10, 2003, 2, 8, box 54, Ovonic Battery Company Advanced Development, 2002–2005, SROP.

55. Eberhard, interview.

56. Forbes Bagatelle-Black, "AC Propulsion, The Quiet Revolutionaries: Tom Gage Talks About the Role His Company Has Played in the Rebirth of the Modern Electric Car," EV World, October 27, 2009, http://evworld.com/article.cfm?storyid=1772; Alan Cocconi, "Electric Car tzero 0-60 3.6 Sec Faster than Tesla Roadster," accessed May 7, 2019, https://www.youtube.com/watch?v=gb9E222QsM0andt=138s.

57. This seems to have been a reference to Eberhart's struggles to tame lithium chemistry. In the plot of *Dark Star*, a spaceship preparing the galaxy for human colonization drops sentient bombs on unstable planets until a glitch prevents a bomb from detaching from the craft. This prompts Lieutenant Doolittle to engage in a philosophical debate with the bomb in an ultimately unsuccessful attempt to convince it not to explode.

58. Eberhard, interview; Martin Eberhard, "Lotus Position," July 25, 2006, https://www.tesla.com/en_GB/blog/lotus-position?redirect=no.

59. Vance, *Elon Musk*, 148–158.

60. Rippel, interview.

61. Eberhard, interview; Vance, *Elon Musk*, 166–167.

62. Vance, *Elon Musk*, 160.

63. Phenix, "A Recharged Th!nk;" Green Car Congress, "Tesla Battery Supply Deal for Th!nk Scuttled," November 2, 2017, accessed March 12, 2019, https://www.greencarcongress.com/2007/11/tesla-battery-s.html.

64. Matthew L. Wald, "Zero to 60 in 4 Seconds, Totally from Revving Batteries," *New York Times*, July 19, 2006, C3.

65. Elon Musk, "The Secret Tesla Motors Master Plan (Just Between You and Me)," August 2, 2006; https://www.tesla.com/blog/secret-tesla-motors-master-plan-just-between-you-and-me.

66. Eberhard, "Lotus Position"; Vance, *Elon Musk*, 165, 179.

67. David R. Baker, "Elon Musk: Tesla Was Founded on 2 False Ideas, and Survived Anyway," *SFGate*, May 31, 2016, https://www.sfgate.com/business/article/Elon-Musk-Tesla-was-founded-on-2-false-ideas-7955528.php.

68. Cocconi, interview.

CHAPTER 11

1. John O'Dell, "GM Turns to a Top Guru in Industry," *Los Angeles Times*, August 3, 2001, https://www.latimes.com/archives/la-xpm-2001-aug-03-fi-30054-story.html.

2. Eberhard, interview.

3. Alan Ohnsman, "Toyota Says It's Now Turning a Profit on the Hybrid Prius," *Los Angeles Times*, December 19, 2001, https://www.latimes.com/archives/la-xpm-2001-dec-19-hy-prius19-story.html.

4. *Energy Policy Act of 2005*, Public Law 109–58, *US Statutes at Large* 119 (2005): 1043, 1047, 1049.

5. Molly F. Sherlock, "The Plug-in Electric Vehicle Tax Credit," *Congressional Research Service*, May 14, 2019, https://fas.org/sgp/crs/misc/IF11017.pdf; *Green Car Congress*, "Worldwide Prius Cumulative Sales Top 2M Mark; Toyota Reportedly Plans Two New Prius Variants for the US by the End of 2012," October 7, 2010, http://www.greencarcongress.com/2010/10/worldwide-prius-cumulative-sales-top-2m-mark-toyota-reportedly-plans-two-new-prius-variants-for-the-.html#more.

6. Dennis McClellan, "Victor Wouk, 86; Developed Hybrid Car in '70s," *Los Angeles Times*, June 19, 2005, https://www.latimes.com/archives/la-xpm-2005-jun-19-me-wouk19-story.html; Bradley Berman, "When Old Things Turn into New Again," *New*

York Times, October 24, 2007, https://www.nytimes.com/2007/10/24/automobiles /autospecial/24history.html.

7. US Energy Information Administration (EIA), "Petroleum and Other Liquids: Cushing, OK WTI Spot Price FOB," accessed April 29, 2020, https://www.eia.gov /dnav/pet/hist/RWTCD.htm.

8. Gavin Green, "Interview: Rick Wagoner, General Motors Co." *Motor Trend*, July 25, 2006, https://www.motortrend.com/news/rick-wagoner-general-motors/; Danny Hakim, "Auto Supplier Delphi Files for Bankruptcy, and GM Will Share Some of the Fallout," *New York Times*, October 9, 2005, Section 1, 28.

9. Green, "Interview: Rick Wagoner."

10. Keith Naughton, "Why Toyota Is Becoming the World's Top Carmaker," *Newsweek*, March 11, 2007, https://www.newsweek.com/why-toyota-becoming-worlds -top-carmaker-95469.

11. Naughton, "Why Toyota."

12. For a detailed account of the plug-in enthusiast community, see Boschert, *Plug-in Hybrids*.

13. *The Economist*, "Plugging into the Future," 379, no. 8481 (June 10, 2006): 33.

14. *Green Car Congress*, "GM To Manufacture Volt Packs in US; LG Chem Providing Cells; Partnership With U. Michigan," January 12, 2009, https://www.greencarcongress .com/2009/01/gm-to-manufactu.html.

15. NPR, "Rick Wagoner on the Future of General Motors," January 9, 2007, https:// www.npr.org/templates/story/story.php?storyId=6768710.

16. Robert Babik, "The Chevrolet Volt," US Environmental Protection Agency (EPA), accessed April 30, 2020, https://www.epa.gov/sites/production/files/2015-01 /documents/05102011mstrs_babik.pdf.

17. *Energy Independence and Security Act of 2007*, Public Law 110–140, *US Statutes at Large* 121 (2007): 1492–1801, 1510–1515.

18. *Energy Independence and Security Act of 2007*, Public Law 110–140, 1509.

19. Haruo Ikehara, "Toyota's Plug-in Hybrid: Debut of Prototype Is Near," *Nikkei Business Online*, January 29, 2007, http://business.nikkeibp.co.jp/article/eng/20070129 /117846/; Toyota Motor Corporation, "Toyota Advances Plug-in Hybrid Development with Partnership Program," July 25, 2007, http://pressroom.toyota.com/pr/tms/toyota /TYT2007072552930.aspx.

20. California Air Resources Board (CARB), "Resolution 08–24," last modified March 27, 2008, https://ww2.arb.ca.gov/sites/default/files/barcu/regact/2008/zev2008/zevfsor .pdf; California Air Resources Board (CARB), "Fact Sheet: The Zero Emission Vehicle Program—2008," May 6, 2008; *Wired*, "California Cuts ZEV Mandate in Favor of Plug-In Hybrids," March 27, 2008, http://www.wired.com/autopia/2008/03/the-california/.

21. Nick Bunkley, "Toyota Ahead of GM in 2008 Sales," *New York Times*, January 21, 2009, https://www.nytimes.com/2009/01/21/business/worldbusiness/21iht-22auto.19 564588.html.

22. John Neff, "Wagoner Arrives for Senate Hearing in Volt Mule," *Autoblog*, December 4, 2008, https://www.autoblog.com/2008/12/04/wagoner-arrives-for-senate-hearing-in -volt-mule/?guccounter=1; AP, "GM Chief Arrives at Capitol Hill in Hybrid," December 4, 2008, https://www.youtube.com/watch?v=FNTM3gRSEBE.

23. For examples of national innovation discourse in academic policy literature contemporaneous with the Great Recession, see Gregory Tassey, "Rationales and Mechanisms for Revitalizing US Manufacturing R&D Strategies," *Journal of Technology Transfer* 35, no. 3 (2010): 283–333; William B. Bonvillian and Richard Van Atta, "ARPA-E and DARPA: Applying the DARPA Model to Energy Innovation," *Journal of Technology Transfer* 36, no. 5 (2011): 469–513; William B. Bonvillian, "Reinventing American Manufacturing: The Role of Innovation," *Innovations: Technology, Governance, Globalization* 7, no. 3 (2012): 97–125.

24. According to Brown and Linden, other key factors in the recovery of the American semiconductor sector were currency manipulation through the Plaza Accord of 1985 and antidumping trade sanctions; see Brown and Linden, *Chips and Change*, 19–20.

25. Brent D. Yacobucci, "The Partnership for a New Generation of Vehicles: Status and Issues," *Congressional Research Service Report RS20852*, January 22, 2003, https:// wikileaks.org/wiki/CRS:_The_Partnership_for_a_New_Generation_of_Vehicles:_ Status_and_Issues,_January_22,_2003.

26. Josh Voorhees, "Obama Favors Plug-In Hybrids over Hydrogen Vehicles," *Scientific American*, July 10, 2009, https://www.scientificamerican.com/article/hybrid-cars -plug-in-obama-stimulus-money/.

27. John M. Broder, "Obama to Toughen Rules on Emissions and Mileage," *New York Times*, May 19, 2009, A1.

28. Kevin Krolicki, "Ford, Nissan, Tesla to Get US Technology Loans," *Reuters*, June 23, 2009, https://www.reuters.com/article/us-ford-loans/ford-nissan-tesla-to-get-u-s -technology-loans-idUSTRE55M39120090623.

29. Fletcher, *Bottled Lightning*.

30. Barack H. Obama, "Remarks of President Barack Obama in State of the Union Address, as Prepared for Delivery," January 25, 2011, https://obamawhitehouse .archives.gov/the-press-office/2011/01/25/remarks-president-barack-obama-state -union-address-prepared-delivery.

31. US Council for Automotive Research (USCAR), "USABC Awards $12.5 Million Battery Technology Development Contract to A123 Systems," last modified May 5, 2008, https://uscar.org/guest/article_view.php?articles_id=210.

32. See Yet-Ming Chiang, "Building a Better Battery," *Science* 330 (December 10, 2010): 1485–1486.

33. Lyle Dennis, "A123 Gets $249 Million Government Grant To Build Battery Factory in Michigan," *Green Car Reports*, August 5, 2009, https://www.greencarreports.com/news/1033931_a123-systems-gets-249-million-government-grant-to-build-battery-factory-in-michigan; Kevin Krolicki, "A123 to Sell Fisker Batteries, Takes Stake," *Reuters*, January 14, 2010, https://uk.reuters.com/article/a123/a123-to-sell-fisker-batteries-takes-stake-idUKN1417119620100114; US Department of Energy (DOE), "Department of Energy Announces Closing of $529 Million Loan to Fisker Automotive," last modified April 23, 2010, https://www.energy.gov/articles/department-energy-announces-closing-529-million-loan-fisker-automotive.

34. Angela Hardin, "LG Chem, Argonne Sign Licensing Deal to Make, Commercialize Advanced Battery Material," January 6, 2011, http://www.anl.gov/articles/lg-chem-argonne-sign-licensing-deal-make-commercialize-advanced-battery-material.

35. US Department of Energy (DOE), *Special Report: The Department of Energy's Management of the Award of a $150 Million Recovery Act Grant to LG Chem Michigan Inc.* (Washington, DC: OAS-RA-13–10, 2013); Bernie Woodall, Paul Lienert, and Ben Klayman, "Insight: GM's Volt: The Ugly Math of Low Sales, High Costs," *Reuters*, September 10, 2012, http://www.reuters.com/assets/print?aid=USBRE88904J20120910.

36. Lithium-manganese oxide chemistry traced to Michael Thackeray's research in spinel-based lithium insertion compounds begun in the early 1980s with Goodenough's assistance. When the South African government ended support for this system in the early 1990s, Thackeray decamped to the US and took a position at the Chemical Technology Division of Argonne National Laboratory, where he completed work on a stable lithium-manganese oxide cathode; see Thackeray, "20 Golden Years of Battery R&D."

37. Nissan Motor Company, "Nissan and NEC Joint Venture AESC Starts Operations," last modified May 19, 2008, https://www.nissan-global.com/EN/NEWS/2008/_STORY/0805.

38. Cheryl Jensen, "Nissan Battery Plant Begins Operations in Tennessee," *New York Times*, December 13, 2012, https://wheels.blogs.nytimes.com/2012/12/13/nissan-battery-plan.

39. *Reuters*, "Johnson Sees Hybrid Engines in 5–8 Pct. of Market," January 19, 2007, https://www.reuters.com/article/us-autos-summit-johnson-hybrids-idUSN1120028420060911.

40. *Detroit Free Press*, "A Lab for Hybrids," September 29, 2005, 2C.

41. Chevrolet built the pack at the Brownstown plant outside Detroit; see Chuck Squatriglia, "GM Fires Up Its Chevrolet Volt Battery Factory," *Wired*, January 7, 2010, https://www.wired.com/2010/01/chevrolet-volt-battery-production/.

42. Andre Morris, "Ford Awards Johnson Controls–Saft Battery Contract for Hybrid Car," *EE/Times*, February 4, 2009, https://www.eetimes.com/ford-awards-johnson

-controls-saft-battery-contract-for-hybrid-car/; Sam Abuelsamid, "Ford Picks Johnson Controls-Saft for PHEV Batteries, Adds 7 Utility Partners to Test Program," *AutoBlog .com*, February 3, 2009, https://www.autoblog.com/2009/02/03/ford-picks-johnson -controls-saft-for-phev-batteries-adds-7-util/2/.

43. Johnson, interview.

44. For a discussion of manufacturing processes of different battery form factors, see Kazuo Tagawa and Ralph J. Brodd, "Chapter 8: Production Processes for Fabrication of Lithium-Ion Batteries," in *Lithium-Ion Batteries: Science and Technologies*, eds. Masaki Yoshio, Ralph J. Brodd, and Akiya Kozawa (New York: Springer, 2009), 181–194.

45. See Matthew N. Eisler, "Materials Science, Instrument Knowledge, and the Power Source Renaissance," *Proceedings of the IEEE* 105, no. 12 (2017): 2382–2389.

46. Johnson, interview.

CHAPTER 12

1. Martin Eberhard, "Lotus Position," Tesla Motors, last modified July 25, 2006, https://www.tesla.com/en_GB/blog/lotus-position?redirect=no.

2. Elon Musk, "The Secret Tesla Motors Master Plan (Just Between You and Me)," Tesla Motors, last modified August 2, 2006, https://www.tesla.com/blog/secret-tesla -motors-master-plan-just-between-you-and-me.

3. I derive this definition of the aesthetic from Terry Eagleton. In the context of complex, science-based consumer products, I use this word to refer to those qualities of technology that do not necessarily directly relate to physical operation (optimal or suboptimal as actors perceive in context), but that relate to and inform social acceptance; see Terry Eagleton, *The Ideology of the Aesthetic* (Oxford, UK: Blackwell, 1990).

4. Customer input was an explicit aspect of Tesla marketing; see Tesla Motors, "Form 10-K for the Fiscal Year Ended 31 Dec. 2014," 13.

5. In this respect, Tesla paralleled the open science movement of the early 2010s; on the open science movement, see Philip Mirowski, "The Future(s) of Open Science," *Social Studies of Science* 48, no. 2 (2018): 171–203, on 188.

6. On Sanyo as the source of cells for the Roadster, see Eberhard, interview. A 2006 paper co-authored by Straubel suggested that Tesla Motors had procured cells from several suppliers but did not name them; see Gene Berdichevsky, Kurt Kelty, JB Straubel, and Erik Toomre, "The Tesla Roadster Battery System," last modified December 19, 2007, http://large.stanford.edu/publications/coal/references/docs/tesla .pdf. On Straubel's work on the controller, see Kevin Bullis, "Innovators Under 35, 2008: JB Straubel, 32, Engineering Electric Sports Cars," *MIT Technology Review*, http://www2.technologyreview.com/tr35/profile.aspx?trid=742.

7. *Top Gear*, "Top Gear Tesla Road Test," accessed May 6, 2020, https://www .youtube.com/watch?v=JKtK493sGAk.

8. Chuck Squatriglia, "Tesla Cries Foul on Top Gear's Test," *Wired*, December 16, 2008, https://www.wired.com/2008/12/tesla-cries-fou/.

9. See, for example, "Cambridge or Bust, Pasadena or Bust;" and Perrin, *Solo*, 33–37.

10. See, for example, Troy R. Hawkins, Bhawna Singh, Guillaume Majeau-Bettez, and Anders Hammer Strømman, "Comparative Environmental Life Cycle Assessment of Conventional and Electric Vehicles," *Journal of Industrial Ecology* 17, no. 1 (2012): 53–64.

11. Tesla Motors, "Tesla vs. Top Gear," March 29, 2011, https://www.tesla.com /en_GB/blog/tesla-vs-top-gear?redirect=no.

12. Brooks, interview; Rippel, interview.

13. Vance, *Elon Musk*, 209; Joanne Muller, "Elon Musk's Financial Car Wreck," *Forbes*, May 28, 2010, https://www.forbes.com/2010/05/28/elon-musk-broke-tesla -business-autos-musk.html#3b8519a56d8b.

14. US Department of Energy (DOE), "Tesla," accessed May 12, 2020, https://www .energy.gov/lpo/tesla; Muller, "Elon Musk's Financial Car Wreck."

15. Chuck Squatriglia, "Tesla IPO Raises $226.1 Million, Stock Surges 41 Percent," *Wired*, June 29, 2010, https://www.wired.com/2010/06/tesla-ipo-raises-226-1-million/.

16. Tesla Motors, "Strategic Partnership: Daimler Acquires Stake in Tesla," April 20, 2010, https://www.tesla.com/en_GB/blog/strategic-partnership-daimler-acquires-stake -tesla; Chuck Squatriglia, "Toyota, Tesla Resurrect the Electric RAV4," *Wired*, July 16, 2010, https://www.wired.com/2010/07/toyota-tesla-rav4-ev/.

17. Jim Motavalli, "Electric Car Agreement for Toyota and Tesla," *New York Times*, May 21, 2010, B7.

18. Dana Hull, "2010: Tesla Gets Ready to Take over the Former NUMMI Auto Plant in Fremont," *The Mercury News*, September 16, 2010, https://www.mercurynews.com /2010/09/16/2010-tesla-gets-ready-to-take-over-the-former-nummi-auto-plant-in -fremont/.

19. Yoshifumi Uesaka, "The Company That Helps Tesla Make Look Aluminum Sexy," *Nikkei Asia*, September 12, 2016, https://asia.nikkei.com/Business/Biotechnology/The -company-that-helps-Tesla-make-aluminum-look-sexy.

20. National Research Council, *Review of the Research Program of the Partnership for a New Generation of Vehicles: Fourth Report* (Washington, DC: National Academies Press, 1998), 43–44; National Research Council, *Review of the Research Program of the Partnership for a New Generation of Vehicles: Fifth Report* (Washington, DC: National Academy Press, 1999), 7, 38–39.

21. This was known as the Advanced Technology Development (ATD) program. The ATD program was designed to aid US battery manufacturers, but many PNGV subcontractors were foreign and ATD researchers depended heavily on materials

and cathode chemistries developed by Japanese firms such as Fuji and Hitachi; see Raymond A. Sutula, *Progress Report for the Advanced Technology Development Program* (Washington, DC: US Department of Energy, 2000).

22. See, for example, K. Amine, C. H. Chen, J. Liu, M. Hammond, A. Jansen, D. Dees, et al., "Factors Responsible for Impedance Rise in High Power Lithium Ion Batteries," *Journal of Power Sources* 97–98 (2001): 684–687; Ira Bloom, Scott A. Jones, Vincent S. Battaglia, Gary L. Henriksen, Jon P. Christophersen, Randy B. Wright, et al., "Effect of Cathode Composition on Capacity Fade, Impedance Rise, and Power Fade in High-Power Lithium-Ion Cells," *Journal of Power Sources* 124, no. 2 (2003): 538–550; D. P. Abraham, J. L. Knuth, D. W. Dees, I. Bloom, and J. P. Christophersen, "Performance Degradation of High-Power Lithium-Ion Cells: Electrochemistry of Harvested Electrodes," *Journal of Power Sources* 170, no. 2 (2007): 465–675; see also Shoichiro Watanabe, Masahiro Kinoshita, and Kensuke Nakura, "Capacity Fade of $LiNi_{(1-x-y)}Co_xAl_yO_2$ Cathode for Lithium-Ion Batteries During Accelerated Calendar and Cycle Life Test. I. Comparison Analysis Between $LiNi_{(1-x-y)}Co_xAl_yO_2$ and $LiCoO_2$ Cathodes in Cylindrical Lithium-Ion Cells During Long Term Storage Test," *Journal of Power Sources* 247 (2014): 412–422.

23. Tesla Motors, "Panasonic Presents First Electric Vehicle Battery to Tesla," last modified April 22, 2010, http://www.teslamotors.com/about/press/releases/panasonic -presents; Tesla Motors, "Panasonic Invests $30 Million in Tesla," last modified November 3, 2010, https://www.tesla.com/en_GB/blog/panasonic-invests-30-million -tesla.

24. Tesla Motors, "Panasonic Enters into Supply Agreement with Tesla Motors to Supply Automotive-Grade Battery Cells," last modified October 11, 2011, https:// www.tesla.com/en_GB/blog/panasonic-enters-supply-agreement-tesla-motors-supply -automotivegrade-battery-c.

25. *Division B Energy Improvement and Extension Act of 2008*, Public Law 110–343, *US Statutes at Large* 122 (2008): 3835–3836.

26. Sherlock, "The Plug-in Electric Vehicle Tax Credit."

27. One observer held that the battery pack added 30 percent to the overall cost of battery power; see Maximilian Holland, "Tesla 2018 Annual Shareholder Meeting: Quick Highlights," *EVObsession*, June 5, 2018, https://evobsession.com/tesla-2018 -annual-tesla-shareholder-meeting-quick-highlights/. In 2015, GM broke with industry practice when it announced the cell cost of its all-battery electric Bolt, annoying its cell supplier LG Chem; see Jay Cole, "LG Chem 'Ticked Off' with GM for Disclosing $145/kWh Battery Cell Pricing," *InsideEVs*, October 23, 2015, https://insideevs.com /news/327874/lg-chem-ticked-off-with-gm-for-disclosing-145-kwh-battery-cell-pricing -video/.

28. Cackette, interview.

29. Collantes and Sperling, "The Origins of California's Zero Emission Vehicle Mandate," 1304.

30. Tesla Motors, Inc., "2013 Form 10-K," 77; Tesla Motors, Inc., "2014 Form 10-K," 53, Tesla Motors, Inc., "2015 Form 10-K," 55, Tesla Motors, Inc., "2016 Form 10-K," 44, Tesla Motors, Inc., "2017 Form 10-K," 72, and Tesla Motors, Inc., "2018 Form 10-K," 56; see also California Air Resources Board (CARB), *California's Zero Emission Vehicle Program*, June 2009, https://www.arb.ca.gov/msprog/zevprog/factsheets/zev_tutorial.pdf.

31. California Air Resources Board (CARB), "Zero-Emission Vehicle Credit Balances," accessed May 3, 2020, https://ww2.arb.ca.gov/our-work/programs/advanced-clean-cars-program/zev-program/zero-emission-vehicle-credit-balances. According to the journalist Jerry Hirsch, zero emission credits and state and federal buyer incentives generated up to $45,000 for each Model S sold; see Jerry Hirsch, "Tesla Drives California Environmental Credits to the Bank," *Los Angeles Times*, May 5, 2013, https://www.latimes.com/business/autos/la-fi-electric-cars-20130506-story.html.

32. The actual value of the credits fluctuated according to supply and demand; see Trefis Team, "Tesla's Lucrative ZEV Credits May Not Be Sustainable," *Forbes*, September 1, 2017, https://www.forbes.com/sites/greatspeculations/2017/09/01/teslas-lucrative-zev-credits-may-not-be-sustainable/#7eeafe976ed5.

33. Tesla Motors, "Tesla Model S: Full Battery Swap Event," last modified June 21, 2013, https://www.youtube.com/watch?v=H5V0vL3nnHYandt=36s.

34. Phil Kerpen, "Tesla and Its Subsidies," *National Review Online*, January 26, 2015, https://www.nationalreview.com/2015/01/tesla-and-its-subsidies-phil-kerpen/.

35. Justin Berkowitz, "Following Coda and Fisker, Spring of EV Carnage Claims Israeli Start-up Better Place," *Car and Driver*, May 30, 2013, https://www.caranddriver.com/news/a15370868/following-coda-and-fisker-spring-of-ev-carnage-claims-israeli-startup-better-place-analysis/; Wayne Cunningham, "Tesla Battery Swap a Dead End," *CNET*, June 21, 2013, https://www.cnet.com/roadshow/news/tesla-battery-swap-a-dead-end/.

36. Tesla Motors, "Tesla Motors Launches Revolutionary Supercharger Enabling Convenient Long-Distance Driving," last modified September 24, 2012, https://ir.tesla.com/press-release/tesla-motors-launches-revolutionary-supercharger-enabling.

37. Tesla, "Tesla Gigafactory," accessed May 3, 2020, https://www.tesla.com/en_GB/gigafactory.

38. Sandra Chereb, "Nevada Gives $1.3 Billion Tax Break to Electric Car Maker Tesla," *Scientific American*, September 12, 2014, https://www.scientificamerican.com/article/nevada-gives-1-3-billion-tax-break-to-electric-car-maker-tesla/.

39. In mainstream automaking, a production run of a commercial automobile typically constituted several hundred thousand units per year. Between 2008 and 2012, Tesla built around 2,500 Roadsters, and between 2012 and 2018, it built 200,000

units of the Model S. Tesla did not consistently record annual production figures for the Model S. In 2016, the company aggregated production figures of the Model S and the Model X sports utility vehicle (SUV); see Tesla Motors Inc., "2012 Form 10-K," 6, Tesla Motors, Inc., "2013 Form 10-K," 5; Tesla Motors, Inc., "2014 Form 10-K," 5, Tesla Motors, Inc., "2015 Form 10-K," 5, and Tesla, Inc., "2016 Form 10-K," 35.

40. Angus MacKenzie, "2013 Motor Trend Car of the Year: Model S," *Motor Trend*, December 10, 2012, https://www.motortrend.com/news/2013-motor-trend-car-of -the-year-tesla-model-s/.

41. "2013 Tesla Model S," *Consumer Reports*, accessed November 30, 2018, https:// www.consumerreports.org/cars/tesla/model-s/2013/overview.

42. Peter Valdes-Dapena, "New Tesla Earns Perfect Score from Consumer Reports," *CNN*, August 27, 2015, https://money.cnn.com/2015/08/27/autos/consumer-reports -tesla-p85d/index.html.

43. Andrew J. Hawkins, "Two Guys Did a Coast-to-Coast 'Run' in a Tesla Model S for a New Record," *The Verge*, July 9, 2017, https://www.theverge.com/2017/7/9 /15938028/tesla-model-s-cannonball-run-record.

44. Tesla Motors, "Form 10-K for the Fiscal Year Ended 31 Dec. 2015," 3–4.

45. Mark Rechtin, "Tesla Reliability Doesn't Match Its High Performance," *Consumer Reports*, October 20, 2015, https://www.consumerreports.org/cars-tesla-reliability -doesnt-match-its-high-performance/. One owner of a Model S recounted an inci- dent when the car was disabled after hitting a large pothole that flattened two tires. Within an hour, held the owner, Tesla technicians recovered the car on a flatbed trailer, and they repaired it the same day; Daniel, interview by author, May 30, 2015.

46. John M. Broder, "Stalled Out on Tesla's Electric Highway," *New York Times*, Feb- ruary 8, 2013, https://www.nytimes.com/2013/02/10/automobiles/stalled-on-the-ev -highway.html.

47. Tesla Motors, "Claim Form," March 29, 2011, file:///Users/mne/Downloads/ tesla_-_claim_form_claimants_copy_29_03_11.pdf.

48. Elon Musk, "A Most Peculiar Test Drive," Tesla Motors, last modified February 13, 2013, https://www.tesla.com/blog/most-peculiar-test-drive; John M. Broder, "That Tesla Data: What It Says and What It Doesn't," *New York Times*, February 14, 2013, https://wheels.blogs.nytimes.com/2013/02/14/that-tesla-data-what-it-says-and-what-it -doesnt/.

49. Deepa Seetharaman, "Tesla Shares Drop 6 Percent After Report of Model S Fire," *Reuters*, October 3, 2013, https://www.reuters.com/article/us-autos-tesla-crash/tesla -shares-drop-6-percent-after-report-of-model-s-fire-idUSBRE99200020131003.

50. See, for example, John Croft, "Lithium Battery Rules Could Get Safety Over- haul," *Aviation Week and Space Technology*, October 2, 2015; US Department of Transportation (DOT), "Guidance on Testing and Installation of Rechargeable Lithium

Battery and Battery Systems on Aircraft," October 15, 2015, https://www.faa.gov /documentLibrary/media/Advisory_Circular/AC_20-184_Final_proof.pdf.

51. Mark Osborne and Rex Sakamoto, "'West Wing' Actress Calls Out Tesla After Husband's Car Bursts into Flames," *ABC News*, June 17, 2018, https://abcnews.go .com/US/west-wing-actress-calls-tesla-husbands-car-bursts/story?id=55953027; Elizabeth Puckett, "Tesla Model S Bursts into Flames Two Separate Times on Same Day Following Tire Issue," *The Drive*, December 21, 2018, http://www.thedrive.com/news /25603/tesla-model-s-bursts-into-flames-two-separate-times-on-same-day-following -tire-issue.

52. Tesla Motors offered a limited eight-year "infinite mile" warranty on the battery and drivetrain of its cars that pointedly did not cover the battery "charging capacity." This language suggested that the company would replace a battery only if it completely failed, not if it lost significant capacity over time while under warranty; see Tesla Motors, "Form 10-K for the Fiscal Year Ended December 31, 2016," "Service and Warranty," 17.

53. John Voelcker, "Should I Buy a Used Nissan Leaf (or Another Electric Car?)," *Green Car Reports*, June 15, 2015, https://www.greencarreports.com/news/1098554_ should-i-buy-a-used-nissan-leaf-or-another-electric-car.

54. Tesla Motors, "Resale Value Guarantee," accessed May 3, 2020, https://www .tesla.com/sites/default/files/pdfs/rvg/RVG_Agreement_R20160318_en_US.pdf; Alex Davies, "Used Teslas Are More About Saving You Time than Money," *Wired*, May 5, 2015, https://www.wired.com/2015/05/used-teslas.

55. Travis Hoium, "Competitors Made $1.5 Billion from Tesla Motors' Success," *Motley Fool*, October 31, 2015, https://www.fool.com/investing/general/2015/10/31 /how-competitors-made-15-billion-tesla-motors-succe.aspx.

56. For historical data on Tesla's share price since 2010, see Yahoo!finance, "Tesla, Inc. (TSLA), accessed February 17, 2022, https://finance.yahoo.com/quote/TSLA /history?period1=1277856000&period2=1645056000&interval=1mo&filter=history &frequency=1mo&includeAdjustedClose=true.

CHAPTER 13

1. On the role of concurrency in the F-35 stealth fighter, see Valerie Insinna, "Inside America's Dysfunctional Trillion-Dollar Fighter-Jet Program," *New York Times Magazine*, August 21, 2019, https://www.nytimes.com/2019/08/21/magazine/f35-joint -strike-fighter-program.html.

2. Jordan Golson, "Tesla Ends 'Resale Value Guarantee' On New Vehicle Purchases," *The Verge*, July 13, 2016, https://www.theverge.com/2016/7/13/12173310/tesla -model-s-resale-value-guarantee-ending; Fred Lambert, "Tesla Officially Announces End of Unlimited Free Supercharging, New 'Supercharging Credit Program' Starts in

2017," *Electrek*, November 7, 2016, https://electrek.co/2016/11/07/tesla-end-of-free
-supercharging-new-supercharging-credit-program-2017/.

3. Alexandria Sage, "Tesla Owner Lawsuit Claims Software Update Fraudulently
Cut Battery Capacity," *Reuters*, August 8, 2019, https://www.reuters.com/article/us
-tesla-battery/tesla-owner-lawsuit . . . aims-software-update-fraudulently-cut-battery
-capacity-idUSKCN1UY2TW.

4. Jim Collins, "Forget the Autopilot Crash and Recall: Tesla's Daunting Debt Repayment Schedule is the Real Issue," *Forbes*, April 2, 2018, https://www.forbes.com/sites
/jimcollins/2018/04/02/forget-the-autopil . . . las-daunting-debt-repayment-sched-
ule-is-the-real-issue/#284af5f93e94. Unsurprisingly, Martin Eberhard was not a fan
of Tesla's self-driving technology; Eberhard, interview.

5. See, for example, Mark Rechtin, "Tesla Nimbly Updates Model S over the Air,"
Automotive News, January 16, 2013, https://www.autonews.com/article/20130116
/OEM06/130119843/tesla-nimbly-updates-model-s-over-the-air; Lou Shipley, "How
Tesla Sets Itself Apart," *Harvard Business Review*, February 28 2020, https://hbr.org
/2020/02/how-tesla-sets-itself-apart.

6. Terry Ericsen, "The Second Electronic Revolution (It's All About Control)," *IEEE
Transactions on Industry Applications* 46, no. 5 (September/October 2010): 1778–1786.

7. See Moore, "The Cost Structure of the Semiconductor Industry," "Are We Really
Ready for VLSI^2?" and "Solid-State;" Ian King, "Chips: Off Quarters for a Hot Company
Highlight Wider Concerns," *Bloomberg Businessweek*, November 19, 2018–January 6,
2019, 30; Ian King, "Tesla Shifts to Intel from Nvidia for Infotainment," *Bloomberg*,
September 26, 2017, https://www.bloomberg.com/news/articles/2017-09-26/tesla-is
-said-to-shift-to-intel-from-nvidia-for-infotainment; Georgina Prodhan, "Car Industry
Players Diverge on Timescale for Self-Driving Cars," *Reuters*, March 16, 2017, https://
www.reuters.com/article/us-autos-autonomous-idUSKBN16N2NF.

8. Noel Randewich, "Tesla Becomes Most Valuable US Car Maker, Edges out
GM," *Reuters*, April 11, 2017, https://www.reuters.com/article/us-usa-stocks-tesla
-idUSKBN17C1XF.

9. Alexandria Sage, "Build Fast, Fix Later: Speed Hurts Quality at Tesla, Some
Workers Say," *Reuters*, November 29, 2017, https://uk.reuters.com/article/us-tesla
-quality-insight/build-fast-fix-later-speed-hurts-quality-at-tesla-some-workers-say
-idUKKBN1DT0N3.

10. Drew Harwell, "Tesla Hits 5000-a-Week Model 3 Production Goal," *Washington Post*, July 2, 2018, https://www.washingtonpost.com/business/economy/tesla
-hits-5000-a-week-model-3-production-goal/2018/07/02/a3306ca0-7e48-11e8-b660
-4d0f9f0351f1_story.html.

11. Dana Hull, John Lippert, and Sarah Gardner, "The Future of Tesla Hinges on
This Gigantic Tent," *Bloomberg News*, June 25, 2018, https://www.bloomberg.com
/news/articles/2018-06-25/the-future-of-tesla-hinges-on-this-gigantic-tent.

12. Pavel Alpeyev, Yuki Furukawa, and Masatsugu Horie, "Panasonic Says Gigafactory Profit in Sight as Tesla Ramps Output," *Bloomberg*, November 1, 2018, https://www.bloomberg.com/news/articles/2018-11-01/panasonic-says-gigafactory-profit-in-sight-as-tesla-ramps-output.

13. Linette Lopez, "Insiders Describe a World of Chaos and Waste at Panasonic's Massive Battery-Making Operation for Tesla," *Business Insider*, April 16, 2019, https://www.businessinsider.com/panasonic-battery-cell-operations-tesla-gigafactory-chaotic-2019-4?r=USandIR=T; Linette Lopez, "Internal Documents Reveal Tesla Is Blowing Through an Insane Amount of Raw Material and Cash to Make Model 3s, and Production Is Still a Nightmare," *Business Insider*, June 4, 2018, https://www.businessinsider.com/tesla-model-3-scrap-waste-high-gigafactory-2018-5?r=USandIR=T; Lora Kolodny, "Tesla Whistleblower Tweets Details About Allegedly Flawed Cars, Scrapped Parts," *CNBC*, August 16, 2018, https://www.cnbc.com/2018/08/15/tesla-whistleblower-tweets-details-about-flawed-cars-scrapped-parts.html.

14. Graham Rapier, "Wall Street Analysts Tore Down a Tesla Model 3 and Found 'Significant Fit and Finish Issues,'" *Business Insider*, August 27, 2018, https://markets.businessinsider.com/news/stocks/tesla-model-3-wall-street-analysts-find-significant-fit-and-finish-issues-2018-8-1027480762.

15. *Consumer Reports*, "Consumer Reports: Tesla Model 3, Chrysler 300 Among Cars No Longer 'Recommended' Based on New Reliability Findings," last updated February 21, 2019, https://www.consumerreports.org/media-room/press-releases/2019/02/consumer_reports_tesla_model_3_chrysler_300_among_cars_no_longer_recommended_based_on_new_reliability_findings/.

16. Eberhard, interview.

17. Charley Grant, "Is Tesla Abandoning the Mass Market? Elon Musk's About-Face on Model 3 Pricing is a Warning Sign for the Stock," *Wall Street Journal*, May 21, 2018, https://www.wsj.com/articles/is-tesla-abandoning-the-mass-market-1526917239.

18. Russ Mitchell, "As Tesla Tax Credits Disappear, Will Model 3 Deposit-Holders Stick Around?," *Los Angeles Times*, July 3, 2018, https://www.latimes.com/business/autos/la-fi-hy-tesla-tax-credit-subsidy-20180703-story.html.

19. Jae Hyun Lee, Scott J. Hardman, and Gil Tal, "Who Is Buying Electric Vehicles in California? Characterizing Early Adopter Heterogeneity and Forecasting Market Diffusion," *Energy Research and Social Science* 55 (2019): 218–226; Sherlock, "The Plug-in Electric Vehicle Tax Credit."

20. Fred Lambert, "Tesla Model 3 Battery Packs Have Capacities of ~ 50 kWh and ~70 kWh, Says Elon Musk," *Electrek*, August 8, 2017, https://electrek.co/2017/08/08/tesla-model-3-battery-packs-50-kwh-75-kwh-elon-musk/.

21. Tesla, "Select Your Car," accessed February 18, 2022, https://3.tesla.com/model3/design#battery.

22. Elon Musk, "Tweet," January 9, 2019, https://twitter.com/elonmusk/status/1083141248872075265?lang=en.

23. Fred Lambert, "Tesla Removes Any Mention of Standard Model 3 Battery from Website; Fans Panic," *Electrek*, February 18, 2019, https://electrek.co/2019/02/18/tesla-standard-model-3-battery-website-fans-panic/.

24. Arash Massoudi and Richard Waters, "Saudi Arabia Slashes Exposure to Tesla Via Hedging Deal," *Financial Times*, January 28, 2019, https://www.ft.com/content/d501c670-2307-11e9-b329-c7e6ceb5ffdf.

25. Carl O'Donnell and Ross Kerber, "Investors Query Funding Costs at a Private Tesla," *Reuters*, August 17, 2018, https://www.reuters.com/article/us-tesla-musk-board-analysis/investors-query-funding-costs-at-a-private-tesla-idUSKBN1L21XN.

26. David Gelles and Peter Eavis, "Elon Musk Wants to Take Tesla Private; Can He Make the Math Work?" *New York Times*, August 23, 2018, https://www.nytimes.com/2018/08/23/business/dealbook/tesla-investors-elon-musk.html.

27. David Gelles, "Why Elon Musk Reversed Course on Taking Tesla Private," *New York Times*, August 25, 2018, https://www.nytimes.com/2018/08/25/business/elon-musk-tesla-private.html?action=clickandmodule=Top%20Storiesandpgtype=Homepage.

28. William D. Cohan, "Tesla's Biggest Problem Isn't Elon Musk," *New York Times*, September 20, 2018, https://www.nytimes.com/2018/09/20/opinion/tesla-elon-musk.html.

29. Sean O'Kane, "The Court Has Approved Elon Musk's New Agreement to Let Lawyers Oversee His Tesla Tweets," *The Verge*, April 30, 2019. https://www.theverge.com/2019/4/26/18484751/elon-musk-sec-fraud-tesla-tweets-contempt-agreement.

30. Alison Frankel, "Hedge Funds Step up to Lead Shareholder Suit Against Tesla," *Reuters*, October 10, 2018, https://uk.reuters.com/article/legal-us-otc-tesla/hedge-funds-step-up-to-lead-shareholder-suit-against-tesla-idUKKCN1MK2HY; Lora Kolodny, "Tesla and Elon Musk Face Dozens of Lawsuits and Investigations Far Beyond the SEC Court Fight," *CNBC*, March 19, 2019, https://www.cnbc.com/2019/03/19/tesla-and-elon-musk-lawsuits-overview.html.

31. Elon Musk, interview by Lesley Stahl, *60 Minutes*, CBS, December 9, 2018, https://www.cbsnews.com/news/tesla-ceo-elon-musk-the-2018-60-minutes-interview/.

32. Kadhim Shubber, "SEC Endorses Investor View of Elon Musk's Indispensable Role at Tesla," *Financial Times*, October 2, 2018, https://search.proquest.com/docview/2115551960?accountid=14116.

33. Kadhim Shubber, "Musk Mocks SEC in Tweet Only Days After Settling with Regulator," *Financial Times*, October 4, 2018, https://search.proquest.com/docview/2116253761?accountid=14116.

34. Chrissie Thompson, "Chevy Volt Was Going to Save Detroit: Now Its Workers Are Losing Jobs," *Detroit Free Press*, November 27, 2018, https://eu.freep.com/story/money/business/2018/11/27/chevy-volt-donald-trump-general-motors/2120687002/.

35. Mike Monticello, "Volt vs. Prius: Chevrolet's Plug-In Takes on Toyota's Hybrid," *Consumer Reports*, April 22, 2016, https://www.consumerreports.org/hybrids-evs/volt-vs-prius-review/?EXTKEY=AGTS004.

36. *Fortune*, "Another One out the Door: Now It's Tesla's Supply Chain Chief Who's Leaving," September 21, 2018, http://fortune.com/2018/09/21/tesla-supply-chain -executive-departure/.

37. *CNBC*, "Tesla Starts Offering Leases for Model 3," April 12, 2019, https://www .cnbc.com/2019/04/12/tesla-begins-offering-leases-for-model-3.html.

38. Alan Ohnsman, "Elon Alone: Longtime Tesla Tech Chief Straubel's Exit Leaves Musk as Sole Remaining Cofounder," *Forbes*, July 24, 2019, https://www.forbes.com /sites/alanohnsman/2019/07/24/elon-alone-long- . . . a-tech-chief-straubels-exit-leaves -musk-as-sole-remaining-cofounder/.

39. Norway used its vast oil wealth to fund a national decarbonization strategy that supported a thriving market for electrics. In the late 2010s, the country had a higher proportion of electrics in its light duty fleet than any other country and was a tempting place for US leasing companies to sell compliance cars after using them for a few years; see Alister Doyle, "From California to Oslo: Foreign Subsidies Fuel Norway's E-Car Boom, for Now," *Reuters*, March 21, 2019, https://uk.reuters.com /article/uk-autos-norway-insight-idUKKCN1R20J4.

40. See Lorenzo Totaro and Daniele Lepido, "Fiat to Pool Cars with Tesla to Meet EU Emissions Targets on CO2," *Bloomberg*, April 7, 2019, https://www.bloomberg .com/news/articles/2019-04-07/fiat-chrysler-teams-with-tesla-to-meet-eu-emissions -targets.

41. Sonam Rai and Jasmine I.S. Bengaluru, "Musk Not Worried About Tesla Model 3 Demand But Wall Street Thinks Otherwise," *Reuters*, January 31, 2019, https://uk .reuters.com/article/uk-tesla-results-stocks/musk-not-worried-about-tesla-model-3 -demand-but-wall-street-is-idUKKCN1PP1TL.

42. Kenneth Rapoza, "Here's Why Tesla CEO Elon Musk Was Dancing in China," *Forbes*, January 13, 2020, https://www.forbes.com/sites/kenrapoza/2020/01/13/heres -why-tesla-ceo-elon-musk-was-dancing-in-china/.

43. Daishi Chiba and Itsuro Fujino, "Tesla and Panasonic Freeze Spending on $4.5 Billion Gigafactory," *Nikkei Asia*, April 11, 2019, https://asia.nikkei.com/Business /Companies/Tesla-and-Panasonic-freeze-spending-on-4.5bn-Gigafactory.

44. Lawrence Ulrich, "Is Elon Musk Back in 'Production Hell' with Tesla's 4680 Battery?" *IEEE Spectrum*, September 1, 2021, https://spectrum.ieee.org/tesla-4680 -battery; Daniel Harrison and Christopher Ludwig, *Electric Vehicle Battery Supply Chain Analysis: How Battery Demand and Production Are Reshaping the Automotive Industry* (London: Ultima Media, 2021), 56–57; Hyunjoo Jin, "LG Hopes to Make New Battery Cells for Tesla in 2023 in US or Europe," *Reuters*, March 9, 2021, https:// www.reuters.com/article/us-tesla-lg-evs-exclusive-idUSKBN2B12HY.

45. John Voelcker, "Who Sold the Most Plug-in Electric Cars in 2015? (It's Not Tesla or Nissan)," *Green Car Reports*, January 15, 2016, https://www.greencarreports.com

/news/1101883_who-sold-the-most-plug-in-electric-cars-in-2015-its-not-tesla-or
-nissan.

46. Harrison and Ludwig, *Electric Vehicle Battery Supply Chain Analysis*, 27, 64.

47. Bill Vlasic, "Chinese Firm Wins Bid for Auto Battery Maker," *New York Times*, December 9, 2012, https://www.nytimes.com/2012/12/10/business/global/auction -for-a123-systems-won-by-wanxiang-group-of-china.html; Dustin Walsh, "Wanxiang Group Closes Deal to Acquire Assets of A123 Systems," *Crain's Detroit Business*, January 29, 2013, https://www.crainsdetroit.com/article/20130129/NEWS/130129846 /wanxiang-group-closes-deal-to-acquire-assets-of-a123-systems.

48. Nissan Motor Corporation, "Nissan Completes Sale of Battery Business to Envision Group," March 29, 2019, https://newsroom.nissan-global.com/releases/190329 -01-e?la=1anddownloadUrl=%2Freleases%2F190329-01-e%2Fdownload.

49. Noel Randewich, "Tesla's Market Value Zooms Past That of GM and Ford Combined," *Reuters*, January 8, 2020, https://www.reuters.com/article/us-usa-stocks-tesla /teslas-market-value-zooms-past-that-of-gm-and-ford-combined-idUSKBN1Z72MU.

50. Elon Musk, "All Our Patent Are Belong To You," Tesla Motors, last modified June 12, 2014, https://www.tesla.com/en_GB/blog/all-our-patent-are-belong-you.

51. Daniel Bell, "The Axial Age of Technology Foreword: 1999," in *The Coming of Post-Industrial Society: A Venture in Social Forecasting* (Basic Books: New York, 1999), ix–lxxxvi, on xxxiv–xlv.

52. See, for example, Jon Gertner, "The Risk of a New Machine," *Fastcompany.com* (April 2012): 104–133; and Shipley, "How Tesla Sets Itself Apart."

53. James C. Collins and Jerry I. Porras, *Built to Last: Successful Habits of Visionary Companies* (London: Random House, 2004), 55.

54. Phil Rosenzweig, *The Halo Effect and the Eight Other Business Delusions That Deceive Managers* (New York: Free Press, 2007).

CHAPTER 14

1. Perrin, *Solo.*

2. This represented about 32 percent of the world total of 1.25 million electric passenger cars; see International Energy Agency, *Global EV Outlook 2016*, 34.

3. Heidi Gjøen and Mikael Hård, "Cultural Politics in Action: Developing User Scripts in Relation to the Electric Vehicle," *Science, Technology, and Human Values* 27, no. 2 (2002): 262–281.

4. Reid R. Heffner, Kenneth S. Kurani, and Thomas S. Turrentine, "Symbolism in California's Early Market for Hybrid Electric Vehicles," *Transportation Research Part D: Transport and Environment* 12, no. 6 (2007): 396–413.

5. Ozaki et al., "The Coproduction of 'Sustainability.'"

6. Christopher W. Wells, *Car Country: An Environmental History* (University of Washington Press: Seattle, 2012), xxiii.

7. Mimi Sheller, "Automotive Emotions: Feeling the Car," *Theory, Culture, and Society* 21, nos. 4–5 (2004): 221–242.

8. Battery researchers long struggled to predict and prove the useful lifetime of rechargeable batteries, an enterprise central to the economics of the all-battery electric vehicle. In the 2010s, pioneering work on this problem was conducted by a group at Dalhousie University led by the materials scientist Jeff Dahn and funded by Tesla. Dahn and his collaborators held that the problem with empirical testing regimes was that they did not approximate battery behavior over time and the way that electric vehicles were driven in real-world circumstances. Electrolyte and charged electrode materials constantly undergo parasitic or side reactions whether batteries are being cycled or not. The solution, argued Dahn and his team, was the ultrahigh-precision charger, a device that used high-rate cycling to "beat the clock" (an expression they attributed to Saft researcher Phillippe Biensan) on problematic side reactions. This instrument technology was developed by Dalhousie and Kyoto University and transferred to instrument manufacturers in the 2010s, a classic example of how academic research and development sometimes addressed innovation in niche fields that established industry had hitherto ignored; see Jeff R. Dahn, J. Christopher Burns, and David A. Stevens, "Importance of Coulombic Efficiency Measurements in R&D Efforts to Obtain Long-Lived Li-Ion Batteries," *Electrochemical Society Interface* 25, no. 3 (2016): 75–78; and Eisler, "Materials Science."

9. Ozaki et al., "The Coproduction of 'Sustainability,'" 530; and Ritsuko Ozaki and Katerina Sevastyanova, "Going Hybrid: An Analysis of Consumer Purchase Motivations," *Energy Policy* 39, no. 5 (2011): 2217–2227, on 2223.

10. Heffner et al., "Symbolism in California's Early Market," 407.

11. See Perrin, *Solo*, and Gjøen and Hård, "Cultural Politics in Action," 264.

12. Tom, interview by author, May 25, 2015.

13. Perrin, *Solo*.

14. Tom, interview.

15. Assem. Bill 475, 2011-2012 Reg. Sess., ch. 274, 2011 Cal Stat., accessed February 19, 2022, https://leginfo.legislature.ca.gov/faces/billTextClient.xhtml?bill_id=201120120AB 475l; *Plug In America*, "Why We're Asking the Governor to Veto AB 475," last modified August 26, 2011, https://pluginamerica.org/why-were-asking-governor-veto-ab -475/.

16. Perrin, *Solo*, 56.

17. Morton, interview by author, May 26, 2015.

18. Phenix, "A Recharged Th!nk Contemplates Its Comeback."

19. Felicia and Peter, interview by author, May 27, 2015.

20. Daniel, interview by author, May 30, 2015.

21. Daniel, interview; and Eberhard, interview.

22. Harvey, interview by author, June 7, 2015. On separate spheres advertising, see Scharff, *Taking the Wheel*, 35–50.

23. See, for example, William J. Mitchell, Christopher E. Borroni-Bird, and Lawrence D. Burns, *Reinventing the Automobile: Personal Urban Mobility for the 21st Century* (Cambridge, MA: MIT Press, 2010); Philip E. Ross, "Ford: Robotaxis in 2021, Self-Driving Cars for Consumer 2025," *IEEE Spectrum*, September 12, 2016, https://spectrum.ieee .org/cars-that-think/transportation/self-driving/ford-robotaxis-in-2021-selfdriving -cars-for-consumer-2025; Philip E. Ross, "Robocars and Electricity: A Match Made in Heaven," *IEEE Spectrum*, June 1, 2017, https://spectrum.ieee.org/cars-that-think /transportation/self-driving/why-robocars-will-run-on-electricity.

24. Raoul, interview by author, July 29, 2016.

25. Maarten Vinkhuyzen, "Nissan's Long Strange Trip with Leaf Batteries," *Clean-Technica*, September 29, 2018, https://cleantechnica.com/2018/09/29/nissans-long -strange-trip-with-leaf-batteries/.

26. Farhad Manjoo, "I've Seen a Future Without Cars, and It's Amazing," *New York Times*, July 9, 2020, https://www.nytimes.com/2020/07/09/opinion/ban-cars -manhattan-cities.html; Micah Toll, "Here's Why Electric Bike Sales Have Skyrocketed During the Coronavirus Lockdown," *Electrek*, May 1, 2020, https://electrek.co /2020/05/01/electric-bike-sales-skyrocket-during-lockdown/.

CONCLUSION

1. Zu and Li, "Thermodynamic Analysis," 2615.

2. See Tesla , "Insane vs. Ludicrous," accessed June 10, 2020, https://forums.tesla .com/forum/forums/insane-vs-ludicrous.

3. *Green Car Congress*, "Trend to Heavier, More Powerful Hybrids Eroding the Technology's Fuel Consumption Benefit," last modified March 28, 2007, https://www .greencarcongress.com/2007/03/trend_to_heavie.html. Analyses by the US Environmental Protection Agency (EPA) that correlated the historical trend in increasing vehicle weight with increasing carbon dioxide emissions looked only at vehicles powered by ICEs, and considered only tailpipe emissions, not emissions from the point of primary energy conversion; see US EPA, *The 2020 EPA Automotive Trends Report: Greenhouse Gas Emissions, Fuel Economy, and Technology Since 1975* (EPA-420-R-21–003, January 2021).

4. Wouk, "Hybrids: Then and Now," 16.

5. Abdulkadir Bedir, Noel Crisostomo, Jennifer Allen, Eric Wood, and Clément Rames, *California Plug-in Electric Vehicle Infrastructure Projections, 2017–2025: Future*

Infrastructure Needs for Reaching the State's Zero-Emission-Vehicle Deployment Goals (California Energy Commission, March 2018), https://www.nrel.gov/docs/fy18osti /70893.pdf.

6. Brad Templeton, "Competing Electric Car Charging Standards Can Be Easily Fixed," *Forbes*, December 19, 2019, https://www.forbes.com/sites/bradtempleton/2019 /12/19/competing-electric-car-charging-standards-can-be-easily-fixed/#6e5ae3f3f40d; Nick Chambers, "Power Politics: Competing Charging Standards Could Threaten Adoption of Electric Vehicles," *Scientific American*, July 5, 2011, https://www.scientific american.com/article/fast-charging-electric-vehicle-standards/.

7. Jae Hyun Lee, Debapriya Chakraborty, Scott J. Hardman, and Gil Tal, "Exploring Electric Vehicle Charging Patterns: Mixed Usage of Charging Infrastructure," *Transportation Research Part D: Transport and Environment* 79 (2020): 102249, 2, 12.

8. Alexander Ritschel and Greg P. Smestad, "Energy Subsidies in California's Electricity Market Deregulation," *Energy Policy* 31 (2003): 1379–1391; Ghazal Razeghi, Brendan Shaffer, and Scott Samuelsen, "Impact of Electricity Deregulation in the State of California," *Energy Policy* 103 (2017): 105–115.

9. For a review of vehicle-to-grid research and development projects, see Adrene Briones, James Francfort, Paul Heitmann, Michael Schey, Steven Schey, and John Smart, *Vehicle-to-Grid (V2G) Power Flow Regulations and Building Codes Review by the AVTA* (Idaho Falls: Idaho National Laboratory, US Department of Energy, 2012), 40–45. See also Jonathan Coignard, Samveg Saxena, Jeffery Greenblatt, and Dai Wang, "Clean Vehicles as an Enabler for a Clean Electricity Grid," *Environmental Research Letters* 13 (2018): 054031; Willett Kempton, Keith Decker, and Li Liao, "Vehicle to Grid Demonstration Project DE-FC26–08NT01905: Final Report," May 7, 2011; Fermata Energy, "Our Story," accessed October 1, 2020, https://www.fermataenergy .com/our-story; Enel X, "Home Page," accessed October 1, 2020, https://www.enelx .com/uk/en.

10. A key problem with distributed generation systems was the phenomenon of "islanding," wherein distributed generation continued to function following grid crash, upsetting the balance between generation, load, voltage, and frequency and creating safety hazards for utility personnel; see Zhang Kai, Liu Kexue, Yao Naipeng, Jia Yuhong, Li Wenjun, and Qin Lihan, "The Impact of Distributed Generation and Its Parallel Operation on Distribution Power Grid," *2015 5th International Conference on Electric Utility Deregulation and Restructuring and Power Technologies* (2015): 2041–2045; Kari Mäki, Anna Kulmala, Sami Repo, and Pertti Järventausta, "Problems Related to Islanding Protection of Distributed Generation in Distribution Network," *2007 IEEE Lausanne Power Tech* (2007): 467–472. David Hawkins, an engineer with the California Independent System Operator (CAISO) who advised Brooks during the ACP experiment, participated in another experiment in the early 2000s involving the simulated use of electric vehicle storage batteries in the regulation application that revealed an adverse reaction by grid control computers responsible for

automatically signaling generators to power up or down. The computers were programmed to control large thermal and hydro plants that gradually responded to signals; when the computers interfaced with storage batteries, they took advantage of the propensity of batteries to respond nearly instantaneously to commands to draw or supply power, leaving the devices completely charged or discharged in a matter of minutes with no spare capacity to regulate the grid; Hawkins, interview.

11. Alec Brooks, interview by author, August 18, 2020.

12. Fermata Energy founder David Slutzky, interview by author, November 12, 2020.

13. Hawkins, interview.

14. Slutzky, interview.

15. Kevin Bullis, "Making Electric Vehicles Practical," *MIT Technology Review*, November 29, 2006, https://www.technologyreview.com/2006/11/29/227392/making-electric -vehicles-practical-2/.

16. Battery University, "BU-205: Types of Lithium-Ion," accessed January 16, 2021, https://batteryuniversity.com/learn/article/types_of_lithium_ion.

17. Battery longevity was a favorite topic of online electric vehicle user forums; see, for example, https://www.reddit.com/r/volt/comments/km2fs5/battery_longevity/ and https://www.reddit.com/r/teslamotors/comments/ckow1k/you_want_a_little_battery _longevity_update_2012/, both accessed January 15, 2021.

18. Harrison and Ludwig, *Electric Vehicle Battery Supply Chain Analysis*, 25, 27.

19. Jay Cole, "LG Chem Says It's Ready to Supply 300 Mile, 120 kWh Batteries," *InsideEVs*, May 28, 2015, https://insideevs.com/news/326517/lg-chem-says-its-ready -to-supply-300-mile-120-kwh-batteries/.

20. Noel Randewich, "Tesla's Market Value Zooms Past That of GM and Ford Combined," *Reuters*, January 8, 2020, https://www.reuters.com/article/us-usa-stocks-tesla /teslas-market-value-zooms-past-that-of-gm-and-ford-combined-idUSKBN1Z72MU.

21. Annual emissions of carbon monoxide and nitrogen oxide from highway vehicles and volatile organic compounds from the transportation sector fell from around 163 million tons, 12.6 million tons, and 18.5 million tons, respectively, in 1970 to 16.2 million tons, 2.4 million tons, and 2.4 million tons in 2020; see US EPA, "Criteria Pollutants National Tier 1 for 1970–2020."

22. Cackette, interview.

23. CARB spokesperson Dave Clegern admitted as much to *Vox* reporter Umair Irfan in 2019; see Irfan, "Trump's Fight with California over Vehicle Emissions Rules Has Divided Automakers," *Vox*, November 5, 2019, https://www.vox.com/policy-and -politics/2019/11/5/20942457/california-trump-fuel-economy-auto-industry.

24. US Environmental Protection Agency (EPA), "Trump Administration Announces One National Program Rule on Federal Preemption of State Fuel Economy Standards,"

September 19, 2019, https://www.epa.gov/newsreleases/trump-administration
-announces-one-national-program-rule-federal-preemption-state-fuel; Jeff Tollefson,
"Trump's Decision to Block California Vehicle Emissions Rules Could Have a Wide
Impact," *Nature*, September 18, 2019, https://www.nature.com/articles/d41586-019
-02812-0.

25. Coral Davenport, "Automakers Drop Efforts to Derail California Climate Rules,"
New York Times, February 2, 2021, https://www.nytimes.com/2021/02/02/climate
/automakers-climate-change.html; US Environmental Protection Agency (EPA),
"Notice of Reconsideration of a Previous Withdrawal of a Waiver for California's
Advanced Clean Car Program (Light-Duty Vehicle Greenhouse Gas Emission Standards and Zero Emission Vehicle Requirements," accessed July 21, 2021, https://
www.epa.gov/regulations-emissions-vehicles-and-engines/notice-reconsideration
-previous-withdrawal-waiver.

26. See, for example, Rachael Nealer, David Reichmuth, and Don Anair, *Cleaner Cars
from Cradle to Grave: How Electric Cars Beat Gasoline Cars on Lifetime Global Warming
Emissions* (Cambridge, MA: Union of Concerned Scientists, 2015), 2–3, 18, https://www
.ucsusa.org/sites/default/files/attach/2015/11/Cleaner-Cars-from-Cradle-to-Grave-full
-report.pdf; and Hawkins, Singh, Majeau-Bettez, and Strømman, "Comparative Environmental Life Cycle Assessment," 53–64; Troy R. Hawkins, Ola Moa Gausen, Anders
Hammer Strømman, "Environmental Impacts of Hybrid and Electric Vehicles—A
Review," *International Journal of Life Cycle Assessment* 17, no. 8 (2012): 997–1014.

27. See Nealer et al., *Cleaner Cars*, 18. In their 2012 review, Hawkins et al. held that
one exception to the one-battery assumption in the lifecycle assessment literature
was a 2010 paper by Notter et al., that acknowledged a requirement for battery
replacement if the life of the notional vehicle they selected for analysis was extended
beyond 240,000 kilometers; see Dominic A. Notter, Marcel Gauch, Rolf Widmer,
Patrick Wäger, Anna Stamp, Rainer Zah, and Hans-Jörg Althaus, "Contribution of
Li-Ion Batteries to the Environmental Impact of Electric Vehicles," *Environmental Science and Technology* 44, no. 17 (2010): 6550–6556, on 6551. On the energy costs of
lightweight vehicle structures, see Sujit Das, Diane Graziano, Venkata K. K. Upadhyayula, Eric Masanet, Matthew Riddle, and Joe Cresko, "Vehicle Lightweighting Energy
Use Impacts in US Light-Duty Vehicle Fleet," *Sustainable Materials and Technologies* 8
(2016): 5–13, on 11; Hyung Chul Kim and Timothy J. Wallington, "Life-Cycle Energy
and Greenhouse Gas Emission Benefits of Lightweighting in Automobiles: Review and
Harmonization," *Environmental Science and Technology* 47, no. 12 (2013): 6089–6097.

28. Mark Vaughn, "What's Going to Happen to All Those Electric Car Batteries
Anyway?" *Autoweek*, March 11, 2021, https://www.autoweek.com/news/green-cars
/a35803612/battery-recycling/.

29. Paul Lienert, "Ex-Tesla Exec Straubel Aims to Build World's Top Battery Recycler," *Reuters*, October 7, 2020, https://www.reuters.com/article/us-batteries-redwood
-recycling-idUSKBN26S3IU.

30. See, for example, Anna Boyden, Vi Kie Soo, and Matthew Doolan, "The Environmental Impacts of Recycling Portable Lithium-Ion Batteries," *Procedia CRIP* 48 (2016): 188–193; Rong Deng, Nathan L. Chang, Zi Ouyang, and Chee Mun Chong, "A Techno-Economic Review of Silicon Photovoltaic Module Recycling," *Renewable and Sustainable Energy Reviews* 109 (2019): 532–550.

31. Cocconi believed that permanent magnet motors offered only marginal performance improvement over induction motors; Cocconi, interview.

32. Laura Millan Lombrana, "Bolivia's Almost Impossible Lithium Dream," *Bloomberg*, December 3, 2018, https://www.bloomberg.com/news/features/2018-12-03/bolivia-s-almost-impossible-lithium-dream; C. J. Atkins, "Bolivia Coup Against Morales Opens Opportunity for Multinational Mining Companies," *People's World*, November 11, 2019, https://www.peoplesworld.org/article/bolivia-coup-against-morales-opens-opportunity-for-multinational-mining-companies/.

33. Tim Treadgold, "Palladium Heads for $2000 An Ounce; Lock up Your Prius!" *Forbes*, December 16, 2019, https://www.forbes.com/sites/timtreadgold/2019/12/16/palladium-heads-for-2000-an-ounce-lock-up-your-prius/; Steve Gorman, "As Hybrid Cars Gobble Rare Metals, Shortage Looms," *Reuters*, August 31, 2009, https://www.reuters.com/article/us-mining-toyota/as-hybrid-cars-gobble-rare-metals-shortage-looms-ionlusidUSTRE57U02B20090831.

34. See US Energy Information Agency (EIA), *August 2020 Monthly Energy Review*, 128. Some environmental analysts held that the use of carbon as a processing agent in steelmaking meant that carbon dioxide emissions from steel fabrication were among the most difficult to abate; see Marian Flores-Granobles and Mark Saeys, "Minimizing CO_2 Emissions with Renewable Energy: A Comparative Study of Emerging Technologies in the Steel Industry," *Energy and Environmental Science* 13 (2020): 1923–1932.

BIBLIOGRAPHY

Abbott, Allan, and Alec N. Brooks. "Flying Fish, The First Human-Powered Hydrofoil to Sustain Flight." Accessed January 8, 2020. https://flyingfishhydrofoil.com/.

Abraham, D. P., J. L. Knuth, D. W. Dees, I. Bloom, and J. P. Christophersen. "Performance Degradation of High-Power Lithium-Ion Cells: Electrochemistry of Harvested Electrodes." *Journal of Power Sources* 170, no. 2 (2007): 465–475.

Abuelsamid, Sam. "Ford Picks Johnson Controls-Saft for PHEV Batteries, Adds 7 Utility Partners to Test Program." *AutoBlog.com*, February 3, 2009. https://www.autoblog .com/2009/02/03/ford-picks-johnson-controls-saft-for-phev-batteries-adds-7-util/2/.

AC Propulsion (ACP). "About Us." Accessed November 8, 2018. https://www .acpropulsion.com/index.php/about-us/management.

Advanced Research Projects Agency (ARPA). "Summary of Proposal of Research on Energy Conversion," February 6, 1961. Project Lorraine, Energy Conversion, 1958–1966 Official Correspondence Files, Materials Sciences Office, ARPA. National Archives and Records Administration, College Park, MD.

AeroVironment and Howard G. Wilson. *Final Study Report and Proposal: The Electric Vehicle—Time for a New Look.* Monrovia, California: AeroVironment, Inc., 1988.

Åhman, Max. "Government Policy and the Development of Electric Vehicles in Japan." *Energy Policy* 34, no. 4 (2006): 433–443.

Air Quality Act of 1967. Public Law 90–148. *US Statutes at Large* 81 (1967): 485–506.

Alpeyev, Pavel Yuki Furukawa, and Masatsugu Horie. "Panasonic Says Gigafactory Profit in Sight as Tesla Ramps Output." *Bloomberg*, November 1, 2018. https://www .bloomberg.com/news/articles/2018-11-01/panasonic-says-gigafactory-profit-in -sight-as-tesla-ramps-output.

Amine, K., C. H. Chen, J. Liu, M. Hammond, A. Jansen, D. Dees, et al. "Factors Responsible for Impedance Rise in High Power Lithium Ion Batteries." *Journal of Power Sources* 97–98 (2001): 684–687.

Amster, Morton. Letter to Chester T. Kamin, February 28, 1996. Robert C. Stempel Papers. Bentley Historical Library, University of Michigan at Ann Arbor.

Ancker-Johnson, Betsy, and Bruce MacDonald. "Robert C. Stempel, 1933-2011." In *Memorial Tributes: National Academy of Engineering, Volume 16*, 309-313. Washington, DC: National Academies Press, 2012.

Andrew, M. R., W. J. Gressler, J. K. Johnson, R. T. Short, and K. R. Williams. "A Fuel-Cell/Lead-Acid Battery Hybrid Car." In *Fuel Cell Technology for Vehicles*, edited by Richard Stobart. Warrendale, PA: Society of Automotive Engineers, 2001.

AP. "GM Chief Arrives at Capitol Hill in Hybrid." December 4, 2008. https://www.youtube.com/watch?v=FNTM3gRSEBE.

Appleby, A. J. "Issues in Fuel Cell Commercialization." *Journal of Power Sources* 58, no. 2 (1996): 153–176.

Arribart, Hervé, and Bernadette Bensaude-Vincent. "Beta-Alumina." Caltech Library. Last modified February 16, 2001. http://authors.library.caltech.edu/5456/1/hrst.mit.edu/hrs/materials/public/Beta-alumina.htm.

Asanuma, Banri. "Manufacturer-Supplier Relationships in Japan and the Concept of the Relation-Specific Skill." *Journal of the Japanese and International Economies* 3, no. 1 (1989): 1–30.

Asner, Glen R. "The Linear Model, the US Department of Defense, and the Golden Age of Industrial Research." In *The Science-Industry Nexus: History, Policy, Implications*, edited by Karl Grandin, Nina Wormbs, and Sven Widmalm, 3–30. Sagamore Beach, MA: Science History Publications/USA, 2004.

Aspray, William, ed. *Chasing Moore's Law: Information Technology Policy in the United States*. Raleigh, NC: SciTech Publishing, 2004.

Assem. Bill 475. 2011-2012 Reg. Sess., ch. 274, 2011 Cal Stat. Accessed February 19, 2022. https://leginfo.legislature.ca.gov/faces/billTextClient.xhtml?bill_id=201120120AB475l.

Atkins, C. J. "Bolivia Coup Against Morales Opens Opportunity for Multinational Mining Companies." *People's World*, November 11, 2019. https://www.peoplesworld.org/article/bolivia-coup-against-morales-opens-opportunity-for-multinational-mining-companies/.

AutoWeek. "World News This Week." 348, no. 2 (1998): 2–3.

Babik, Robert. "The Chevrolet Volt." US Environmental Protection Agency (EPA). Accessed April 30, 2020. https://www.epa.gov/sites/production/files/2015-01/documents/05102011mstrs_babik.pdf.

Bagatelle-Black, Forbes. "AC Propulsion, The Quiet Revolutionaries: Tom Gage Talks About the Role His Company has Played in the Rebirth of the Modern Electric Car." *EV World*, October 27, 2009. http://evworld.com/article.cfm?storyid=1772.

Bain, Dominique M., and Thomas L. Acker. "Hydropower Impacts on Electrical System Production Costs in the Southwest United States." *Energies* 11, no. 2 (2018): 368.

Baker, David R. "Elon Musk: Tesla Was Founded on 2 False Ideas, and Survived Anyway." *SFGate*, May 31, 2016. https://www.sfgate.com/business/article/Elon-Musk-Tesla-was-founded-on-2-false-ideas-7955528.php.

Baker, William O. "The National Role of Materials Research and Development." In *Properties of Crystalline Solids: ASTM Special Technical Publication No. 283*, 1–7. Baltimore, MD: American Society for Testing Materials, 1961.

Baker, William O. "Advances in Materials Research and Development." In *Advancing Materials Research*, edited by Peter A. Psaras and H. Dale Langford, 3–22. Washington, DC: National Academies Press, 1987.

Baldwin, Elizabeth, Valerie Rountree, and Janet Jock. "Distributed Resources and Distributed Governance: Stakeholder Participation in Demand Side Management Governance." *Energy Research and Social Science* 39 (2018): 37–45.

Ball, Jeffrey. "DaimlerChrysler Unveils Prototype Car Using Fuel Cell, Seeks Sales in 5 Years." *Wall Street Journal*, March 18, 1999, B2.

Ball, Jeffrey. "Fuel-Cell Makers Get a Big Boost from Bush's Auto Subsidy Plan." *Wall Street Journal*, January 10, 2002, B2.

Ball, Jeffrey. "Evasive Maneuvers: Detroit Again Tries to Dodge Pressures for a 'Greener' Fleet." *Wall Street Journal*, January 28, 2002, A1.

Balogh, Brian. *Chain Reaction: Expert Debate and Public Participation in American Commercial Nuclear Power, 1945–1975*. Cambridge: Cambridge University Press, 1991.

Barker, Brent. "Electric Power Research Institute: Born in a Blackout." *EPRI Journal* (Summer 2012): 14–17.

Bassett, Ross Knox. *To the Digital Age: Research Labs, Start-up Companies, and the Rise of MOS Technology*. Baltimore: Johns Hopkins University Press, 2002.

Battery University. "BU-205: Types of Lithium-Ion." Accessed January 16, 2021. https://batteryuniversity.com/learn/article/types_of_lithium_ion.

Beder, Sharon. *Global Spin: The Corporate Assault on Environmentalism*. Devon, UK: Green Books, 2002.

Bedir, Abdulkadir, Noel Crisostomo, Jennifer Allen, Eric Wood, and Clément Rames. *California Plug-in Electric Vehicle Infrastructure Projections: 2017–2025: Future Infrastructure Needs for Reaching the State's Zero-Emission-Vehicle Deployment Goals*. California Energy Commission, March 2018. https://www.nrel.gov/docs/fy18osti/70893.pdf.

Bell, Daniel. "The Axial Age of Technology Foreword: 1999." In *The Coming of Post-Industrial Society: A Venture in Social Forecasting*, ix–lxxxvi. New York: Basic Books, 1999.

Bensaude-Vincent, Bernadette. "The Construction of a Discipline: Materials Science in the United States." *Historical Studies in the Physical and Biological Sciences* 31, no. 2 (2001): 223–248.

Benson, Etienne S. "Infrastructural Invisibility: Insulation, Interconnection, and Avian Excrement in the Southern California Power Grid." *Environmental Humanities* 6, no. 1 (2015): 103–130.

Benson, Etienne S. *Surroundings: A History of Environments and Environmentalisms*. Chicago: University of Chicago Press, 2020.

Berdichevsky, Gene, Kurt Kelty, JB Straubel, and Erik Toomre. "The Tesla Roadster Battery System." Tesla Motors. Last modified December 19, 2007. http://large.stanford.edu/publications/coal/references/docs/tesla.pdf

Berkowitz, Justin. "Following Coda and Fisker, Spring of EV Carnage Claims Israeli Start-up Better Place." *Car and Driver*, May 30, 2013. https://www.caranddriver.com/news/a15370868/following-coda-and-fisker-spring-of-ev-carnage-claims-israeli-startup-better-place-analysis/.

Berlin, Leslie. *The Man Behind the Microchip: Robert Noyce and the Invention of Silicon Valley*. Oxford: Oxford University Press, 2005.

Berman, Bradley. "When Old Things Turn into New Again." *New York Times*, October 24, 2007. https://www.nytimes.com/2007/10/24/automobiles/autospecial/24history .html.

Blackbourn, David. *The Conquest of Nature: Water, Landscape, and the Making of Modern Germany*. New York: Norton, 2006.

Block, Fred. "Swimming Against the Current: The Rise of a Hidden Developmental State in the United States." *Politics and Society* 36, no. 2 (2008): 169–206.

Bloom, Ira, Scott A. Jones, Vincent S. Battaglia, Gary L. Henriksen, Jon P. Christophersen, Randy B. Wright, et al. "Effect of Cathode Composition on Capacity Fade, Impedance Rise, and Power Fade in High-Power Lithium-Ion Cells." *Journal of Power Sources* 124, no. 2 (2003): 538–550.

Bockris, John, and A. J. Appleby. "The Hydrogen Economy: An Ultimate Economy? A Practical Answer to the Problem of Energy Supply and Pollution." *Environment This Month: The International Journal of Environmental Science* 1, no. 1 (July 1972): 29–35.

Bogost, Ian. "The Tesla Model 3 Is Still a Rich Person's Car." *The Atlantic*, April 7, 2016. https://www.theatlantic.com/technology/archive/2016/04/tesla-model-3-/477243/m.

Bonner, James. "Arie Jan Haagen-Smit, 1900–1977: A Biographical Memoir." In *Biographical Memoirs* 58, 189–216. Washington, DC: National Academies Press, 1989.

Bonvillian, William B. "Reinventing American Manufacturing: The Role of Innovation." *Innovations: Technology, Governance, Globalization* 7, no. 3 (2012): 97–125.

Bonvillian, William B., and Richard Van Atta. "ARPA-E and DARPA: Applying the DARPA Model to Energy Innovation." *Journal of Technology Transfer* 36, no. 5 (2011): 469–513.

Borroni-Bird, Christopher E. "Fuel Cell Commercialization Issues for Light-Duty Vehicle Applications." *Journal of Power Sources* 61, nos. 1–2 (1996): 33–48.

Boschert, Sherry. *Plug-in Hybrids: The Cars that Will Recharge America*. Gabriola Island, Canada: New Society Publishers, 2006.

Boyden, Anna, Vi Kie Soo, and Matthew Doolan. "The Environmental Impacts of Recycling Portable Lithium-Ion Batteries." *Procedia CRIP* 48 (2016): 188–193.

Breitschwerdt, Dirk, Andreas Cornet, Sebastian Kempf, Lukas Michor, and Martin Schmidt. *The Changing Aftermarket Game and How Automotive Suppliers Can Benefit From Arising Opportunities*. McKinsey & Company, 2017.

Briones, Adrene, James Francfort, Paul Heitmann, Michael Schey, Steven Schey, and John Smart. *Vehicle-to-Grid (V2G) Power Flow Regulations and Building Codes Review by the AVTA*. Idaho Falls: Idaho National Laboratory, US Department of Energy, 2012.

Brock, David C., and Christophe Lécuyer. "Digital Foundations: The Making of Silicon-Gate Manufacturing Technology." *Technology and Culture* 53, no. 3 (2012): 561–597.

Brodd, Ralph J. "Chapter 1: Synopsis of the Lithium-Ion Battery Markets." In *Lithium-Ion Batteries: Science and Technologies*, edited by Masaki Yoshio, Ralph J. Brodd, and Akiya Kozawa, 1–7. New York: Springer, 2009.

Brodd, Ralph J. *Factors Affecting US Production Decisions: Why Are There No Volume Lithium-Ion Battery Manufacturers in the United States?* Gaithersburg, MD: National Institute of Standards and Technology, 2005.

Broder, John M. "Obama to Toughen Rules on Emissions and Mileage." *New York Times*, May 19, 2009, A1.

Broder, John M. "Stalled Out on Tesla's Electric Highway." *New York Times*, February 8, 2013. https://www.nytimes.com/2013/02/10/automobiles/stalled-on-the-ev-highway .html.

Broder, John M. "That Tesla Data: What It Says and What It Doesn't." *New York Times*, February 14, 2013. https://wheels.blogs.nytimes.com/2013/02/14/that-tesla -data-what-it-says-and-what-it-doesnt/.

Brooks, Alec N. *Final Report: Vehicle-to-Grid Demonstration Project: Grid Regulation Ancillary Service with a Battery Electric Vehicle: Prepared for the California Air Resources Board and the California Environmental Protection Agency, Contract Number 01–313.* San Dimas, CA: AC Propulsion, 2002.

Brown, Nik. "Hope Against Hype: Accountability in Biopasts, Presents and Futures." *Science Studies* 16, no. 2 (2003): 3–21.

Brown, Stuart F. "The Automakers' Big-Time Bet on Fuel Cells." *Fortune*, March 30, 1998. 122[D].

Brown, Clair, and Greg Linden. "Offshoring in the Semiconductor Industry: A Historical Perspective." *Brookings Trade Forum* (2005): 279–322.

Brown, Clair, and Greg Linden. *Chips and Change: How Crisis Reshapes the Semiconductor Industry.* Cambridge, MA: MIT Press, 2009.

Brown, Clair, Greg Linden, and Jeffrey T. Macher. "Offshoring in the Semiconductor Industry: A Historical Perspective." In *Brookings Trade Forum: Offshoring White-Collar Work*, 279–333. Washington, DC: Brookings Institution Press, 2005.

Brown, Nik, and Mike Michael. "A Sociology of Expectations: Retrospecting Prospects and Prospecting Retrospects." *Technology Analysis and Strategic Management* 15, no. 1 (2003): 3–18.

Buchner, H., and R. Povel. "The Daimler-Benz Hydride Vehicle Project." *International Journal of Hydrogen Energy* 7, no. 3 (1982): 259–266.

Bullis, Kevin. "Making Electric Vehicles Practical." *MIT Technology Review*, November 29, 2006. https://www.technologyreview.com/2006/11/29/227392/making-electric -vehicles-practical-2/.

Bullis, Kevin. "Innovators Under 35, 2008: JB Straubel, 32, Engineering Electric Sports Cars." *MIT Technology Review*, no date. http://www2.technologyreview.com /tr35/profile.aspx?trid=742.

Burke, Andrew. "The First Modern Hybrid Car, HTV-1, 1978–1982." Accessed December 9, 2016. https://www.youtube.com/watch?v=p_pqT21eLdI.

Business Wire. "Solectria Unveils United States' First Mass Producible All Composite Ground-Up Electric Vehicle." December 2, 1994. http://www.sunrise-ev.com /SolectriaPR1994.htm.

Business Wire. "Texaco to Acquire General Motors' Share of GM-Ovonic Battery Joint Venture." October 10, 2000. https://www.proquest.com/docview/445865036?accountid=14116.

California Air Resources Board (CARB). "Proposed Regulations for Low-Emissions Vehicles and Clean Fuels: Technical Support Document." August 13, 1990.

California Air Resources Board (CARB). *"1998 Zero-Emission Vehicle Biennial Program Review."* July 6, 1998.

California Air Resources Board (CARB). "Resolution 08–24." Last modified March 27, 2008. https://ww2.arb.ca.gov/sites/default/files/barcu/regact/2008/zev2008/zevfsor.pdf.

California Air Resources Board (CARB). *"2000 Zero Emission Vehicle Program."*

California Air Resources Board (CARB). "Fact Sheet: The Zero Emission Vehicle Program-2008." May 6, 2008.

California Air Resources Board (CARB). "California's Zero Emission Vehicle Program." June 2009. https://www.arb.ca.gov/msprog/zevprog/factsheets/zev_tutorial.pdf.

California Air Resources Board (CARB). "Staff Report: Public Hearing to Consider Proposed Amendments to the Low Emission Vehicle III Greenhouse Gas Emission Regulation." August 7, 2018. https://ww2.arb.ca.gov/sites/default/files/barcu/regact/2018/leviii2018/leviiiisor.pdf.

California Air Resources Board (CARB). "Zero-Emission Vehicle Credit Balances." https://ww2.arb.ca.gov/our-work/programs/advanced-clean-cars-program/zev-program/zero-emission-vehicle-credit-balances.

California Air Resources Board (CARB). "History." https://ww2.arb.ca.gov/about/history.

California Environmental Protection Agency. *2000 Emission Vehicle Program Biennial Review.* August 7, 2000. https://ww2.arb.ca.gov/2000-mailouts-list.

Carse, Ashley. "Nature as Infrastructure: Making and Managing the Panama Canal Watershed." *Social Studies of Science* 42, no. 4 (2012): 539–563.

Carse, Ashley. *Beyond the Big Ditch: Politics, Ecology, and Infrastructure at the Panama Canal.* Cambridge, MA: MIT Press, 2014.

Casten, Sean, Peter Teagan, and Richard Stobart. "Fuels for Fuel Cell-Powered Vehicles." In *Fuel Cell Technology for Vehicles,* edited by Richard Stobart, 61–62. Warrendale, PA: Society of Automotive Engineers, 2001.

Ceruzzi, Paul E. *A History of Modern Computing.* Cambridge, MA: MIT Press, 2003.

Chalk, Steven G., James F. Miller, and Fred W. Wagner. "Challenges for Fuel Cells in Transport Applications." *Journal of Power Sources* 86, nos. 1–2 (2000): 40–51.

Chalk, Steven G., Pandit G. Patil, and S. R. Venkateswaran. "The New Generation of Vehicles: Market Opportunities for Fuel Cells." *Journal of Power Sources* 61, nos. 1–2 (1996): 7–13.

Chambers, Nick. "Power Politics: Competing Charging Standards Could Threaten Adoption of Electric Vehicles." *Scientific American,* July 5, 2011. https://www.scientificamerican.com/article/fast-charging-electric-vehicle-standards/.

Chang, Shiuan, Kwo-hsiung Young, Jean Nei, and Cristian Fierro. "Reviews on the US Patents Regarding Nickel/Metal Hydride Batteries." *Batteries* 2, no. 10 (2016): 1–29.

Chapman, Robert M. *The Machine That Could: PNGV, a Government-Industry Partnership*. Santa Monica, CA: RAND, 1998.

Chereb, Sandra. "Nevada Gives $1.3 Billion Tax Break to Electric Car Maker Tesla." *Scientific American*, September 12, 2014. https://www.scientificamerican.com/article /nevada-gives-1-3-billion-tax-break-to-electric-car-maker-tesla/.

Chiang, Yet-Ming. "Building a Better Battery." *Science* 330 (December 10, 2010): 1485–1486. (2015): 21–42.

Chiba, Daishi, and Itsuro Fujino. "Tesla and Panasonic Freeze Spending on $4.5 Billion Gigafactory." *Nikkei Asia*, April 11, 2019. https://asia.nikkei.com/Business /Companies/Tesla-and-Panasonic-freeze-spending-on-4.5bn-Gigafactory.

Chicago Tribune. "Ford Unplugs Electric Vans After Two Fires." June 6, 1994. https:// www.chicagotribune.com/news/ct-xpm-1994-06-06-9406060018-story.html.

Choi, Hyungsub. "The Boundaries of Industrial Research: Making Transistors at RCA, 1948–1960." *Technology and Culture* 48, no. 4 (2007): 758–782.

Choi, Hyungsub, and Brit Shields. "A Place for Materials Science: Laboratory Buildings and Interdisciplinary Research at the University of Pennsylvania." *Minerva* 53, no. 1

Cioc, Mark. *The Rhine: An Eco-Biography, 1815–2000*. Seattle: University of Washington Press, 2002.

Clarke, Deborah. *Driving Women: Fiction and Automobile Culture in Twentieth-Century America*. Baltimore: Johns Hopkins University Press, 2007.

Clarke, Sally H. *Trust and Power: Consumers, the Modern Corporation, and the Making of the United States Automobile Market*. Cambridge: Cambridge University Press, 2007.

Clarkson, Diana, and John T. Middleton. "The California Control Program for Motor Vehicle Created Air Pollution." *Journal of the Air Pollution Control Association* 12, no. 1 (1962): 22–28.

Clinton, William J., and Albert Gore Jr. *Technology for America's Economic Growth, A New Direction to Build Economic Strength*. February 22, 1993. The White House: Office of the Press Secretary.

Clinton, William J., and Albert Gore Jr. "High Technology Policy Initiatives." Filmed February 22, 1993, *C-SPAN, San Jose, California*. https://www.c-span.org/video /?38171-1/high-technology-policy-initiatives.

CNBC. "Tesla Starts Offering Leases for Model 3." April 12, 2019. https://www.cnbc .com/2019/04/12/tesla-begins-offering-leases-for-model-3.html.

Cocconi, Alan. "Electric Car tzero 0-60 3.6 Sec Faster than Tesla Roadster." https:// www.youtube.com/watch?v=gb9E222QsM0andt=138s. Accessed May 7, 2019.

Coetzer, Johan. "A New High Energy Density Battery System." *Journal of Power Sources* 18, no. 4 (1986): 377–380.

Cohan, William D. "Tesla's Biggest Problem Isn't Elon Musk." *New York Times*, September 20, 2018. https://www.nytimes.com/2018/09/20/opinion/tesla-elon-musk.html.

Cohen, Morris A., Narendra Agrawal, and Vipul Agrawal. "Winning in the Aftermarket." *Harvard Business Review*, May 2006. https://hbr.org/2006/05/winning-in-the-aftermarket.

Cohn, Ernst M. "The Growth of Fuel Cell Systems," August 1965. Propulsion, Auxiliary Power: Fuel Cells, 1961–1999. NASA Headquarters Archive, Washington, DC.

Coignard, Jonathan, Samveg Saxena, Jeffery Greenblatt, and Dai Wang. "Clean Vehicles as an Enabler for a Clean Electricity Grid." *Environmental Research Letters* 13 (2018): 054031.

Cole, Jay. "LG Chem Says It's Ready to Supply 300 Mile, 120 kWh Batteries." *InsideEVs*, May 28, 2015. https://insideevs.com/news/326517/lg-chem-says-its-ready-to-supply-300-mile-120-kwh-batteries/.

Cole, Jay. "LG Chem 'Ticked Off' with GM for Disclosing $145/kWh Battery Cell Pricing." *InsideEVs*, October 23, 2015. https://insideevs.com/news/327874/lg-chem-ticked-off-with-gm-for-disclosing-145-kwh-battery-cell-pricing-video/.

Collantes, Gustavo, and Daniel Sperling. "The Origin of California's Zero Emission Vehicle Mandate." *Transportation Research Part A: Policy and Practice* 42, no. 10 (2008): 1302–1313.

Collins, Jim, and Jerry I. Porras. *Built to Last: Successful Habits of Visionary Companies*. London: Random House, 2004.

Collins, Jim. "Forget the Autopilot Crash and Recall: Tesla's Daunting Debt Repayment Schedule is the Real Issue." *Forbes*, April 2, 2018. https://www.forbes.com/sites/jimcollins/2018/04/02/forget-the-autopil . . . las-daunting-debt-repayment-schedule-is-the-real-issue/#284af5f93e94.

Condon, John E. "Practical Values of Space Exploration," October 10, 1962. Technology Utilization, Addresses, Speeches. NASA Headquarters Archive, Washington, DC.

Consumer Reports. "2013 Tesla Model S." Accessed November 30, 2018. https://www.consumerreports.org/cars/tesla/model-s/2013/overview.

Consumer Reports. "Consumer Reports: Tesla Model 3, Chrysler 300 Among Cars No Longer 'Recommended' Based on New Reliability Findings." Last updated February 21, 2019. https://www.consumerreports.org/media-room/press-releases/2019/02/consumer_reports_tesla_model_3_chrysler_300_among_cars_no_longer_recommended_based_on_new_reliability_findings/.

Corrigan, Dennis A. Letter to Jim Metzger, "Subject: OBC Licensee Royalty Opportunities," December 10, 2003. Stanford R. Ovshinsky Papers. Bentley Historical Library, University of Michigan at Ann Arbor.

Corrigan, Dennis A. Letter to Michael A. Fetcenko, June 15, 1990. Stanford R. Ovshinsky Papers. Bentley Historical Library, University of Michigan at Ann Arbor.

Corrigan, Dennis A. Letter to Stanford Ovshinsky and Robert Stempel, "Subject: HEV Presentations to GM," March 21, 1997. Robert C. Stempel Papers. Bentley Historical Library, University of Michigan at Ann Arbor.

Corrigan, Dennis A. Letter to Stanford Ovshinsky, Robert Stempel, and Subhash Dhar, "Subject: Response from USABC Regarding HEV Funding," May 7, 1997. Robert C. Stempel Papers. Bentley Historical Library, University of Michigan at Ann Arbor.

Crandall, Robert W. "The Effects of US Trade Protection for Autos and Steel." *Brookings Papers on Economic Activity* 1 (1987): 271–288.

Cringely, Robert X. "Safety Last." *New York Times,* September 1, 2006. https://www.nytimes.com/2006/09/01/opinion/01cringely.html?p.

Croft, John. "Lithium Battery Rules Could Get Safety Overhaul." *Aviation Week and Space Technology,* October 2, 2015.

Cronon, William. "Foreword: Revisiting Origins: Questions That Won't Go Away." In *Conservation in the Progressive Era: Classic Texts,* edited by David Stradling, vii–ix. Seattle: University of Washington Press, 2004.

Cunningham, Wayne. "Tesla Battery Swap a Dead End." *C/NET,* June 21, 2013. https://www.cnet.com/roadshow/news/tesla-battery-swap-a-dead-end/.

Cushman, John H. Jr. "Intense Lobbying Against Global Warming Treaty." *New York Times,* December 7, 1997, Section 1, 28.

Dahn, Jeff R., J. Christopher Burns, and David A. Stevens. "Importance of Coulombic Efficiency Measurements in R&D Efforts to Obtain Long-Lived Li-Ion Batteries." *Electrochemical Society Interface* 25, no. 3 (2016): 75–78.

Daimler AG. "Electric Motors as an Alternative to Combustion Engines." Last modified November 9, 2007. http://media.daimler.com/marsMediaSite/en/instance/ko/Electric-motors-as-an-alternative-to-combustion-engines.xhtml?oid=9274529.

Darlington, Thomas, Jon Heuss, Dennis Kahlbaum, and George Wolff. "Bibliography of Information Relevant to NHTSA's Reconsideration of the 2022–2025 Model Year GHG Standards." January 24, 2018.

Das, Sujit, Diane Graziano, Venkata K. K. Upadhyayula, Eric Masanet, Matthew Riddle, and Joe Cresko. "Vehicle Lightweighting Energy Use Impacts in US Light-Duty Vehicle Fleet." *Sustainable Materials and Technologies* 8 (2016): 5–13.

Davies, Alex. "Used Teslas Are More About Saving You Time than Money." *Wired,* May 5, 2015. https://www.wired.com/2015/05/used-teslas.

Davis, Mike. *Ecology of Fear: Los Angeles and the Imagination of Disaster.* New York: Vintage Books, 1998.

Dean, Paul. "It's a Bird . . . It's a Plane; It's Weird—But It Can Fly." *Los Angeles Times,* February 16, 1986. https://www.latimes.com/archives/la-xpm-1986-02-16-vw-8846-story.html.

DeCicco, John M. *Fuel Cell Vehicles: Technology, Market and Policy Issues.* Warrendale, PA: Society of Automotive Engineers, 2001.

Deng, Rong, Nathan L. Chang, Zi Ouyang, and Chee Mun Chong, "A Techno-Economic Review of Silicon Photovoltaic Module Recycling." *Renewable and Sustainable Energy Reviews* 109 (2019): 532–550.

Dennis, Michael Aaron. "'Our First Line of Defense:' Two University Laboratories in the Postwar American State." *Isis* 85, no. 3 (1994): 427–455.

Demuro, Doug. "The RAV4 EV Has Had Two Obscure Generations." *Autotrader,* May 17, 2019. https://www.autotrader.com/car-news/toyota-RAV4EV-has-had-two -obscure-generations-281474979930553.

Dennis, Lyle. "A123 Gets $249 Million Government Grant To Build Battery Factory in Michigan." *Green Car Reports,* August 5, 2009. https://www.greencarreports.com /news/1033931_a123-systems-gets-249-million-government-grant-to-build-battery -factory-in-michigan.

Dennison, James T. "Contributions of Aerospace Research to the Business Economy," September 26, 1963. Technology Utilization, Addresses, Speeches. NASA Headquarters Archive, Washington, DC.

Depalma, Anthony. "Ford Joins in a Global Alliance to Develop Fuel-Cell Auto Engines." *New York Times,* December 16, 1997, D1.

Detroit Free Press. "A Lab for Hybrids." September 29, 2005, 2C.

Dhar, Subhash K. Letter to Stanford Ovshinsky. "Subject: Hybrid Electric Vehicle," March 21, 1997. Robert C. Stempel Papers. Bentley Historical Library, University of Michigan at Ann Arbor.

Dhar, Subhash. K. Letter to Stanford Ovshinsky, "EV1 Test Results with Ovonic Batteries," April 17, 1997. Robert C. Stempel Papers. Bentley Historical Library, University of Michigan at Ann Arbor.

Dhar, Subhash K., Stanford R. Ovshinsky, Paul R. Gifford, Dennis A. Corrigan, Michael A. Fetcenko, and Srinivasan Venkatesan. "Nickel/Metal Hydride Technology for Consumer and Electric Vehicle Batteries—A Review and Update." *Journal of Power Sources* 65, nos. 1–2 (1997): 1–7.

Dixon, Chris. "Lots of Zoom, with Batteries." *New York Times,* September 19, 2003, F1.

Doing, Park. *Velvet Revolution at the Synchrotron: Biology, Physics, and Change in Science.* Cambridge, MA: MIT Press, 2009.

Doyle, Jack. *Taken for a Ride: Detroit's Big Three and the Politics of Pollution.* New York and London: Four Walls Eight Windows, 2000.

Doyle, Alister. "From California to Oslo: Foreign Subsidies Fuel Norway's E-Car Boom, for Now." *Reuters,* March 21, 2019. https://uk.reuters.com/article/uk-autos -norway-insight-idUKKCN1R20J4.

Dunn, Richard Chase, and Ann Johnson. "Chasing Molecules: Chemistry and Technology for Automotive Emissions Control." In *Toxic Airs: Body, Place, Planet in Historical Perspective,* edited by James Rodger Fleming and Ann Johnson, 109–126. Pittsburgh: University of Pittsburgh Press, 2014.

Division B: Energy Improvement and Extension Act of 2008. Public Law 110–343. *US Statutes at Large* 122 (2008): 3807–3861.

Eagleton, Terry. *The Ideology of the Aesthetic.* Oxford, UK: Blackwell, 1990.

Ecklund, E. Eugene. "Federal Hydrogen Energy Activities in the United States of America." In *Hydrogen Energy Progress IV: Proceedings of the Fourth World Hydrogen Energy Conference, Pasadena, California, USA., June 13–17, 1982*, vol. 4, edited by T. N. Veziroglu, W. D. Van Vorst, and J. H. Kelley, 1431–1434. Oxford, UK: Pergamon Press, 1982.

The Economist. "Science and Technology: Hybrid Vigour"? 354, no. 8155 (January 29, 2000): 94–95.

The Economist. "Plugging into the Future," 379, no. 8481 (June 10, 2006): 33.

Eberhard, Martin. "Lotus Position." Tesla Motors. Last modified July 25, 2006. https://www.tesla.com/en_GB/blog/lotus-position?redirect=no.

Edwards, Owen. "The Death of the EV-1: Fans of a Battery-Powered Emissions Free Sedan Mourn Its Passing." *Smithsonian Magazine*, June 2006. https://www.smithsonianmag.com/science-nature/the-death-of-the-ev-1-118595941/.

Egan, Timothy "Tapes Show Enron Arranged Plant Shutdown." *New York Times*, February 4, 2005, A12.

Eisler, Matthew N. "'A Modern Philosopher's Stone': Techno-Analogy and the Bacon Cell." *Technology and Culture* 50, no. 2 (2009): 345–365.

Eisler, Matthew N. "Energy Innovation at Nanoscale: Case Study of an Emergent Industry." *Science Progress*, May 23, 2011. http://scienceprogress.org/2011/05/innovation-case-study-nanotechnology-and-clean-energy/.

Eisler, Matthew N. *Overpotential: Fuel Cells, Futurism, and the Making of a Power Panacea.* New Brunswick, NJ: Rutgers University Press, 2012.

Eisler, Matthew N. "'The Ennobling Unity of Science and Technology:' Materials Sciences and Engineering, the Department of Energy, and the Nanotechnology Enigma." *Minerva* 51, no. 2 (2013): 225–251.

Eisler, Matthew N. "Exploding the Black Box: Personal Computing, the Notebook Battery Crisis, and Postindustrial Systems Thinking." *Technology and Culture* 58, no. 2 (2017): 368–391.

Eisler, Matthew N. "Materials Science, Instrument Knowledge, and the Power Source Renaissance." *Proceedings of the IEEE* 105, no. 12 (2017): 2382–2389.

Eisler, Matthew N. "Vehicle-to-Grid and the Energy Conversion Imaginary." In *Rethinking Electric History: From Esoteric Knowledge to Invisible Infrastructure to Fragile Networks*, edited by W. Bernard Carlson and Erik M. Conway. Forthcoming, University of Virginia Press, 2023.

Elkind, Sarah S. *How Local Politics Shape Federal Policy: Business, Power, and the Environment in Twentieth-Century Los Angeles.* Chapel Hill: University of North Carolina Press, 2011.

Engineering and Science. "Cambridge or Bust, Pasadena or Bust: Both Teams in the Great Electric Car Race Made It—and Busted Too." 32, no. 1 (October 1968): 10–17.

Electric and Hybrid Vehicle Research, Development, and Demonstration Act of 1976. Public Law 94–413. *US Statutes at Large* 90 (1976): 1260–1272.

Enel X. "Home Page." Accessed October 1, 2020. https://www.enelx.com/uk/en.

Energy Conversion Devices (ECD). "Energy Conversion Devices and Its Subsidiary, Ovonic Battery Company, Announce Battery Agreement with a Major Japanese Automobile Manufacturer," September 19, 1991. Stanford R. Ovshinsky Papers. Bentley Historical Library, University of Michigan at Ann Arbor.

Energy Conversion Devices (ECD). "ECD/OBC Announces Agreement with Honda," January 4, 1994. Stanford R. Ovshinsky Papers. Bentley Historical Library, University of Michigan at Ann Arbor.

Energy Conversion Devices (ECD). "Potential Settlement Plan," January 28, 1997. Stanford R. Ovshinsky Papers. Bentley Historical Library, University of Michigan at Ann Arbor.

Energy Conversion Devices (ECD). "Annual Meeting 1997." Stanford R. Ovshinsky Papers. Bentley Historical Library, University of Michigan at Ann Arbor.

Energy Conversion Devices (ECD). "ECD/Ovonic NiMH Battery Update," no date. Robert C. Stempel Papers. Bentley Historical Library, University of Michigan at Ann Arbor.

Energy Conversion Devices (ECD). "News Release: MBI Litigation Concluded in Favor of ECD," January 5, 1998. Robert C. Stempel Papers. Bentley Historical Library, University of Michigan at Ann Arbor.

Energy Conversion Devices (ECD). "News Release: ECD Announces Ovonic Battery Is Filing a Lawsuit Against Matsushita Battery," March 6, 2001. Stanford R. Ovshinsky Papers. Bentley Historical Library, University of Michigan at Ann Arbor.

Energy Conversion Devices (ECD). "Form 8-K," July 7, 2004.

Energy Conversion Devices (ECD), Ovonic Battery Company (OBC), and Texaco. "Waiver (Re: Chevron Merger)," July 17, 2001. Robert C. Stempel Papers. Bentley Historical Library, University of Michigan at Ann Arbor.

Energy Independence and Security Act of 2007. Public Law 110–140. *US Statutes at Large* 121 (2007): 1492–1801.

Energy Policy Act of 2005. Public Law 109–58. *US Statutes at Large* 119 (2005): 594–1143.

Energy Security Act. Public Law 96–294. *US Statutes at Large* 94 (1980): 611–779.

Ericsen, Terry. "The Second Electronic Revolution (It's All About Control)." *IEEE Transactions on Industry Applications* 46, no. 5 (September/October 2010): 1778–1786.

EVAmerica and the US Department of Energy (DOE). "1998 Ford Ranger EV."

EVAmerica and the US Department of Energy (DOE). "1999 Ford Ranger EV."

Esso Research and Engineering Company. "Proposal for the Continuation of Government Contract Research on Fuel Cells." Project Lorraine, Energy Conversion, 1958–1966 Official Correspondence Files—Materials Sciences Office. National Archives and Records Administration, College Park, MD.

Farmer, Richard, Dennis Zimmerman, and Gail Cohen. *Causes and Lessons of the California Electricity Crisis.* Congressional Budget Office, September 2001.

Fermata Energy. "Our Story." Accessed October 1, 2020. https://www.fermataenergy.com/our-story.

Ferreira, Fernanda. "The Great Big Headache of 1968." *MIT Technology Review*, February 26, 2020. https://www.technologyreview.com/2020/02/26/905991/the-great-big-headache-of-1968/.

Fetcenko, M. A., S. R. Ovshinsky, B. Reichman, K. Young, C. Fierro, J. Koch, et al. "Recent Advances in NiMH Battery Technology." *Journal of Power Sources* 165, no. 2 (2007): 544–551.

Fialka, John J. *Car Wars: The Rise, the Fall, and the Resurgence of the Electric Car*. New York: Thomas Dunne Books, 2015.

Fisher, Lawrence M. "California Is Backing off Mandate for Electric Car: Board Finds Shortcomings in Technology." *New York Times*, December 26, 1995, A14.

Fisher, Lawrence M. "GM, in a First, Will Sell a Car Designed for Electric Power This Fall." *New York Times*, January 5, 1996, A10.

Fletcher, Seth. *Bottled Lightning: Superbatteries, Electric Cars, and the New Lithium Economy*. New York: Hill and Wang, 2011.

Flores-Granobles, Marian, and Mark Saeys. "Minimizing CO_2 Emissions with Renewable Energy: A Comparative Study of Emerging Technologies in the Steel Industry." *Energy and Environmental Science* 13 (2020): 1923–1932.

Ford Motor Company. "Th!nk City Electric Vehicle Demonstration Program: Final Project Report, June 2005." June 18, 2004, Award DE-FG26-O1ID14048.

Fortun, Michael. "Mediated Speculations in the Genomics Future Markets." *New Genetics and Society* 20, no. 2 (2001): 139–156.

Fortun, Michael. *Promising Genomics: Iceland and DeCODE Genetics in a World of Speculation*. Berkeley: University of California Press, 2008.

Fortune. "Another One out the Door: Now It's Tesla's Supply Chain Chief Who's Leaving." September 21, 2018. http://fortune.com/2018/09/21/tesla-supply-chain-executive-departure/.

Frankel, Alison. "Hedge Funds Step up to Lead Shareholder Suit Against Tesla." *Reuters*, October 10, 2018. https://uk.reuters.com/article/legal-us-otc-tesla/hedge-funds-step-up-to-lead-shareholder-suit-against-tesla-idUKKCN1MK2HY.

Freiwald, D. A., and W. J. Barattino. "Technical Note: Alternative Transportation Vehicles for Military-Base Operations." *International Journal of Hydrogen Energy* 6, no. 6 (1981): 631–636.

Freund, Peter, and George Martin. *The Ecology of the Automobile*. Montreal: Black Rose Books, 1993.

Fri, Robert W., William Agnew, Peter D. Blair, Ralph Cavanagh, Uma Chowdhry, Linda R. Cohen, et al. *Energy Research at DOE: Was It Worth It? Energy Efficiency and Fossil Energy Research, 1978 to 2000*. Washington, DC: National Academy Press, 2001.

Gaither, Chris, and Dawn C. Chmielewski, "Fears of Dot-Com Crash, Version 2.0." *Los Angeles Times*. July 16, 2006, https://www.latimes.com/archives/la-xpm-2006-jul-16-fi-overheat16-story.html.

Gartman, David. *Auto Opium: A Social History of American Automobile Design*. London: Routledge, 1994.

Gartman, David. "Three Ages of the Automobile: The Cultural Logics of the Car." *Theory, Culture and Society* 21, nos. 4–5 (2004): 169–195.

Geller, Marc. "Wired Blogger Takes on Nissan Leaf." *Plug and Cars*, January 25, 2010. https://plugsandcars.blogspot.com/2010_01_25_archive.html.

Gelles, David, and Peter Eavis. "Elon Musk Wants to Take Tesla Private; Can He Make the Math Work?" *New York Times*, August 23, 2018. https://www.nytimes.com/2018/08/23/business/dealbook/tesla-investors-elon-musk.html.

Gelles, David. "Why Elon Musk Reversed Course on Taking Tesla Private." *New York Times*, August 25, 2018. https://www.nytimes.com/2018/08/25/business/elon-musk-tesla-private.html?action=clickandmodule=Top%20Storiesandpgtype=Homepage.

General Motors (GM). "Firebird III." Accessed February 21, 2022. https://www.gmheritagecenter.com/docs/gm-heritage-archive/historical-brochures/1958-firebird-III/1958_Firebird_III_Brochure.pdf.

General Motors (GM). "Impact's Aluminum Frame Provides Lightweight Support." Undated photo. Robert C. Stempel Papers. Bentley Historical Library, University of Michigan at Ann Arbor.

General Motors (GM). "GM-Ovonic Forms Management Team, Names Board of Managers." September 9, 1994. Stanford R. Ovshinsky Papers. Bentley Historical Library, University of Michigan at Ann Arbor.

General Motors (GM). "Manufacturing EV1 and S-10 Electric NiMH Batteries." December 3, 1998. Robert C. Stempel Papers. Bentley Historical Library, University of Michigan at Ann Arbor.

General Motors (GM). "Form 8-K, Harry J. Pearce," February 1, 2000.

Gertner, Jon. *The Idea Factory: Bell Labs and the Great Age of American Innovation*. New York: The Penguin Press, 2012.

Gertner, Jon. "The Risk of a New Machine." *Fastcompany.com* (April 2012): 104–133.

Gifford, Paul R. Letter to Stanford Ovshinsky, "Subject: Meeting with GM Hybrid Vehicle Team," March 21, 1997. Robert C. Stempel Papers. Bentley Historical Library, University of Michigan at Ann Arbor.

Gifford, Paul R., John Adams, Dennis Corrigan, and Srinivasan Venkatesan. "Development of Advanced Nickel/Metal Hydride Batteries for Electric and Hybrid Vehicles." *Journal of Power Sources* 80, nos. 1–2 (1999): 157–163.

Gjøen, Heidi, and Mikael Hård. "Cultural Politics in Action: Developing User Scripts in Relation to the Electric Vehicle." *Science, Technology, and Human Values* 27, no. 2 (2002): 262–281.

GM-Ovonic. "Slide 4 [GM-Ovonic organizational tree]." April 26, 1995. Stanford R. Ovshinsky Papers. Bentley Historical Library, University of Michigan at Ann Arbor.

GM-Ovonic. "Production Status" and "Cost Projections." April 26, 1995. Stanford R. Ovshinsky Papers. Bentley Historical Library, University of Michigan at Ann Arbor.

Godin, Benoît. "The Linear Model of Innovation: The Historical Construction of an Analytical Framework." *Science, Technology and Human Values* 31, no. 6 (2006): 639–667.

Goldstein, Bernard R. "Saving the Phenomena: The Background to Ptolemy's Planetary Theory." *Journal for the History of Astronomy* 28 (1997): 1–12.

Golson, Jordan. "Tesla Ends 'Resale Value Guarantee' On New Vehicle Purchases." *The Verge*, July 13, 2016. https://www.theverge.com/2016/7/13/12173310/tesla -model-s-resale-value-guarantee-ending.

Goodenough, John B. "Rechargeable Batteries: Challenges Old and New." *Journal of Solid-State Electrochemistry* 16, no. 6 (2012): 2019–2029.

Goodenough, John B., K. Mizushima, and T. Takeda. "Solid-Solution Oxides for Storage-Battery Electrodes." *Japanese Journal of Applied Physics* 19 (1980): 305–313.

Goodenough, John B., and Youngsik Kim. "Challenges for Rechargeable Li Batteries." *Chemistry of Materials Review* 22, no. 3 (2010): 587–603.

Gordin, Michael D., Helen Tilley, and Gyan Prakash. "Utopia and Dystopia Beyond Space and Time." In *Utopia/Dystopia: Conditions of Historical Possibility*, edited by Michael D. Gordin, Helen Tilley, and Gyan Prakash, 1–17. Princeton, NJ: Princeton University Press, 2010.

Gorman, Steve. "As Hybrid Cars Gobble Rare Metals, Shortage Looms." *Reuters*, August 31, 2009. https://www.reuters.com/article/us-mining-toyota/as-hybrid-cars -gobble-rare-metals-shortage-looms-ionlusidUSTRE57U02B20090831.

Gottlieb, Robert. *Forcing the Spring: The Transformation of the American Environmental Movement*. Washington, DC: Island Press, 2005.

Gow, Phil. Letter to Stanford Ovshinsky, Robert Stempel, and Subhash Dhar, "Subject: AeroVironment Testing of OBC Prototype Hybrid Battery Module," April 4, 1997. Robert C. Stempel Papers. Bentley Historical Library, University of Michigan at Ann Arbor.

Grandin, Karl, Nina Wormbs, and Sven Widmalm, eds. *The Science-Industry Nexus: History, Policy, Implications*. Sagamore Beach, MA: Science History Publications, 2005.

Grant, Charley. "Is Tesla Abandoning the Mass Market? Elon Musk's About-Face on Model 3 Pricing is a Warning Sign for the Stock." *Wall Street Journal*, May 21, 2018. https://www.wsj.com/articles/is-tesla-abandoning-the-mass-market-1526917239.

Green, Gavin. "Interview: Rick Wagoner, General Motors Co." *Motor Trend*, July 25, 2006. https://www.motortrend.com/news/rick-wagoner-general-motors/.

Green Car Congress. "Trend to Heavier, More Powerful Hybrids Eroding the Technology's Fuel Consumption Benefit." March 28, 2007. https://www.greencarcongress .com/2007/03/trend_to_heavie.html.

Green Car Congress. "GM To Manufacture Volt Packs in US; LG Chem Providing Cells; Partnership With U. Michigan." January 12, 2009. https://www.greencarcongress .com/2009/01/gm-to-manufactu.html.

Green Car Congress. "Worldwide Prius Cumulative Sales Top 2M Mark; Toyota Reportedly Plans Two New Prius Variants for the US by the End of 2012." October 7, 2010. http://www.greencarcongress.com/2010/10/worldwide-prius-cumulative-sales -top-2m-mark-toyota-reportedly-plans-two-new-prius-variants-for-the-.html#more.

Green Car Congress. "Tesla Battery Supply Deal for Th!nk Scuttled." November 2, 2017, https://www.greencarcongress.com/2007/11/tesla-battery-s.html.

Greenberg, Daniel S. *The Politics of Pure Science.* Chicago: University of Chicago Press, 1967.

Greenberg, Daniel S. *Science, Money, and Politics: Political Triumph and Ethical Erosion.* Chicago: University of Chicago Press, 2001.

Greenpeace. "Th!nk Again: Ford Does a U-Turn." September 17, 2004. https://web .archive.org/web/20060609043839/http://www.greenpeace.org/international/news /th-nk-again-ford-does-a-u-tur#.

Gregory, D.P., P.J. Anderson, R.J. Dufour, R.H. Elkins, W.J.D. Escher, R.B. Foster, et al. *A Hydrogen-Energy System.* Institute of Gas Technology/American Gas Association, 1973.

Gross, Benjamin. *The TVs of Tomorrow: How RCA's Flat-Screen Dreams Led to the First LCDs.* Chicago: University of Chicago Press, 2018.

Grove, William Robert. "On a Gaseous Voltaic Battery." *Philosophical Magazine and Journal of Science* 21, S.3 (December 1842): 417.

Gultom, Yohanna M. L. "Governance Structures and Efficiency in the US Electricity Sector After the Market Restructuring and Deregulation." *Energy Policy* 129 (2019): 1008–1019.

Haagen-Smit, Arie Jan. "Chemistry and Physiology of Los Angeles Smog." *Industrial and Engineering Chemistry* 44, no. 6 (1952): 1342–1346.

Haagen-Smit, Zus. Interview by Shirley K. Cohen. Pasadena, California, March 16 and 20, 2000. Oral History Project, California Institute of Technology Archives. https://oralhistories.library.caltech.edu/42/.

Hacker, Barton C., and James M. Grimwood. *On the Shoulders of Titans: A History of Project Gemini.* Washington, DC: National Aeronautics and Space Administration (NASA), 1977.

Hackett, Edward J., Olga Amsterdamska, Michael Lynch, and Judy Wajcman. "Introduction." In *Handbook of Science and Technology Studies,* edited by Edward J. Hackett, Olga Amsterdamska, Michael Lynch, and Judy Wajcman, 1–8. Cambridge, MA: MIT Press, 2008.

Hakim, Danny. "Automakers Drop Suits on Air Rules." *New York Times,* August 12, 2003, A1.

Hakim, Danny. "Auto Supplier Delphi Files for Bankruptcy, and GM Will Share Some of the Fallout." *New York Times,* October 9, 2005, Section 1, 28.

Halpern, Megan K. "Negotiations and Love Songs: Integration, Fairness, and Balance in an Art–Science Collaboration." In *Routledge Handbook of Art, Science, and*

Technology Studies, edited by Hannah Star Rogers, Megan K. Halpern, Dehlia Hannah, and Kathryn de Ridder-Vignone, 319-334. Routledge, 2021.

Hardin, Angela. "LG Chem, Argonne Sign Licensing Deal to Make, Commercialize Advanced Battery Material." January 6, 2011. http://www.anl.gov/articles/lg-chem -argonne-sign-licensing-deal-make-commercialize-advanced-battery-material.

Harrar, George. "Technology: The 'Concept Car' Pushes Change." *New York Times*, July 1, 1990, Section 3, p. 5.

Harrison, Daniel, and Christopher Ludwig. *Electric Vehicle Battery Supply Chain Analysis: How Battery Demand and Production Are Reshaping the Automotive Industry*. London: Ultima Media, 2021.

Harwell, Drew. "Tesla Hits 5000-a-Week Model 3 Production Goal." *Washington Post*, July 2, 2018. https://www.washingtonpost.com/business/economy/tesla-hits -5000-a-week-model-3-production-goal/2018/07/02/a3306ca0-7e48-11e8-b660 -4d0f9f0351f1_story.html.

Haskins, Harold. Letter to Dennis Corrigan, May 6, 1997. Robert C. Stempel Papers. Bentley Historical Library, University of Michigan at Ann Arbor.

Hawken, Paul, Amory B. Lovins, and L. Hunter Lovins. *Natural Capitalism: Creating the Next Industrial Revolution*. New York: Little, Brown, and Company, 1999.

Hawkins, Andrew J. "Two Guys Did a Coast-to-Coast 'Cannonball Run' in a Tesla Model S for a New Record." *The Verge*, July 9, 2017. https://www.theverge.com/2017 /7/9/15938028/tesla-model-s-cannonball-run-record.

Hawkins, Troy R., Ola Moa Gausen, and Anders Hammer Strømman. "Environmental Impacts of Hybrid and Electric Vehicles—A Review." *International Journal of Life Cycle Assessment* 17, no. 8 (2012): 997–1014.

Hawkins, Troy R., Bhawna Singh, Guillaume Majeau-Bettez, and Anders Hammer Strømman. "Comparative Environmental Life Cycle Assessment of Conventional and Electric Vehicles." *Journal of Industrial Ecology* 17, no.1 (2012): 53–64.

Hayes, Paul R. "Auto Facts: 200 Motorists to Give Chrysler Turbine 'Ride-and-Drive' Test for Year." *Philadelphia Inquirer*, May 20, 1963, p. 24.

Hays, Samuel P. *Beauty, Health, and Permanence: Environmental Politics in the United States, 1955–1985*. Cambridge: Cambridge University Press, 1987.

Heavner, Brad. *Pollution Politics 2000: California Political Expenditures of the Automobile and Oil Industries, 1997–2000*. Santa Barbara, CA: California Public Interest Research Group Charitable Trust, 2000.

Heffner, Reid R., Kenneth S. Kurani, and Thomas S. Turrentine. "Symbolism in California's Early Market for Hybrid Electric Vehicles." *Transportation Research Part D: Transport and Environment* 12, no. 6 (2007): 396–413.

Heizer, Robert F. "The Background of Thomsen's Three-Age System." *Technology and Culture* 3, no. 3 (1962): 259–266.

Hilgartner, Stephen. *Science on Stage: Expert Advice as Public Drama*. Stanford, CA: Stanford University Press, 2000.

Hirsch, Jerry. "Tesla Drives California Environmental Credits to the Bank." *Los Angeles Times*, May 5, 2013. https://www.latimes.com/business/autos/la-fi-electric-cars-20130506-story.html.

Hobsbawm, Eric. *Industry and Empire: From 1750 to the Present Day*. London: Pelican, 1968.

Hoddeson, Lillian, and Peter Garrett. *The Man Who Saw Tomorrow: The Life and Inventions of Stanford R. Ovshinsky*. Cambridge, MA: MIT Press, 2018.

Hoium, Travis. "Competitors Made $1.5 Billion from Tesla Motors' Success." *Motley Fool*, October 31, 2015. https://www.fool.com/investing/general/2015/10/31/how-competitors-made-15-billion-tesla-motors-succe.aspx.

Holland, Maximilian. "Tesla 2018 Annual Shareholder Meeting: Quick Highlights," *EVObsession*. June 5, 2018. https://evobsession.com/tesla-2018-annual-tesla-shareholder-meeting-quick-highlights/.

Honda Motor Company. "Honda Electric Vehicle Program Enters Next Phase." April 26, 1999. Stanford R. Ovshinsky Papers. Bentley Historical Library, University of Michigan at Ann Arbor.

Honda Motor Company. "Honda EV Plus: The Dream of an Electric Vehicle/1988." Accessed July 4, 2020. https://global.honda/heritage/episodes/1988evplus.html.

Horton, Peter. "Peter Buys an Electric Car." *Los Angeles Times*, June 8, 2003. https://www.latimes.com/archives/la-xpm-2003-jun-08-tm-ev123-story.html.

Huff, James R., and John C. Orth. "The USAMECOM-MERDC Fuel Cell Electric Power Generation Program." In *Fuel Cell Systems II: 5th Biennial Fuel Cell Symposium Sponsored by the Division of Fuel Chemistry at the 154th Meeting of the American Chemical Society, Chicago, Illinois, September 12–14, 1967*, 323–326. Washington, DC: American Chemical Society, 1969.

Hughes, Thomas P. *Networks of Power: Electrification in Western Society, 1880–1930*. Baltimore: Johns Hopkins University Press, 1983.

Hull, Dana. "2010: Tesla Gets Ready to Take over the Former NUMMI Auto Plant in Fremont." *The Mercury News*, September 16, 2010. https://www.mercurynews.com/2010/09/16/2010-tesla-gets-ready-to-take-over-the-former-nummi-auto-plant-in-fremont/.

Hull, Dana, John Lippert, and Sarah Gardner. "The Future of Tesla Hinges on This Gigantic Tent." *Bloomberg News*, June 25, 2018. https://www.bloomberg.com/news/articles/2018-06-25/the-future-of-tesla-hinges-on-this-gigantic-tent.

Hydrogen Future Act of 1996. Public Law 104–271. *US Statutes at Large* 110 (1996): 3304–3308.

Ikehara, Haruo. "Toyota's Plug-in Hybrid: Debut of Prototype Is Near." *Nikkei Business Online*, January 29, 2007. http://business.nikkeibp.co.jp/article/eng/20070129/117846/.

Ingram, Frederick C. "Delphi Automotive Systems Corporation." *International Directory of Company Histories:Encyclopedia.com*. January 24, 2022. https://www.encyclopedia.com/books/politics-and-business-magazines/delphi-automotive-systems-corporation.

Insinna, Valerie. "Inside America's Dysfunctional Trillion-Dollar Fighter-Jet Program." *New York Times Magazine*, August 21, 2019. https://www.nytimes.com/2019/08/21/magazine/f35-joint-strike-fighter-program.html.

International Battery Materials Association (IBA). "Special Symposium to Honor Michael Thackeray." Last modified March 10, 2013. http://congresses.icmab.es/iba2013/images/stories/PDF/mt.pdf.

International Energy Agency (IEA). *Global EV Outlook 2016: Beyond One Million Electric Cars. Paris: OECD/International Energy Agency*, 2016. https://www.iea.org/reports/global-ev-outlook-2016.

International Energy Agency (IEA). *Global EV Outlook 2021: Accelerating Ambitions Despite The Pandemic. Accessed February 21, 2022.* https://www.iea.org/reports/global-ev-outlook-2021/trends-and-developments-in-electric-vehicle-markets.

Irfan, Umair. "Trump's Fight with California over Vehicle Emissions Rules Has Divided Automakers." *Vox*, November 5, 2019. https://www.vox.com/policy-and-politics/2019/11/5/20942457/california-trump-fuel-economy-auto-industry.

Irvin, Robert W. "The Revival of Electric Vehicles: Passim [sic] Fancy or Car of the Future?" *New York Times*, April 7, 1974.

Israel, Paul B. *Edison: A Life of Invention*. New York: John Wiley, 1998.

Israel, Paul B. "Inventing Industrial Research: Thomas Edison and the Menlo Park Laboratory." *Endeavour* 26, no. 2 (2002): 48–54.

Itazaki, Hideshi. *The Prius That Shook the World: How Toyota Developed the World's First Mass-Production Hybrid Vehicle*. Tokyo: Nikkan Kogyo Shimbun, 1999.

Jack Morton Company. "Highlights of GM Advanced Technology Vehicles Press Conference, North American International Auto Show," January 4, 1998. Stanford R. Ovshinsky Papers. Bentley Historical Library, University of Michigan at Ann Arbor.

Jacobs, Chip, and William J. Kelly. *Smogtown: The Lung-Burning History of Pollution in Los Angeles*. New York: Overlook, 2008.

Jasanoff, Sheila. "Future Imperfect: Science, Technology, and Imaginations of Modernity." In *Dreamscapes of Modernity: Sociotechnical Imaginaries and the Fabrication of Power*, edited by Sheila Jasanoff and Sang-Hyun Kim, 1–33. Chicago: University of Chicago Press, 2015.

Jensen, Cheryl. "Nissan Battery Plant Begins Operations in Tennessee." *New York Times*, December 13, 2012. https://wheels.blogs.nytimes.com/2012/12/13/nissan-battery-plan.

Jensen, Johs, Jorgen Lundsgaard, and Carol M. Perram. *Electric Vehicles for Urban Transport: A Preliminary Investigation into the Possibilities for Introduction of Electric Buses and Other Electric Vehicles in Odense, Denmark*. Odense, Denmark: Odense University Press, 1980.

Jin, Hyunjoo. "LG Hopes to Make New Battery Cells for Tesla in 2023 in US or Europe." *Reuters*, March 9, 2021. https://www.reuters.com/article/us-tesla-lg-evs-exclusive-idUSKBN2B12HY.

Johnson, Ann. "The End of Pure Science: Science Policy from Bayh-Dole to the NNI." In *Discovering the Nanoscale*, edited by Davis Baird and Alfred Nordmann, 217–230. Amsterdam: IOS Press, 2004.

Johnson, Chalmers. *MITI and the Japanese Miracle: The Growth of Industrial Policy, 1925–1975*. Stanford: Stanford University Press, 1982.

Johnson, Lyndon B. "Special Message to the Congress: Protecting Our Natural Heritage." Speech, Washington, DC, January 30, 1967. The American Presidency Project, University of California at Santa Barbara. https://www.presidency.ucsb.edu /documents/special-message-the-congress-protecting-our-natural-heritage.

Jones, Lawrence W. "Hydrogen: A Fuel to Run Our Engines in Clean Air." *Saturday Evening Post*, Spring 1972, 34.

Jordan, Kent A. "Civil Action No. 96–101: Matsushita Battery Industrial Co., Ltd., versus Energy Conversion Devices, Inc., and Ovonic Battery Company, Inc." February 28, 1996. Robert C. Stempel Papers. Bentley Historical Library, University of Michigan at Ann Arbor.

Jørgensen, Dolly. "Mixing Oil and Water: Naturalizing Offshore Oil Platforms in Gulf Coast Aquariums." *Journal of American Studies* 46, no. 2 (2012): 461–480.

Jorgenson, Dale W., and Charles W. Wessner. "Preface." In *Productivity and Cyclicality in Semiconductors: Trends, Implications, and Questions; Report of a Symposium*, edited by Dale W. Jorgenson and Charles W. Wessner, xiii–xviii. Washington, DC: National Academies Press, 2004.

Joskow, Paul L. "Markets for Power in the United States: An Interim Assessment." *Energy Journal* 27, no. 1 (2006): 1–36.

Justi, Eduard. *Leitungsmechanismus und Energieumwandlung in Festkörpern*. Göttingen, Germany: Vandenhoeck and Ruprecht, 1965.

Kahn, Alfred E. "Surprises of Airline Deregulation." *American Economic Review* 78, no. 2 (1988): 316–322.

Kahn, Alfred E. *Whom the Gods Would Destroy, or How Not to Deregulate*. Washington, DC: AEI Press, 2001.

Kai, Zhang, Liu Kexue, Yao Naipeng, Jia Yuhong, Li Wenjun, and Qin Lihan. "The Impact of Distributed Generation and Its Parallel Operation on Distribution Power Grid." *2015 5th International Conference on Electric Utility Deregulation and Restructuring and Power Technologies* (2015): 2041–2045.

Kamin, Chester T. Letter to Morton Amster, February 26, 1996. Robert C. Stempel Papers. Bentley Historical Library, University of Michigan at Ann Arbor.

Kamin, Chester T. Letter to Robert C. Stempel and Stanford R. Ovshinsky, "Re: ChevronTexaco," October 20, 2004. Stanford R. Ovshinsky Papers. Bentley Historical Library, University of Michigan at Ann Arbor.

Kanellos, Michael. "Can Anything Tame the Battery Flames?" *C/NET*, August 16, 2006. http://news.cnet.com/Can-anything-tame-the-battery-flames/2100-11398_3-6105924 .html.

Karpinski, A. P., B. Makovetski, S. J. Russell, J. R. Serenyi, and D. C. Williams. "Silver-Zinc: Status of Technology and Applications." *Journal of Power Sources* 80, nos. 1–2 (1999): 53–60.

Kawauchi, Shosuke. Letter to Stanford Ovshinsky. June 12, 1992. Stanford R. Ovshinsky Papers. Bentley Historical Library, University of Michigan at Ann Arbor.

Keith, David R., Samantha Houston, and Sergey Naumov. "Vehicle Fleet Turnover and the Future of Fuel Economy." *Environmental Research Letters* 14 (2019): 021001.

Kelly, Henry, and Robert H. Williams. "Fuel Cells and the Future of the US Automobile." Unpublished manuscript, December 7, 1992.

Kempton, Willett, and Steven E. Letendre. "Electric Vehicles as a New Power Source for Electric Utilities." *Transportation Research Part D: Transport and Environment* 2, no. 3 (1997): 157–175.

Kempton, Willett, Jasna Tomić, Steven E. Letendre, Alec N. Brooks, and Timothy Lipman. "Vehicle-to-Grid Power: Battery, Hybrid, and Fuel Cell Vehicles as Resources for Distributed Electric Power in California." UC Davis Institute for Transportation Studies, ECD-ITS-RR-01–03, June 2001.

Kempton, Willett, Keith Decker, and Li Liao. "Vehicle to Grid Demonstration Project DE-FC26–08NT01905: Final Report." May 7, 2011.

Kerpen, Phil. "Tesla and Its Subsidies." *National Review Online*, January 26, 2015. https://www.nationalreview.com/2015/01/tesla-and-its-subsidies-phil-kerpen/.

Killian, James R., Jr. *Sputnik, Scientists, and Eisenhower: A Memoir of the First Special Assistant to the President for Science and Technology*. Cambridge, MA: MIT Press, 1977.

Kim, Hyung Chul, and Timothy J. Wallington. "Life-Cycle Energy and Greenhouse Gas Emission Benefits of Lightweighting in Automobiles: Review and Harmonization." *Environmental Science and Technology* 47, no. 12 (2013): 6089–6097.

King, Ian. "Tesla Shifts to Intel from Nvidia for Infotainment." *Bloomberg*, September 26, 2017. https://www.bloomberg.com/news/articles/2017-09-26/tesla-is-said-to-shift-to-intel-from-nvidia-for-infotainment.

King, Ian. "Chips: Off Quarters for a Hot Company Highlight Wider Concerns." *Bloomberg Businessweek*, November 19, 2018–January 6, 2019, 30.

Kingwill, D. G. *The CSIR: The First 40 Years*. Pretoria, South Africa: Scientia Printers/ CSIR, 1990.

Kirk, Andrew G. *Counterculture Green: The Whole Earth Catalog and American Environmentalism*. Lawrence: University Press of Kansas, 2007.

Kirsch, David A. *The Electric Vehicle and the Burden of History*. New Brunswick, NJ: Rutgers University Press, 2000.

Kirsch, David A., and Gijs Mom. "Visions of Transportation: The EVC and the Transition from Service- to Product-Based Mobility." *Business History Review* 76, no. 1 (2002): 75–110.

Kisiel, Ralph. "Chrysler Designs a Mild Hybrid: Small Battery Only Boosts the Diesel, Costs Just $15,000 More." *Automotive News Europe*, February 2, 1998.

Knepper, Mike. "Citicar: Have You Heard the One About the Voltswagen?" *Motor Trend*, November 1976, 60–63.

Kolodny, Lora. "Tesla Whistleblower Tweets Details About Allegedly Flawed Cars, Scrapped Parts." *CNBC*, August 16, 2018. https://www.cnbc.com/2018/08/15/tesla -whistleblower-tweets-details-about-flawed-cars-scrapped-parts.html.

Kolodny, Lora. "Tesla and Elon Musk Face Dozens of Lawsuits and Investigations Far Beyond the SEC Court Fight." *CNBC*, March 19, 2019. https://www.cnbc.com/2019 /03/19/tesla-and-elon-musk-lawsuits-overview.html.

Koppel, Tom. *Powering the Future: The Ballard Fuel Cell and the Race to Change the World*. Toronto: John Wiley and Sons Canada, 1999.

Krain, Leon J. Untitled letter to Stanford R. Ovshinsky, March 4, 1994. Stanford R. Ovshinsky Papers. Bentley Historical Library, University of Michigan at Ann Arbor.

Krolicki, Kevin. "Ford, Nissan, Tesla to Get US Technology Loans." *Reuters*, June 23, 2009. https://www.reuters.com/article/us-ford-loans/ford-nissan-tesla-to-get-u-s -technology-loans-idUSTRE55M39120090623.

Krolicki, Kevin. "A123 to Sell Fisker Batteries, Takes Stake." *Reuters*, January 14, 2010. https://uk.reuters.com/article/a123/a123-to-sell-fisker-batteries-takes-stake -idUKN1417119620100114.

Kummer, Joseph T., and Neill Weber. "A Sodium-Sulfur Secondary Battery." *SAE Transactions* 76 (1968): 1003–1007.

Kutkut, Nasser H., Herman L. N. Wiegman, Deepak M. Divan, and Donald W. Novotny. "Design Considerations for Charge Equalization of an Electric Vehicle Battery System." *IEEE Transactions on Industry Applications* 35, no. 1 (1999): 28–35.

Leary, Warren E. "Use of Hydrogen as Fuel is Moving Closer to Reality." *New York Times*, April 16, 1995, S1, 15.

Laird, Frank N. "Constructing the Future: Advocating Energy Technologies in the Cold War." *Technology and Culture* 44, no. 1 (2003): 27–49.

Lambert, Fred. "Tesla Officially Announces End of Unlimited Free Supercharging, New 'Supercharging Credit Program' Starts in 2017." *Electrek*, November 7, 2016. https://electrek.co/2016/11/07/tesla-end-of-free-supercharging-new-supercharging -credit-program-2017/.

Lambert, Fred. "Tesla Model 3 Battery Packs Have Capacities of ~ 50 kWh and ~70 kWh, Says Elon Musk." *Electrek*, August 8, 2017. https://electrek.co/2017/08/08/tesla -model-3-battery-packs-50-kwh-75-kwh-elon-musk/.

Lambert, Fred. "Tesla Removes Any Mention of Standard Model 3 Battery from Website; Fans Panic." *Electrek*, February 18, 2019. https://electrek.co/2019/02/18/tesla -standard-model-3-battery-website-fans-panic/.

Lamm, Michael. "PM Owners Report: Electric Cars." *Popular Mechanics*, March 1977, 90–93, 137.

Lave, Rebecca, Philip Mirowski, and Samuel Randalls. "Introduction: STS and Neo-liberal Science." *Social Studies of Science* 40, no. 5 (2010): 659–675.

Lawder, David. "GM Names Smith Chairman as Smale Steps Down." *Buffalo News*, December 4, 1995.

Lécuyer, Christophe. *Making Silicon Valley: Innovation and the Growth of High Tech, 1930–1970*. Cambridge, MA: MIT Press, 2005.

Lécuyer, Christophe, and David C. Brock. "The Materiality of Microelectronics." *History and Technology* 22, no. 3 (2006): 301–325.

Lécuyer, Christophe, and David C. Brock. "High Tech Manufacturing." *History and Technology*, 25, no. 3 (2009): 165–171.

Lécuyer, Christophe, and David C. Brock. "From Nuclear Physics to Semiconductor Manufacturing: The Making of Ion Implantation." *History and Technology* 25, no. 3 (2009): 193–217.

Lee, Jae Hyun, Debapriya Chakraborty, Scott J. Hardman, and Gil Tal. "Exploring Electric Vehicle Charging Patterns: Mixed Usage of Charging Infrastructure." *Transportation Research Part D: Transport and Environment* 79 (2020): 102249, 1–13.

Lee, Jae Hyun, Scott J. Hardman, and Gil Tal. "Who Is Buying Electric Vehicles in California? Characterizing Early Adopter Heterogeneity and Forecasting Market Diffusion." *Energy Research and Social Science* 55 (2019): 218–226.

Lesher, Dave. "Midwest Governors Give Wilson's Campaign a Jolt over Electric Cars." *Los Angeles Times*, May 20, 1995.

Leslie, Stuart W. *The Cold War and American Science: The Military-Industrial-Academic Complex at MIT and Stanford*. New York: Columbia University Press, 1993.

Leslie, Stuart W. "Blue Collar Science: Bringing the Transistor to Life in the Lehigh Valley." *Historical Studies in the Physical and Biological Sciences* 32, no. 1 (2001): 71–113.

Letendre, Steven E., Paul Denholm, and Peter Lilienthal. "New Load, or New Resource?" *Public Utilities Fortnightly* 144, no. 12 (2006): 28–33.

Letendre, Steven E., and Willett Kempton. "The V2G Concept: A New Model for Power?" *Public Utilities Fortnightly* 140, no. 4 (2002): 16–26.

Levin, Doron P. "GM to Begin Production of Battery-Powered Car." *New York Times*, April 19, 1990, D5.

Levin, Doron P. "Mr. Pearce's Growing Domain." *New York Times*, November 15, 1992.

Levy, David L., and Sandra Rothenberg. "Heterogeneity and Change in Environmental Strategy: Technological and Political Responses to Climate Change in the Global Automobile Industry." In *Organizations, Policy, and the Natural Environment: Institutional and Strategic Perspectives*, edited by Andrew Hoffman and Marc Ventresca, 173–193. Stanford CA: Stanford University Press, 2002.

Lienert, Paul. "Ex-Tesla Exec Straubel Aims to Build World's Top Battery Recycler." *Reuters*, October 7, 2020. https://www.reuters.com/article/us-batteries-redwood-recycling-idUSKBN26S3IU.

Loeb, Leon S. "Across the USA with MIT's Electric Car." *Popular Mechanics*, November 1968, 52J, 52K, 184,186, 188.

Lombrana, Laura Millan. "Bolivia's Almost Impossible Lithium Dream." *Bloomberg*, December 3, 2018. https://www.bloomberg.com/news/features/2018-12-03/bolivia-s -almost-impossible-lithium-dream.

Lopez, Linette. "Internal Documents Reveal Tesla Is Blowing Through an Insane Amount of Raw Material and Cash to Make Model 3s, and Production Is Still a Nightmare." *Business Insider*, June 4, 2018. https://www.businessinsider.com/tesla -model-3-scrap-waste-high-gigafactory-2018-5?r=USandIR=T.

Lopez, Linette. "Insiders Describe a World of Chaos and Waste at Panasonic's Massive Battery-Making Operation for Tesla." *Business Insider*, April 16, 2019. https:// www.businessinsider.com/panasonic-battery-cell-operations-tesla-gigafactory -chaotic-2019-4?r=USandIR=T.

Lovins, Amory B. "Energy Strategy: The Road Not Taken?" *Foreign Affairs* (Oct. 1976): 65–96.

Lovins, Amory B., and L. Hunter Lovins. *Brittle Power: Energy Strategy for National Security* Andover, MA: Brick House Publishing, 1982.

Luger, Stan. *Corporate Power, American Democracy, and the Automobile Industry*. Cambridge: Cambridge University Press, 2000.

MacArthur, Donald, George Blomgren, and Robert A. Powers. *Lithium and Lithium Ion Batteries, 2000: A Review and Analysis of Technical, Market and Commercial* Developments. Westlake, OH: Robert A. Powers Associates, 2000.

MacCready, Paul B. "Sunraycer Odyssey: Winning the Solar-Powered Car Race Across Australia." *Engineering and Science* (Winter 1988): 3–13.

Macher, Jeffrey T., and David C. Mowery. "Introduction." In *Innovation in Global Industries: US Firms Competing in a New World*, edited by Jeffrey T. Macher and David C. Mowery, 1–18. Washington, DC: National Academies Press, 2008.

MacKenzie, Angus. "2013 Motor Trend Car of the Year: Model S," *Motor Trend*, December 10, 2012. https://www.motortrend.com/news/2013-motor-trend-car-of -the-year-tesla-model-s/.

MacLeod, Miles. "What Makes Interdisciplinarity Difficult? Some Consequences of Domain Specificity in Interdisciplinary Practice." *Synthese* 195 (2018): 697–720.

Magretta, Joan. "The Power of Virtual Integration: An Interview with Dell Computer's Michael Dell." *Harvard Business Review* (Mar-April 1998): 72–84.

Mäki, Kari, Anna Kulmala, Sami Repo, and Pertti Järventausta. "Problems Related to Islanding Protection of Distributed Generation in Distribution Network." *2007 IEEE Lausanne Power Tech* (2007): 467–472.

Manjoo, Farhad. "I've Seen a Future Without Cars, and It's Amazing." *New York Times*, July 9, 2020. https://www.nytimes.com/2020/07/09/opinion/ban-cars-manhattan -cities.html.

Marchetti, Cesare. "From the Primeval Soup to World Government: An Essay on Comparative Evolution." *International Journal of Hydrogen Energy* 2, no. 1 (1977): 1–5.

Mark, Jason. "Cleaning up Cars." *Washington Post*, August 14, 1996, HO2.

Mark, Jason. "Clean Car's Wrong Turn." *New York Times*, October 26, 1997, WK14.

Marks, Craig, Edward A. Rishavy, and Floyd A. Wyczalek. "Electrovan: A Fuel Cell Powered Vehicle." *SAE Transactions* 76 (1968): 992–1002.

Martin, Joseph D. *Solid State Insurrection: How The Science of Substance Made American Physics Matter*. Pittsburgh: University of Pittsburgh Press, 2018.

Massachusetts Institute of Technology (MIT). "MIT Electric Vehicle Team History: The Clean Air Car Race." Accessed February 26, 2022. http://web.mit.edu/evt /CleanAirCarRace.html.

Massoudi, Arash, and Richard Waters. "Saudi Arabia Slashes Exposure to Tesla Via Hedging Deal." *Financial Times*, January 28, 2019. https://www.ft.com/content /d501c670-2307-11e9-b329-c7e6ceb5ffdf.

McCarthy, Tom. *Auto Mania: Cars, Consumers, and the Environment*. New Haven, CT: Yale University Press, 2007.

McClellan, Dennis. "Victor Wouk, 86; Developed Hybrid Car in '70s." *Los Angeles Times*, June 19, 2005. https://www.latimes.com/archives/la-xpm-2005-jun-19-me -wouk19-story.html.

McCormick, J. Byron, and James R. Huff. "The Case for Fuel-Cell–Powered Vehicles." *Technology Review* (August/September 1980): 54–65.

McCray, W. Patrick. *The Visioneers: How a Group of Elite Scientists Pursued Space Colonies, Nanotechnologies, and a Limitless Future*. Princeton, NJ: Princeton University Press, 2013.

Melosi, Martin V. "The Automobile and the Environment in American History." University of Michigan. Accessed February 21, 2022. http://www.autolife.umd.umich .edu/Environment/E_Overview/E_Overview3.htm#:~:text=The%20Automobile%20 and%20the%20Environment%20in%20American%20History.,production%20 of%20motor%20vehicles%20with%20internal%20combustion%20engines.

Mirowski, Philip. "The Future(s) of Open Science." *Social Studies of Science* 48, no. 2 (2018): 171–203.

Mirowski, Philip, and Esther-Mirjam Sent. "The Commercialization of Science and the Response of STS." In *Handbook of Science and Technology Studies*, edited by Edward J. Hackett, Olga Amsterdamska, Michael Lynch, and Judy Wajcman, 635–689. Cambridge, MA: MIT Press, 2008.

Mitchell, Russ. "As Tesla Tax Credits Disappear, Will Model 3 Deposit-Holders Stick Around?" *Los Angeles Times*, July 3, 2018. https://www.latimes.com/business/autos /la-fi-hy-tesla-tax-credit-subsidy-20180703-story.html.

Mitchell, William J., Christopher E. Borroni-Bird, and Lawrence D. Burns. *Reinventing the Automobile: Personal Urban Mobility for the 21st Century*. Cambridge, MA: MIT Press, 2010.

Mitchener, Brandon, and Tamsin Carlisle. "Daimler, Ballard Team to Develop Fuel-Cell Engine." *Wall Street Journal*, April 15, 1997, B, 8:4.

Miwa, Yoshiro, and J. Mark Ramseyer. "The Fable of the Keiretsu." *Journal of Economics and Management Strategy* 11, no. 2 (2002): 169–224.

Mizushima, K., P. C. Jones, P. J. Wiseman, and J. B. Goodenough. "LixCoO$_2$: A New Cathode Material for Batteries of High Energy Density." *Materials Research Bulletin* 15, no. 6 (1980): 783–789.

Mody, Cyrus, C. M. *The Long Arm of Moore's Law: Microelectronics and American Science.* Cambridge, MA: MIT Press, 2016.

Mody, Cyrus, C. M., and Hyungsub Choi. "From Materials Science to Nanotechnology: Interdisciplinary Center Programs at Cornell University, 1960–2000." *Historical Studies in the Natural Sciences* 43, no. 2 (2013): 121–161.

Mom, Gijs. *The Electric Vehicle: Technology and Expectations in the Automobile Age.* Baltimore: Johns Hopkins University Press, 2004.

Monticello, Mike. "Volt vs. Prius: Chevrolet's Plug-In Takes on Toyota's Hybrid." *Consumer Reports*, April 22, 2016. https://www.consumerreports.org/hybrids-evs/volt -vs-prius-review/?EXTKEY=AGTS004.

Moore, Gordon E. "Cramming More Components onto Integrated Circuits." *Electronics* 38, no. 8 (1965): 114–117.

Moore, Gordon E. "Progress in Digital Integrated Electronics." *Technical Digest*, IEEE International Electron Devices Meeting 21 (1975): 11–13.

Moore, Gordon E. "The Cost Structure of the Semiconductor Industry and its Implications for Consumer Electronics." *IEEE Transactions on Consumer Electronics* CE-23, no. 1 (1977): x–xvi.

Moore, Gordon E. "Are We Really Ready For VLSI^2?" *Digest of Technical Papers*, IEEE International Solid-State Circuits Conference (1979): 54–55.

Moore, Gordon E. "VLSI: Some Fundamental Challenges." *IEEE Spectrum* 16, no. 4 (1979): 30.

Morris, Andre. "Ford Awards Johnson Controls–Saft Battery Contract for Hybrid Car." *EE/Times*, February 4, 2009. https://www.eetimes.com/ford-awards-johnson -controls-saft-battery-contract-for-hybrid-car/.

Motavalli, Jim. *Forward Drive: The Race to Build 'Clean' Cars for the Future.* New York: Earthscan, 2001.

Motavalli, Jim. "Electric Car Agreement for Toyota and Tesla." *New York Times*, May 21, 2010, B7.

Mowery, David C. "Collaborative R&D: How Effective Is It?" *Issues in Science and Technology* 15, no. 1 (1998): 37–44.

Mowery, David C., Richard R. Nelson, Bhaven N. Sampat, and Arvids A. Ziedonis. "The Growth of Patenting and Licensing by US Universities: An Assessment of the Effects of the Bayh-Dole Act of 1980." *Research Policy* 30, no. 1 (2001): 99–119.

Mowery, David C., Richard R. Nelson, Bhaven N. Sampat, and Arvids A. Ziedonis. *Ivory Tower and Industrial Innovation: University-Industry Technology Transfer Before and After the Bayh-Dole Act.* Stanford, CA: Stanford University Press, 2004.

Muller, Joanne. "Elon Musk's Financial Car Wreck." *Forbes*, May 28, 2010. https:// www.forbes.com/2010/05/28/elon-musk-broke-tesla-business-autos-musk.html #3b8519a56d8b.

Musk, Elon. "The Secret Tesla Motors Master Plan (Just Between You and Me)." Tesla Motors. Last modified August 2, 2006. https://www.tesla.com/blog/secret-tesla -motors-master-plan-just-between-you-and-me.

Musk, Elon. "A Most Peculiar Test Drive." Tesla Motors. Last modified February 13, 2013. https://www.tesla.com/blog/most-peculiar-test-drive.

Musk, Elon. "All Our Patent Are Belong To You." Tesla Motors. Last modified June 12, 2014. https://www.tesla.com/en_GB/blog/all-our-patent-are-belong-you.

Musk, Elon. Interview by Lesley Stahl. *60 Minutes*, CBS, December 9, 2018. https:// www.cbsnews.com/news/tesla-ceo-elon-musk-the-2018-60-minutes-interview/.

Musk, Elon. "Tweet." January 9, 2019. https://twitter.com/elonmusk/status/108314 1248872075265?lang=en.

Nader, Ralph. "The Management of Environmental Violence: Regulation or Reluctance." In *Environment in Peril*, edited by Anthony N. Wolbarst, 2–25. Washington, DC: Smithsonian Institution Press, 1991.

Nader, Ralph. *Unsafe at Any Speed: The Designed-in Dangers of the American Automobile* New York: Grossman Publishers, 1965.

Nardone, Anthony, Joan A. Casey, Rachel Morello-Frosch, Mahasin Mujahid, John R. Balmes, and Neeta Thakur. "Associations Between Historical Residential Redlining and Current Age-Adjusted Rates of Emergency Department Visits Due to Asthma Across Eight Cities in California: An Ecological Study." *Lancet Planet Health* 4 (2020): 24–31.

National Academy of Sciences. *Materials and Man's Needs, Vol. 1: The History, Scope, and Nature of Materials Science and Engineering.* Washington, DC: National Academy of Sciences, 1975.

National Highway Traffic Safety Administration (NHTSA). "Summary of CAFE Fines Collected." Last modified July 24, 2014. file:///Users/mne/Downloads/cafe_fines-07-2014.pdf.

National Research Council. *Materials Science and Engineering for the 1990s: Maintaining Competitiveness in the Age of Materials.* Washington, DC: National Academies Press, 1989.

National Research Council. *Review of the Research Program of the Partnership for a New Generation of Vehicles: Second Report.* Washington, DC: National Academies Press, 1996.

National Research Council. *Review of the Research Program of the Partnership for a New Generation of Vehicles: Third Report.* Washington, DC: National Academies Press, 1997.

National Research Council. *Review of the Research Program of the Partnership for a New Generation of Vehicles: Fourth Report.* Washington, DC: National Academies Press, 1998.

National Research Council. *Review of the Research Program of the Partnership for a New Generation of Vehicles: Fifth Report.* Washington, DC: National Academies Press, 1999.

National Research Council. *Review of the Research Program of the Partnership for a New Generation of Vehicles: Sixth Report.* Washington, DC: National Academies Press, 2000.

National Research Council. *Condensed-Matter and Materials Physics: The Science of the World Around Us.* Washington, DC: National Academies Press, 2007.

National Science and Technology Council. *Materials Genome Initiative for Global Competitiveness.* Washington, DC: Executive Office of the President of the United States, 2011.

Naughton, Keith. "Why Toyota Is Becoming the World's Top Carmaker." *Newsweek*, March 11, 2007. https://www.newsweek.com/why-toyota-becoming-worlds -top-carmaker-95469.

Nauss, Donald W. "GM's Man Who Bested NBC Helps Rouse Sleeping Giant." *Los Angeles Times*, February 17, 1993. https://www.latimes.com/archives/la-xpm-1993 -02-17-mn-238-story.html.

Nauss, Donald W. "Autos: GM Group Forms Unit to Make, Sell Parts for Electric Vehicles." *Los Angeles Times*, September 22, 1994. https://www.latimes.com/archives /la-xpm-1994-09-22-fi-41631-story.html.

Nauss, Donald W. "GM Rolls Dice with Roll-Out of Electric Car." *Los Angeles Times*, December 5, 1996. https://www.latimes.com/archives/la-xpm-1996-12-05-mn-6000 -story.html.

Nauss, Donald W. "Ford Investing $420 Million for Fuel-Cell-Powered Auto." *Los Angeles Times*, December 16, 1997. https://www.latimes.com/archives/la-xpm-1997 -dec-16-mn-64565-story.html.

Nealer, Rachael, David Reichmuth, and Don Anair. *Cleaner Cars from Cradle to Grave: How Electric Cars Beat Gasoline Cars on Lifetime Global Warming Emissions.* Cambridge, MA: Union of Concerned Scientists, 2015. https://www.ucsusa.org/sites/default/files /attach/2015/11/Cleaner-Cars-from-Cradle-to-Grave-full-report.pdf.

Neff, John. "Wagoner Arrives for Senate Hearing in Volt Mule." *Autoblog*, December 4, 2008. https://www.autoblog.com/2008/12/04/wagoner-arrives-for-senate-hearing -in-volt-mule/?guccounter=1.

New York Times. "Detroit Turns a Corner." January 11, 1998, Section 4, 18.

New York Times. "Spencer Abraham's Dream Car." January 14, 2002, A14.

Nishi, Yoshio. "The Development of Lithium Ion Secondary Batteries." *The Chemical Record* 1 (2001): 406–413.

Nishi, Yoshio. "Lithium Ion Secondary Batteries: Past 10 Years and the Future." *Journal of Power Sources* 100, nos. 1–2 (2001): 101–106.

Nishi, Yoshio. "My Way to Lithium-Ion Batteries." In *Lithium-Ion Batteries: Science and Technologies*, edited by Masaki Yoshio, Ralph J. Brodd, and Akiya Kozawa, v–vii. New York: Springer, 2009.

Nissan Motor Company. "Nissan and NEC Joint Venture AESC Starts Operations." Last modified May 19, 2008. https://www.nissan-global.com/EN/NEWS/2008/_STORY /0805.

Nissan Motor Corporation. "Nissan Completes Sale of Battery Business to Envision Group." Last modified March 29, 2019. https://newsroom.nissan-global.com/releases /190329-01-e?la=1anddownloadUrl=%2Freleases%2F190329-01-e%2Fdownload.

Nivola, Pietro S. *The Politics of Energy Conservation*. Washington, DC: Brookings Institution, 1986.

Noble, David F. *America by Design: Science, Technology, and the Rise of Corporate Capitalism*. New York: Knopf, 1977.

Noble, David F. *The Religion of Technology: The Divinity of Man and the Spirit of Invention*. Penguin, 1999.

North Dakota Office of the Governor. "Harry J. Pearce." August 11, 2004. https:// www.governor.nd.gov/theodore-roosevelt-rough-rider-award/harry-j-pearce.

Notter, Dominic A., Marcel Gauch, Rolf Widmer, Patrick Wäger, Anna Stamp, Rainer Zah, and Hans-Jörg Althaus. "Contribution of Li-Ion Batteries to the Environmental Impact of Electric Vehicles." *Environmental Science and Technology* 44, no. 17 (2010): 6550–6556.

Notter, Dominic A., Marcel Gauch, Rolf Widmer, Patrick Wäger, Anna Stamp, Rainer Zah, and Hans-Jörg Althaus. "Contribution of Li-Ion Batteries to the Environmental Impact of Electric Vehicles." *Environmental Science and Technology* 44, no. 17 (2010): 6550–6556.

NPR. "Rick Wagoner on the Future of General Motors." January 9, 2007. https:// www.npr.org/templates/story/story.php?storyId=6768710.

Nuttall, Nick. "Breathtaking: The Vehicle Powered by Air." *The Times*, May 15, 1996, Home News 7.

Nye, David E. "Technological Prediction: A Promethean Problem." In *Technological Visions: The Hopes and Fears that Shape New Technologies*, edited by Marita Sturken, Douglas Thomas, and Sandra J. Ball-Rokeach, 159–161. Philadelphia: Temple University Press, 2004.

Obama, Barack H. "Remarks of President Barack Obama in State of the Union Address, as Prepared for Delivery." January 25, 2011. https://obamawhitehouse.archives.gov /the-press-office/2011/01/25/remarks-president-barack-obama-state-union-address -prepared-delivery.

O'Dell, John. "GM Turns to a Top Guru in Industry." *Los Angeles Times*, August 3, 2001. https://www.latimes.com/archives/la-xpm-2001-aug-03-fi-30054-story.html.

O'Dell, John. "Car Companies Team up to Fight State's ZEV Rule." *Los Angeles Times*, January 23, 2002. https://www.latimes.com/archives/la-xpm-2002-jan-23-hy -green23-story.html.

O'Donnell, Carl, and Ross Kerber. "Investors Query Funding Costs at a Private Tesla." *Reuters*, August 17, 2018. https://www.reuters.com/article/us-tesla-musk -board-analysis/investors-query-funding-costs-at-a-private-tesla-idUSKBN1L21XN.

Ogiso, Satoshi. "The Story Behind the Birth of the Prius, Part 2." Last modified December 13, 2017. https://newsroom.toyota.co.jp/en/prius20th/challenge/birth/02/.

Ohnsman, Alan. "Toyota Says It's Now Turning a Profit on the Hybrid Prius." *Los Angeles Times*, December 19, 2001. https://www.latimes.com/archives/la-xpm-2001 -dec-19-hy-prius19-story.html.

Ohnsman, Alan. "Elon Alone: Longtime Tesla Tech Chief Straubel's Exit Leaves Musk as Sole Remaining Cofounder." *Forbes*, July 24, 2019. https://www.forbes .com/sites/alanohnsman/2019/07/24/elon-alone-long- . . . a-tech-chief-straubels -exit-leaves-musk-as-sole-remaining-cofounder/.

O'Kane, Sean. "The Court Has Approved Elon Musk's New Agreement to Let Lawyers Oversee His Tesla Tweets." *The Verge*, April 30, 2019. https://www.theverge.com /2019/4/26/18484751/elon-musk-sec-fraud-tesla-tweets-contempt-agreement.

Öniş, Ziya. "The Logic of the Developmental State." *Comparative Politics* 24, no. 1 (1991): 109–126.

Oreskes, Naomi, and Erik M. Conway. *Merchants of Doubt: How a Handful of Scientists Obscured the Truth on Issues from Tobacco Smoke to Global Warming*. New York: Bloomsbury, 2010.

Oreskes, Naomi, Erik M. Conway, and Matthew Shindell. "From Chicken Little to Dr. Pangloss: William Nierenberg, Global Warming, and the Social Deconstruction of Scientific Knowledge." *Historical Studies in the Natural Sciences* 38, no. 1 (2008): 109–152.

Osborne, Mark, and Rex Sakamoto. "'West Wing' Actress Calls Out Tesla After Husband's Car Bursts into Flames." *ABC News*, June 17, 2018. https://abcnews.go.com /US/west-wing-actress-calls-tesla-husbands-car-bursts/story?id=55953027.

Oswald, Larry. Letter to Paul Gifford and Dennis Corrigan, "Re: High Power-to-Energy Ratio Ovonic Batteries," April 17, 1997. Robert C. Stempel Papers. Bentley Historical Library, University of Michigan at Ann Arbor.

Ovonic Battery Company (OBC). "GM-Ovonic Forms Management Team, Readies Manufacturing Facility." August 30, 1994. Stanford R. Ovshinsky Papers. Bentley Historical Library, University of Michigan at Ann Arbor.

Ovonic Battery Company (OBC). "OBC_PLN3.DOC." June 13, 1994. Robert C. Stempel Papers. Bentley Historical Library, University of Michigan at Ann Arbor.

Ovshinsky, Stanford R. Letter to Andrew Ng. December 20, 1996. Stanford R. Ovshinsky Papers. Bentley Historical Library, University of Michigan at Ann Arbor.

Ovshinsky, Stanford R. Letter to Robert C. Stempel, "Subject: ECD as Resource to General Motors," May 15, 1997. Robert C. Stempel Papers. Bentley Historical Library, University of Michigan at Ann Arbor.

Ovshinsky, Stanford R. Letter to Robert C. Stempel, "Subject: GMO 3 Status," July 24, 1998. Stanford R. Ovshinsky Papers. Bentley Historical Library, University of Michigan at Ann Arbor.

Ovshinsky, Stanford R., and Robert C. Stempel. Letter to *The Economist*, undated. Stanford R. Ovshinsky Papers. Bentley Historical Library, University of Michigan at Ann Arbor.

Ovshinsky, Stanford R., Michael A. Fetcenko, and J. Ross. "A Nickel-Metal Hydride Battery for Electric Vehicles." *Science* 260, no. 5105 (April 9, 1993), 176–181.

Owen, David. *The Conundrum: How Scientific Innovation, Increased Efficiency, and Good Intentions Can Make Our Energy and Climate Problems Worse*. New York: Riverhead Books, 2011.

Ozaki, Ritsuko, and Katerina Sevastyanova. "Going Hybrid: An Analysis of Consumer Purchase Motivations." *Energy Policy* 39, no. 5 (2011): 2217–2227.

Ozaki, Ritsuko, Isabel Shaw, and Mark Dodgson. "The Coproduction of 'Sustainability:' Negotiated Practices and the Prius." *Science, Technology, and Human Values* 38, no. 4 (2013): 518–541.

Ozawa, Kazunori. "Lithium-Ion Rechargeable Batteries with $LiCoO_2$ and Carbon Electrodes: The $LiCoO_2$/C System." *Solid-State Ionics* 69, nos. 3–4 (1994): 212–221.

Pae, Peter. "GM Seen to Drop Suits on Zero-Emission-Vehicle Mandate." *Los Angeles Times*, August 12, 2003. https://www.latimes.com/archives/la-xpm-2003-aug-12-fi -zero12-story.html.

Paine, Chris, dir. *Who Killed the Electric Car?* Sony Pictures Classics, 2006.

Paine, Chris, dir. *Revenge of the Electric Car*. WestMidWest Productions/Area 23a Films, 2011.

Panasonic. "PEV: Battery for Pure Electric Vehicles." Accessed January 28, 2018. http://www.evnut.com/rav_battery_data_sheet.html.

Panasonic. "Panasonic Battery History." Accessed June 29, 2020. https://www.panasonic .com/global/consumer/battery/about_us/hist.

Paris Productions."EV1 Funeral: Hollywood Forever Cemetery, Thursday, July 24, 2003." Accessed December 29, 2021. https://www.youtube.com/watch?v=HZHka8KUj74andt =302s and https://www.youtube.com/watch?v=zFsOxZPR-eQ.

Parrish, Michael. "Electric Vehicle Firm Struggles to Go On." *Los Angeles Times*, March 21, 1995.

Patchell, Jerry. "Creating the Japanese Electric Vehicle Industry: The Challenges of Uncertainty and Cooperation." *Environment and Planning A: Economy and Space* 31, no. 6 (1999): 997–1016.

Paterson, Matthew *Automobile Politics: Ecology and Cultural Political Economy*. Cambridge: Cambridge University Press 2007.

Patil, Pandit G. "Prospects for Electric Vehicles." *IEEE AES Systems Magazine* (December 1990): 15–19.

Pecht, Michael G. "Editorial: Re-Thinking Reliability." *IEEE Transactions on Components and Packaging Technologies* 29, no. 4. (2006): 893–894.

Pemberton, Max. *The Iron Pirate: A Plain Tale of Strange Happenings on the Sea*. London: Cassell and Company, 1893.

Perrin, Noel. *Solo: Life with an Electric Car*. New York: W. W. Norton and Company, 1992.

Perry, M. L., and T. F. Fuller. "A Historical Perspective of Fuel Cell Technology in the 20th Century." *Journal of the Electrochemical Society* 149, no. 7 (2002): S59–S67.

Phenix, Matthew. "A Recharged Th!nk Contemplates Its Comeback." *Wired*, September 8, 2007. https://www.wired.com/2007/09/post-ford-a-rec/.

Pinch, Trevor. "Testing—One, Two, Three . . . Testing!" Toward a Sociology of Testing." *Science, Technology and Human Values* 18, no.1 (1993), 27–31.

Pizer, William A. "A Tale of Two Policies: Clear Skies and Climate Change." In *Painting the White House Green: Rationalizing Environmental Policy Inside the Executive Office of the President*, edited by Randall Lutter and Jason F. Shogren, 10–45. Washington, DC: Resources for the Future, 2004.

Plastics News. "Pivco Bankruptcy Takes Car Off Fast Track." November 9, 1998. https://www.plasticsnews.com/article/19981109/NEWS/311099998/pivco-bankruptcy-takes-car-off-fast-track.

Plug In America. "Why We're Asking the Governor to Veto AB 475." Last modified August 26, 2011. https://pluginamerica.org/why-were-asking-governor-veto-ab-475/.

Polakovic, Gary, and John O'Dell. "Injunction Holds Up ZEV Program." *Los Angeles Times*, June 15, 2002. https://www.latimes.com/archives/la-xpm-2002-jun-15-me-emisions15-story.html.

Pollack, Andrew. "America's Answer to MITI." *New York Times*, March 5, 1989, Section 3, 1, 8.

Pollack, Andrew. "Cars and the Environment: Where to Put the Golf Clubs? Right Next to the Hydrogen!" *New York Times*, May 19, 1999, G20.

Pollock, Neil, and Robin Williams. "The Business of Expectations: How Promissory Organizations Shape Technology and Innovation." *Social Studies of Science* 40, no. 4 (2010): 525–548.

Pool, Bob. "Drivers Find Outlet for Grief Over EV1s." *Los Angeles Times*, July 25, 2003, https://www.latimes.com/archives/la-xpm-2003-jul-25-me-funeral25-story.html.

Prater, Keith B. "The Renaissance of the Solid Polymer Fuel Cell." *Journal of Power Sources* 29, nos. 1–2 (1990): 239–250.

Pritchard, Sara B. *Confluence: The Nature of Technology and the Remaking of the Rhône.* Cambridge, MA: Harvard University Press, 2011.

Pritchard, Sara B. "An Envirotechnical Disaster: Nature, Technology, and Politics at Fukushima." *Environmental History* 17, no. 2 (2012): 219–243.

PR Newswire. "Ovonic NiMH Batteries Featured in GM Advanced Technology Vehicle Introduced at North American Auto Show." January 13, 2000, 1.

Prodhan, Georgina. "Car Industry Players Diverge on Timescale for Self-Driving Cars." *Reuters*, March 16, 2017. https://www.reuters.com/article/us-autos-autonomous-idUSKBN16N2NF.

Public Utility Regulatory Policies Act of 1978. Public Law 95–617. *US Statutes at Large 92 (1978): 3117–3173.*

Puckett, Elizabeth. "Tesla Model S Bursts into Flames Two Separate Times on Same Day Following Tire Issue." *The Drive*, December 21, 2018. http://www.thedrive.com /news/25603/tesla-model-s-bursts-into-flames-two-separate-times-on-same-day -following-tire-issue.

Pursell, Carroll. "The Rise and Fall of the Appropriate Technology Movement in the United States, 1965–1985." *Technology and Culture* 34, no. 3 (1993): 629–637.

Rai, Sonam, and Jasmine I.S. Bengaluru. "Musk Not Worried About Tesla Model 3 Demand But Wall Street Thinks Otherwise." *Reuters*, January 31, 2019. https://uk .reuters.com/article/uk-tesla-results-stocks/musk-not-worried-about-tesla-model-3 -demand-but-wall-street-is-idUKKCN1PP1TL.

Rajan, Sudhir Chella. *The Enigma of Automobility: Democratic Politics and Pollution Control*. Pittsburgh: University of Pittsburgh Press, 1996.

Randewich, Noel. "Tesla Becomes Most Valuable US Car Maker, Edges out GM." *Reuters*, April 11, 2017. https://www.reuters.com/article/us-usa-stocks-tesla-idUSKBN17C1XF.

Randewich, Noel. "Tesla's Market Value Zooms Past That of GM and Ford Combined." *Reuters*, January 8, 2020. https://www.reuters.com/article/us-usa-stocks-tesla /teslas-market-value-zooms-past-that-of-gm-and-ford-combined-idUSKBN1Z72MU.

Rapier, Graham. "Wall Street Analysts Tore Down a Tesla Model 3 and Found 'Significant Fit and Finish Issues.'" *Business Insider*, August 27, 2018. https://markets .businessinsider.com/news/stocks/tesla-model-3-wall-street-analysts-find-significant -fit-and-finish-issues-2018-8-1027480762.

Rapoza, Kenneth. "Here's Why Tesla CEO Elon Musk Was Dancing in China." *Forbes*, January 13, 2020. https://www.forbes.com/sites/kenrapoza/2020/01/13/heres -why-tesla-ceo-elon-musk-was-dancing-in-china/.

Razeghi, Ghazal, Brendan Shaffer, and Scott Samuelsen. "Impact of Electricity Deregulation in the State of California." *Energy Policy* 103 (2017): 105–115.

Rechtin, Mark. "Honda Pulls the Plug on EV Plus." *Automotive News*, April 26, 1999. https://www.autonews.com/article/19990426/ANA/904260758/honda-pulls-the -plug-on-ev-plus.

Rechtin, Mark. "Tesla Nimbly Updates Model S over the Air." *Automotive News*, January 16, 2013. https://www.autonews.com/article/20130116/OEM06/130119843 /tesla-nimbly-updates-model-s-over-the-air.

Rechtin, Mark. "Tesla Reliability Doesn't Match Its High Performance." *Consumer Reports*, October 20, 2015. https://www.consumerreports.org/cars-tesla-reliability -doesnt-match-its-high-performance/.

Redmond, Kent C., and Thomas M. Smith. *From Whirlwind to MITRE: The R&D Story of the SAGE Air Defense Computer*. Cambridge, MA: MIT Press, 2000.

Reitman, Valerie. "Toyota to Sell Hybrid Gas-Electric Car: Auto Maker Cites High Efficiency, Low Emissions." *Wall Street Journal*, March 26, 1997, A 12:1.

Reitman, Valerie. "Ford Is Investing in Daimler-Ballard Fuel-Cell Venture." *Wall Street Journal*, December 16, 1997, B8.

Reuters. "Johnson Sees Hybrid Engines in 5–8 Pct. of Market." January 19, 2007. https://www.reuters.com/article/us-autos-summit-johnson-hybrids-idUSN1120028420060911.

Riezenman, Michael J. "EV Candidate for Mass Production Does Boston-to–New York Run On One Charge." *IEEE Spectrum* (December 1997): 68–70.

Riordan, Michael, and Lillian Hoddeson. *Crystal Fire: The Invention of the Transistor and the Birth of the Information Age.* New York: Norton, 1997.

Rishavy, E. A., W. D. Bond, and T. A. Zechin. "Electrovair: A Battery Electric Car." *SAE Transactions* 76 (1968): 981–991, 1023–1028.

Ritschel, Alexander, and Greg P. Smestad. "Energy Subsidies in California's Electricity Market Deregulation." *Energy Policy* 31 (2003): 1379–1391.

Rodrigues de Almeida, Poliana, Akira Luiz Nakamura, and José Ricardo Sodré. "Evaluation of Catalytic Converter Aging for Vehicle Operation with Ethanol." *Applied Thermal Engineering* 71 (2014): 335–341.

Rogers, Hannah Star. *Art, Science, and the Politics of Knowledge.* Cambridge, MA: MIT Press, 2022.

Rosenzweig, Phil. *The Halo Effect and the Eight Other Business Delusions That Deceive Managers.* New York: Free Press, 2007.

Ross, Benjamin. "California's Regulation Debacle." *Dissent* (Spring 2001): 45–47.

Ross, Philip E. "Ford: Robotaxis in 2021, Self-Driving Cars for Consumer 2025." *IEEE Spectrum.* September 12, 2016. https://spectrum.ieee.org/cars-that-think/transportation/self-driving/ford-robotaxis-in-2021-selfdriving-cars-for-consumer-2025.

Ross, Philip E. "Robocars and Electricity: A Match Made in Heaven." *IEEE Spectrum,* June 1, 2017. https://spectrum.ieee.org/cars-that-think/transportation/self-driving/why-robocars-will-run-on-electricity.

Royal Swedish Academy of Science (RSAC). "The Nobel Prize in Chemistry 2019: The Royal Swedish Academy of Sciences Has Decided to Award the Nobel Prize in Chemistry 2019 to John B. Goodenough, M. Stanley Whittingham, and Akira Yoshino for the Development of Lithium-Ion Batteries: They Created a Rechargeable World." Last modified October 9, 2019. https://www.nobelprize.org/uploads/2019/10/press-chemistry-2019-2.pdf.

Russell, Edmund, James Allison, Thomas Finger, John K. Brown, Brian Balogh, and W. Bernard Carlson. "The Nature of Power: Synthesizing the History of Technology and Environmental History." *Technology and Culture* 52, no. 2 (2011): 246–259.

Saft, Michael, Guy Chagnon, Thierry Faugeras, Guy Sarre, and Pierre Morhet. "Saft Lithium-Ion Energy and Power Storage Technology." *Journal of Power Sources* 80, nos. 1–2 (1999): 180–189.

Sage, Alexandria. "Build Fast, Fix Later: Speed Hurts Quality at Tesla, Some Workers Say." *Reuters,* November 29, 2017. https://uk.reuters.com/article/us-tesla-quality-insight/build-fast-fix-later-speed-hurts-quality-at-tesla-some-workers-say-idUKKBN1DT0N3.

Sage, Alexandria. "Tesla Owner Lawsuit Claims Software Update Fraudulently Cut Battery Capacity." *Reuters,* August 8, 2019. https://www.reuters.com/article/us-tesla-battery

/tesla-owner-lawsuit . . . aims-software-update-fraudulently-cut-battery-capacity-idUSKC-N1UY2TW.

Salihi, Jalal T., Paul D. Agarwal, and George J. Spix. "Induction Motor Scheme for Battery-Powered Electric Car (GM Electrovair-I)." *IEEE Transactions on Industry and General Applications* 3, no. 5 (Sept/Oct 1967): 463–469.

Salon, Deborah, Daniel Sperling, and David Friedman. *California's Partial ZEV Credits and LEV II Program: UCTC No. 470.* Berkeley: University of California Transportation Center, 2001.

Saxenian, AnnaLee. *Regional Advantage: Culture and Competition in Silicon Valley and Route 128.* Cambridge, MA: Harvard University Press, 1994.

Schallenberg, Richard H. *Bottled Energy: Electrical Engineering and the Evolution of Chemical Energy Storage.* Philadelphia: American Philosophical Society, 1982.

Scharff, Virginia. *Taking the Wheel: Women and the Coming of the Motor Age.* Albuquerque: University of New Mexico Press, 1991.

Schumacher, E. F. *Small is Beautiful: A Study of Economics as if People Mattered.* London: Blond and Briggs, 1973.

Sedgwick, David. "Battery Maker Faces High Cost, Low Demand." *Automotive News Europe,* June 23, 1997. http://europe.autonews.com/article/19970623/ANE/706230848?template=printartANE.

Seetharaman, Deepa. "Tesla Shares Drop 6 Percent After Report of Model S Fire." *Reuters,* October 3, 2013. https://www.reuters.com/article/us-autos-tesla-crash/tesla-shares-drop-6-percent-after-report-of-model-s-fire-idUSBRE99200020131003.

Segal, Howard P. *Technological Utopianism in American Culture.* Chicago: University of Chicago Press, 1985.

Shapin, Steven, and Simon Schaffer. *Leviathan and the Air-Pump: Hobbes, Boyle and the Experimental Life.* Princeton, NJ: Princeton University Press, 1985.

Sheller, Mimi. "Automotive Emotions: Feeling the Car." *Theory, Culture, and Society* 21, nos. 4–5 (2004): 221–242.

Sheller, Mimi, and John Urry. "The City and the Car." *International Journal of Urban and Regional Research* 24, no. 4 (2000): 737–757.

Sherlock, Molly F. "The Plug-in Electric Vehicle Tax Credit." *Congressional Research Service,* May 14, 2019. https://fas.org/sgp/crs/misc/IF11017.pdf.

Sherwin, Chalmers W., and Raymond S. Isenson. "Project Hindsight." *Science* (New Series) 156, no. 3782 (1967): 1571–1577.

Shipley, Lou. "How Tesla Sets Itself Apart." *Harvard Business Review,* February 28 2020. https://hbr.org/2020/02/how-tesla-sets-itself-apart.

Shnayerson, Michael. *The Car That Could: The Inside Story of GM's Revolutionary Electric Vehicle.* New York: Random House, 1996.

Shubber, Kadhim. "SEC Endorses Investor View of Elon Musk's Indispensable Role at Tesla." *Financial Times,* October 2, 2018. https://search.proquest.com/docview/2115551960?accountid=14116.

Shubber, Kadhim. "Musk Mocks SEC in Tweet Only Days After Settling with Regulator." *Financial Times*, October 4, 2018. https://search.proquest.com/docview /2116253761?accountid=14116.

Slayton, Rebecca. "From a 'Dead Albatross' to Lincoln Labs: Applied Research and the Making of a Normal Cold War University." *Historical Studies in the Natural Sciences* 42, no. 4 (2012): 255–282.

Smil, Vaclav. *Energy in Nature and Society: General Energetics of Complex Systems*. Cambridge, MA: MIT Press, 2008.

Smith. Roger B. "A Message from the Chairman." Introduction to *Sunraycer*, by Bill Tuckey, 9. Hornsby, Australia: Chevron Publishing Group, 1989.

Sorkin, Andrew Ross, and Neela Banerjee. "Chevron Agrees to Buy Texaco for Stock Valued at $36 Billion: Deal Creates World's 4th-Largest Oil Company." *New York Times*, October 16, 2000, A1.

Spark M. Matsunaga Hydrogen Research, Development, and Demonstration Program Act of 1990. Public Law 101–566. *US Statutes at Large* 104 (1990): 2797–2801.

Sperling, Daniel. *Future Drive: Electric Vehicles and Sustainable Transportation*. Washington, DC: Island Press, 1995.

Sperling, Daniel. "Public-Private Technology R&D Partnerships: Lessons from US Partnership for a New Generation of Vehicles." *Transport Policy* 8, no. 4 (2001): 247–256.

Sperling, Daniel, and Deborah Gordon. *Two Billion Cars: Driving Towards Sustainability*. New York: Oxford University Press, 2009.

Squatriglia, Chuck. "Tesla Cries Foul on Top Gear's Test." *Wired*, December 16, 2008. https://www.wired.com/2008/12/tesla-cries-fou/.

Squatriglia, Chuck. "GM Fires up Its Chevrolet Volt Battery Factory." *Wired*, January 7, 2010. https://www.wired.com/2010/01/chevrolet-volt-battery-production/.

Squatriglia, Chuck. "Tesla IPO Raises $226.1 Million, Stock Surges 41 Percent." *Wired*, June 29, 2010. https://www.wired.com/2010/06/tesla-ipo-raises-226-1-million/.

Squatriglia, Chuck. "Toyota, Tesla Resurrect the Electric RAV4." *Wired*, July 16, 2010. https://www.wired.com/2010/07/toyota-tesla-rav4-ev/.

Star, Susan Leigh. "The Structures of Ill-Structured Solutions: Boundary Objects and Distributed Heterogeneous Problem Solving." In *Readings in Distributed Artificial Intelligence*, edited by M. Huhns and L. Glasser, 37–54. San Mateo, CA: Morgan Kauffmann.

Star, Susan Leigh. "This Is Not a Boundary Object: Reflections on the Origin of a Concept." *Science, Technology, and Human Values* 35, no. 5 (2010): 601–617.

Stempel, Robert C. Letter to Stanford R. Ovhsinsky. September 9, 1994. Stanford R. Ovshinsky Papers. Bentley Historical Library, University of Michigan at Ann Arbor.

Stempel, Robert C. Letter to Stanford R. Ovsinsky. November 26, 1994. Stanford R. Ovshinsky Papers. Bentley Historical Library, University of Michigan at Ann Arbor.

Stempel, Robert C. "Challenge for Tomorrow: Forging the Road Ahead," December 7, 1994. Stanford R. Ovshinsky Papers. Bentley Historical Library, University of Michigan at Ann Arbor.

Stempel, Robert C. Letter to Alex J. Trotman. February 21, 1995. Stanford R. Ovshinsky Papers. Bentley Historical Library, University of Michigan at Ann Arbor.

Stempel, Robert C. Hand notes on meeting with John Adams. February 9, 1996. Robert C. Stempel Papers. Bentley Historical Library, University of Michigan at Ann Arbor.

Stempel, Robert C. Letter to Rich Piellisch. March 6, 1996. Robert C. Stempel Papers. Bentley Historical Library, University of Michigan at Ann Arbor.

Stempel, Robert C. Letter to Harry J. Pearce. April 14, 1997. Robert C. Stempel Papers. Bentley Historical Library, University of Michigan at Ann Arbor.

Stempel, Robert C. Letter to Robert C. Purcell. April 14, 1997. Robert C. Stempel Papers. Bentley Historical Library, University of Michigan at Ann Arbor.

Stempel, Robert C. Letter to Kenneth R. Baker, April 23, 1997. Robert C. Stempel Papers. Bentley Historical Library, University of Michigan at Ann Arbor.

Stempel, Robert C. "Trio Report: Summit of 8 Meetings, Denver, Colorado, June 20, 1997," June 30, 1997. Robert C. Stempel Papers. Bentley Historical Library, University of Michigan at Ann Arbor.

Stempel, Robert C. Handwritten notes, "H. J. Pierce [sic] Office, 9:30 AM." August 13, 1997. Robert C. Stempel Papers. Bentley Historical Library, University of Michigan at Ann Arbor.

Stempel, Robert C. Letter to Kenneth R. Baker, September 15, 1997. Robert C. Stempel Papers. Bentley Historical Library, University of Michigan at Ann Arbor

Stempel, Robert C., Stanford R. Ovshinsky, Paul R. Gifford, and Dennis A. Corrigan. "Nickel-Metal Hydride: Ready to Serve." *IEEE Spectrum* 35, no. 11 (1998): 29–34.

Stephenson, R. Rhoads, N.R. Moore, G.J. Nunz, S.P. DeGrey, H.C. Vivian, G.J. Klose, et al. *Should We Have a New Engine? An Automobile Power Systems Evaluation Volume II, Technical Reports.* Pasadena: Jet Propulsion Laboratory, California Institute of Technology, 1975.

Stobart, Richard K. "Fuel Cell Power for Passenger Cars: What Barriers Remain?" In *Fuel Cell Technology for Vehicles*, edited by Richard Stobart, 14. Warrendale, PA: Society of Automotive Engineers, 2001.

Stone, Charles, and Anne E. Morrison. "From Curiosity to 'Power to Change the World®.'" *Solid-State Ionics* 152–153 (2002): 1–13.

Strand, Dave. Email to Stanford R. Ovshinsky, "Subject: Discussion with Takeo Ohta," April 22, 2001. Stanford R. Ovshinsky Papers. Bentley Historical Library, University of Michigan at Ann Arbor.

Sturken, Marita, and Douglas Thomas. "Introduction: Technological Visions and the Rhetoric of the New." In *Technological Visions: The Hopes and Fears that Shape New Technologies*, edited by Marita Sturken, Douglas Thomas, and Sandra J. Ball-Rokeach, 1–18. Philadelphia: Temple University Press, 2004.

Sudworth, J. L. "Zebra Batteries." *Journal of Power Sources* 51, nos. 1–2 (1994): 105–114.

Suris, Oscar. "Daimler-Benz Unveils Electric Vehicle, Claiming a Breakthrough on Fuel Cells." *Wall Street Journal*, April 14, 1994, B2.

Sutula, Raymond A. *Progress Report for the Advanced Technology Development Program.* Washington, DC: US Department of Energy, 2000.

Sweet, William, and Elizabeth A. Bretz. "How to Make Deregulation Work: Alfred E. Kahn, the Father of Airline Deregulation, Firmly Defends it in an Interview with IEEE Spectrum but is Less Sanguine About the Effect on Electricity and Communications." *IEEE Spectrum* 39, no. 1 (2002): 50–56.

Tagawa, Kazuo, and Ralph J. Brodd. "Chapter 8: Production Processes for Fabrication of Lithium-Ion Batteries." In *Lithium-Ion Batteries Science and Technologies*, edited by Masaki Yoshio, Ralph J. Brodd, and Akiya Kozawa, 181–194. New York: Springer, 2009.

Taniguchi, Akihiro, Noriyuki Fujioka, Munehisa Ikoma, and Akira Ohta. "Development of Nickel/Metal-Hydride Batteries for EVs and HEVs." *Journal of Power Sources* 100, nos. 1–2 (2001): 117–124.

Tassey, Gregory. "Rationales and Mechanisms for Revitalizing US Manufacturing R&D Strategies." *Journal of Technology Transfer* 35, no. 3 (2010): 283–333.

Taylor, Alex, III. "The Star-Crossed Career of a Fallen GM CEO." *Fortune*, May 11, 2011. https://archive.fortune.com/2011/05/10/autos/gm_ceo_robert_stempel.fortune/index.htm.

Taylor, Barbara E. *The Lost Cord: The Storyteller's History of the Electric Car.* Columbus, OH: Greyden Press, 1995.

Team, Trefis. "Tesla's Lucrative ZEV Credits May Not Be Sustainable." *Forbes*, September 1, 2017. https://www.forbes.com/sites/greatspeculations/2017/09/01/teslas-lucrative-zev-credits-may-not-be-sustainable/#7eeafe976ed5.

Templeman, John. "Daimler's New Driver Won't Be Making Sharp Turns." *Bloomberg*, July 4, 1994. https://www.bloomberg.com/news/articles/1994-07-03/daimlers-new-driver-wont-be-making-sharp-turns.

Templeton, Brad. "Competing Electric Car Charging Standards Can Be Easily Fixed." *Forbes*, December 19, 2019. https://www.forbes.com/sites/bradtempleton/2019/12/19/competing-electric-car-charging-standards-can-be-easily-fixed/#6e5ae3f3f40d.

Tesla Motors. "Strategic Partnership: Daimler Acquires Stake in Tesla." Last modified April 20, 2010. https://www.tesla.com/en_GB/blog/strategic-partnership-daimler-acquires-stake-tesla.

Tesla Motors. "Panasonic Presents First Electric Vehicle Battery to Tesla." Last modified April 22, 2010. http://www.teslamotors.com/about/press/releases/panasonic-presents.

Tesla Motors. "Panasonic Invests $30 Million in Tesla." Last modified November 3, 2010. https://www.tesla.com/en_GB/blog/panasonic-invests-30-million-tesla.

Tesla Motors. "Claim Form." March 29, 2011. file:///Users/mne/Downloads/tesla_-_claim_form_claimants_copy_29_03_11.pdf.

Tesla Motors. "Tesla vs. Top Gear." Last modified March 29, 2011. https://www.tesla.com/en_GB/blog/tesla-vs-top-gear?redirect=no.

Tesla Motors. "Panasonic Enters into Supply Agreement with Tesla Motors to Supply Automotive-Grade Battery Cells." Last modified October 11, 2011. https://www .tesla.com/en_GB/blog/panasonic-enters-supply-agreement-tesla-motors-supply -automotivegrade-battery-c.

Tesla Motors. "Tesla Motors Launches Revolutionary Supercharger Enabling Convenient Long-Distance Driving." Last modified September 24, 2012. https://ir.tesla .com/press-release/tesla-motors-launches-revolutionary-supercharger-enabling.

Tesla Motors. "Tesla Model S: Full Battery Swap Event." Last modified June 21, 2013. https://www.youtube.com/watch?v=H5V0vL3nnHYandt=36s.

Tesla Motors. "2013 Form 10-K."

Tesla Motors. "2014 Form 10-K."

Tesla Motors. "2015 Form 10-K."

Tesla. "2016 Form 10-K."

Tesla. "2017 Form 10-K."

Tesla. "2018 Form 10-K."

Tesla. "Tesla Gigafactory." Accessed May 3, 2020. https://www.tesla.com/en_GB /gigafactory.

Tesla. "Insane vs. Ludicrous." Accessed June 10, 2020. https://forums.tesla.com/forum /forums/insane-vs-ludicrous.

Tesla. "Select Your Car." Accessed February 18, 2022. https://3.tesla.com/model3 /design#battery.

Texaco. "Texaco Response to Proposed US National Energy Policy," May 17, 2001. Robert C. Stempel Papers. Bentley Historical Library, University of Michigan at Ann Arbor.

Thackeray, Michael. "20 Golden Years of Battery R&D at CSIR, 1974–1994." *South African Journal of Chemistry* 64 (2011): 61–66.

Therivel, Riki, and Bill Ross. "Cumulative Effects Assessment: Does Scale Matter?" *Environmental Impact Assessment Review* 27 (2007): 365–385.

Thompson, Chrissie. "Chevy Volt Was Going to Save Detroit: Now Its Workers Are Losing Jobs." *Detroit Free Press*, November 27, 2018. https://eu.freep.com/story/money /business/2018/11/27/chevy-volt-donald-trump-general-motors/2120687002/.

Tikhonov, Victor. "Simple Analog BMS for the Tinkerer: Part 1." *Current Events* 44, no. 12 (2012): 1, 34–37.

Toll, Micah. "Here's Why Electric Bike Sales Have Skyrocketed During the Coronavirus Lockdown." *Electrek*, May 1, 2020. https://electrek.co/2020/05/01/electric-bike -sales-skyrocket-during-lockdown/.

Tollefson, Jeff. "Trump's Decision to Block California Vehicle Emissions Rules Could Have a Wide Impact." *Nature* September 18, 2019. https://www.nature.com/articles /d41586-019-02812-0.

Top Gear "Top Gear Tesla Road Test." Accessed May 6, 2020. https://www.youtube.com/watch?v=JKtK493sGAk.

Totaro, Lorenzo, and Daniele Lepido. "Fiat to Pool Cars with Tesla to Meet EU Emissions Targets on CO2." *Bloomberg*, April 7, 2019. https://www.bloomberg.com/news/articles/2019-04-07/fiat-chrysler-teams-with-tesla-to-meet-eu-emissions-targets.

Toyota Motor Corporation. "Toyota RAV4 Electric Vehicle." Last modified August 1999. https://media.toyota.co.uk/wp-content/uploads/sites/5/1324550080rav4_ev_whole.pdf.

Toyota Motor Corporation. "Toyota Advances Plug-in Hybrid Development with Partnership Program," July 25, 2007. http://pressroom.toyota.com/pr/tms/toyota/TYT2007072552930.aspx.

Treadgold, Tim. "Palladium Heads for $2000 An Ounce; Lock up Your Prius!" *Forbes*, December 16, 2019. https://www.forbes.com/sites/timtreadgold/2019/12/16/palladium-heads-for-2000-an-ounce-lock-up-your-prius/.

Trotman, Alex J. Speech, National Press Club, Washington DC, October 27, 1997. https://www.c-span.org/video/?94065-1/business-environment.

Tuckey, Bill. *Sunraycer*. Hornsby, Australia: Chevron Publishing Group, 1989.

Uesaka, Yoshifumi. "The Company That Helps Tesla Make Look Aluminum Sexy." *Nikkei Asia*, September 12, 2016. https://asia.nikkei.com/Business/Biotechnology/The-company-that-helps-Tesla-make-aluminum-look-sexy.

Ulrich, Lawrence. "Is Elon Musk Back in 'Production Hell' with Tesla's 4680 Battery?" *IEEE Spectrum*, September 1, 2021. https://spectrum.ieee.org/tesla-4680-battery.

United Nations Framework Convention on Climate Change (UNFCCC). "Report of the Conference of the Parties on its Second Session, Held at Geneva from 8 to 19 July 1996, Part Two: Action Taken by the Conference of the Parties at its Second Session." https://unfccc.int/sites/default/files/resource/docs/cop2/15a01.pdf.

University of Texas. "UT Austin's John B. Goodenough Wins Engineering's Highest Honor for Pioneering Lithium-Ion Battery." Last modified January 6, 2014. https://news.utexas.edu/2014/01/06/ut-austins-john-b-goodenough-wins-engineerings-highest-honor-for-pioneering-lithium-ion-battery/.

US Congress. Senate. Committee on Commerce and the Subcommittee on Air and Water Pollution of the Committee on Public Works. *Electric Vehicles and Other Alternatives to the Internal Combustion Engine: Joint Hearings Before the Committee on Commerce and the Subcommittee on Air and Water Pollution of the Committee on Public Works*. 90th Cong., 1st sess., March 14–17 and April 10, 1967. https://books.google.co.uk/books?id=zR83AQAAIAAJ&printsec=frontcover&source=gbs_ge_summary_r&cad=0#v=onepage&q&f=false.

US Council for Automotive Research (USCAR). "USCAR as Umbrella for Big Three Research." Robert C. Stempel Papers. Bentley Historical Library, University of Michigan at Ann Arbor.

US Council for Automotive Research (USCAR). "USABC Awards $12.5 Million Battery Technology Development Contract to A123 Systems." Last modified May 5, 2008. https://uscar.org/guest/article_view.php?articles_id=210.

US Council for Automotive Research (USCAR). "Who We Are." Accessed June 23, 2013. https://www.uscar.org/guest/history.php.

US Department of Energy (DOE). "Department of Energy Announces Closing of $529 Million Loan to Fisker Automotive." Last modified April 23, 2010. https://www.energy.gov/articles/department-energy-announces-closing-529-million-loan-fisker-automotive.

US Department of Energy (DOE). *Special Report: The Department of Energy's Management of the Award of a $150 Million Recovery Act Grant to LG Chem Michigan Inc.* Washington, DC: OAS-RA-13–10, 2013.

US Department of Energy (DOE). "Tesla." Accessed May 12, 2020. https://www.energy.gov/lpo/tesla.

US Department of Transportation (DOT). "Summary of Fuel Economy Performance." December 15, 2014. https://www.nhtsa.gov/sites/nhtsa.gov/files/performance-summary-report-12152014-v2.pdf.

US Department of Transportation (DOT). "Guidance on Testing and Installation of Rechargeable Lithium Battery and Battery Systems on Aircraft." October 15, 2015. https://www.faa.gov/documentLibrary/media/Advisory_Circular/AC_20-184_Final_proof.pdf.

US Department of Transportation (DOT). "Number of US Aircraft, Vehicles, Vessels, and Other Conveyances." Bureau of Transportation Statistics, 2021.

US Energy Information Administration (EIA). *August 2020 Monthly Energy Review.*

US Energy Information Administration (EIA). "Petroleum and Other Liquids: Cushing, OK WTI Spot Price FOB." Accessed April 29, 2020. https://www.eia.gov/dnav/pet/hist/RWTCD.htm.

US Environmental Protection Agency (EPA). Air Pollutant Emissions Trends Data. "Criteria Pollutants National Tier 1 for 1970–2020." https://www.epa.gov/air-emissions-inventories/air-pollutant-emissions-trends-data.

US Environmental Protection Agency (EPA). *Light-Duty Automotive Technology, Carbon Dioxide Emissions, and Fuel Economy Trends: 1975 Through 2016.* EPA-420-R-16–010, November 2016.

US Environmental Protection Agency (EPA). "Trump Administration Announces One National Program Rule on Federal Preemption of State Fuel Economy Standards." September 19, 2019. https://www.epa.gov/newsreleases/trump-administration-announces-one-national-program-rule-federal-preemption-state-fuel.

US Environmental Protection Agency (EPA). *The 2020 EPA Automotive Trends Report: Greenhouse Gas Emissions, Fuel Economy, and Technology Since 1975.* EPA-420-R-21–003, January 2021.

US Environmental Protection Agency (EPA). *Inventory of US Greenhouse Gas Emissions and Sinks, 1990–2019.* EPA 430-R-21–005, April 14, 2021.

US Environmental Protection Agency (EPA). "Notice of Reconsideration of a Previous Withdrawal of a Waiver for California's Advanced Clean Car Program (Light-Duty Vehicle Greenhouse Gas Emission Standards and Zero Emission Vehicle Requirements."

Accessed July 21, 2021. https://www.epa.gov/regulations-emissions-vehicles-and-engines/notice-reconsideration-previous-withdrawal-waiver.

US President, Proclamation. "The Energy Emergency." *Federal Register* 9, no. 45 (November 12, 1973): 1312–1318.

Urry, John. "The 'System' of Automobility." *Theory, Culture and Society* 21, nos. 4–5 (2004): 25–39.

Valdes-Dapena, Peter. "New Tesla Earns Perfect Score from Consumer Reports." *CNN*, August 27, 2015. https://money.cnn.com/2015/08/27/autos/consumer-reports-tesla-p85d/index.html.

Van Vorst, William D., J. H. Kelley, and T. N. Veziroglu. "WHEC-IV." *International Journal of Hydrogen Energy* 8, no. 11–12 (1983): 858–859.

Vance, Ashlee. *Elon Musk: How the Billionaire CEO of SpaceX and Tesla Is Shaping Our Future*. HarperCollins, 2015.

Vaughn, Mark. "What's Going to Happen to All Those Electric Car Batteries Anyway?" *Autoweek*, March 11, 2021. https://www.autoweek.com/news/green-cars/a35803612/battery-recycling/.

Venkatesan, Srinivasan, Michael Fetcenko, Benny Reichman, and Kuochih C. Hong. "Development of Ovonic Rechargeable Metal Hydride Batteries." *Proceedings of the 24th Intersociety Energy Conversion Engineering Conference* 3 (1989): 1659–1664.

Venkatesan, Srinivasan, Subhash K. Dhar, Stanford R. Ovshinsky, and Michael Fetcenko. "Ovonic Nickel-Metal Hydride Batteries for Industrial and Electric Vehicle Applications." *Proceedings of the Sixth Annual Battery Conference on Applications and Advances* (1991): 59–73.

Venkatesan, Srinivasan, Matt van Kirk, Lynn Taylor, and Jim Strebe. Letter to Stanford R. Ovshinsky, Subhash K. Dhar, Michael A. Fetcenko, Dennis A. Corrigan, and Paul R. Gifford. May 6, 1996. Robert C. Stempel Papers. Bentley Historical Library, University of Michigan at Ann Arbor.

Verne, Jules. *The Mysterious Island*. Hetzel 1874.

Vinkhuyzen, Maarten. "Nissan's Long Strange Trip with Leaf Batteries." *CleanTechnica*, September 29, 2018, https://cleantechnica.com/2018/09/29/nissans-long-strange-trip-with-leaf-batteries/.

Vlasic, Bill. "Chinese Firm Wins Bid for Auto Battery Maker." *New York Times*, December 9, 2012. https://www.nytimes.com/2012/12/10/business/global/auction-for-a123-systems-won-by-wanxiang-group-of-china.html.

Voelcker, John. "Should I Buy a Used Nissan Leaf (or Another Electric Car?)." *Green Car Reports*, June 15, 2015. https://www.greencarreports.com/news/1098554_should-i-buy-a-used-nissan-leaf-or-another-electric-car.

Voelcker, John. "Who Sold the Most Plug-in Electric Cars in 2015? (It's Not Tesla or Nissan)." *Green Car Reports*, January 15, 2016. https://www.greencarreports.com/news/1101883_who-sold-the-most-plug-in-electric-cars-in-2015-its-not-tesla-or-nissan.

Voorhees, Josh. "Obama Favors Plug-In Hybrids over Hydrogen Vehicles." *Scientific American*, July 10, 2009. https://www.scientificamerican.com/article/hybrid-cars -plug-in-obama-stimulus-money/.

Wald, Matthew L. "A Tough Sell for Electric Cars: Technology Lagging as Markets Emerge." *New York Times*, November 26, 1991, D1.

Wald, Matthew L. "Government Dream Car: Washington and Detroit Pool Resources to Devise a New Approach to Technology." *New York Times*, September 30, 1993, A1.

Wald, Matthew L. "Expecting a Fizzle, GM Puts Electric Car to Test." *New York Times*, January 28, 1994, D4.

Wald, Matthew L. "Company News: Electric Car Venture Set with Itochu." *New York Times*, June 10, 1994, D0000.3.

Wald, Matthew L. "Ford Plans Zero-Emission Fuel Cell Car." *New York Times*, April 22, 1997, D2.

Wald, Matthew L. "Three Guesses: The Fuel of the Future Will be Gas, Gas, or Gas." *New York Times*, October 16, 1997, G16.

Wald, Matthew L. "In a Step Toward a Better Electric Car, Company Uses Fuel Cell to Get Energy from Gasoline." *New York Times*, October 21, 1997, A14.

Wald, Matthew L. "Zero to 60 in 4 Seconds, Totally From Revving Batteries." *New York Times*, July 19, 2006, C3.

Wallace, Harold D., Jr. "Fuel Cells: A Challenging History." *Substantia* 3, no. 2 (2019): 83–97.

Wall Street Journal. "Ford, Chrysler Win Auto Fuel-Cell Work." July 13, 1994, B2.

Walsh, Dustin. "Wanxiang Group Closes Deal to Acquire Assets of A123 Systems." *Crain's Detroit Business*, January 29, 2013. https://www.crainsdetroit.com/article /20130129/NEWS/130129846/wanxiang-group-closes-deal-to-acquire-assets-of-a123 -systems.

Wartzman, Rick. "GM Unveils Electric Car with Lots of Zip But Also a Battery of Unsolved Problems." *Wall Street Journal*, January 4, 1990, A1.

Watanabe, Shoichiro, Masahiro Kinoshita, and Kensuke Nakura. "Capacity Fade of $LiNi(1-x-y)Co_xAl_yO_2$ Cathode for Lithium-Ion Batteries During Accelerated Calendar and Cycle Life Test. I. Comparison Analysis Between $LiNi(1-x-y)Co_xAl_yO_2$ and $LiCoO_2$ Cathodes in Cylindrical Lithium-Ion Cells During Long Term Storage Test." *Journal of Power Sources* 247 (2014): 412–422.

Wei, Haiqiao, Tianyu Zhu, Gequn Shu, Linlin Tan, and Yuesen Wang. "Gasoline Engine Exhaust Gas Recirculation: A Review." *Applied Energy* 99 (2012): 534–544.

Wells, Christopher W. *Car Country: An Environmental History*. University of Washington Press: Seattle, 2012.

Wenner, Jann S., and Will Dana. "Al Gore: The Rolling Stone Interview." *Rolling Stone*, November 9, 2000. https://www.rollingstone.com/feature/al-gore-the-rolling -stone-interview-62074/.

Westbrook, Michael H. *The Electric Car: Development and Future of Battery, Hybrid and Fuel-Cell Cars.* London: Institution of Electrical Engineers, 2001.

Westfall, Catherine. "Retooling for the Future: Launching the Advanced Light Source at Lawrence's Laboratory, 1980-1986." *Historical Studies in the Natural Sciences* 38, no. 4 (2008): 569–609.

Westwick, Peter J. *The National Labs: Science in an American System, 1947–1974.* Cambridge, MA: Harvard University Press, 2003.

White, Gregory L. "GM Stops Making Electric Car, Holds Talks with Toyota." *Wall Street Journal,* January 12, 2000, A14.

White, Joseph P. "GM Says It Plans an Electric Car, But Details Are Spotty." *Wall Street Journal,* April 19, 1990, B1.

White, Richard. *The Organic Machine: The Remaking of the Columbia River.* New York: Hill and Wang, 1995.

Whittingham, M. S. "Electrical Energy Storage and Intercalation Chemistry." *Science* 192, no. 4244 (1976): 1126–1127.

Williams, James C. *Energy and the Making of Modern California.* Akron, OH: University of Akron Press, 1997.

Williams, Robert H. "Fuel Cells, Their Fuels, and the US Automobile." In *Proceedings: First Annual World Car 2001 Conference, June 21–24, 1993, 73–75.* California Institute of Technology, 1993.

Winner, Langdon. "Do Artifacts Have Politics?" *Daedalus* 109, no. 1 (1980): 121–136.

Wired. "California Cuts ZEV Mandate in Favor of Plug-In Hybrids." March 27, 2008. http://www.wired.com/autopia/2008/03/the-california/.

Wise, David W. "The Tides of Deregulation." *Public Utilities Fortnightly,* 124, no. 5 (1989): 39–40.

Womack, James P., Daniel T. Jones, and Daniel Roos. *The Machine That Changed the World: How Lean Production Revolutionized the Global Car Wars.* London: Simon and Schuster, 2007.

Wong, Andrea. "The Untold Story Behind Saudi Arabia's 41-Year Debt Secret." *Bloomberg News,* May 30, 2016. http://www.bloomberg.com/news/features/2016-05 -30/the-untold-story-behind-saudi-arabia-s-41-year-u-s-debt-secret.

Woo-Cumings, Meredith, ed. *The Developmental State.* Ithaca, NY: Cornell University Press, 1999.

Woodall, Bernie, Paul Lienert, and Ben Klayman. "Insight: GM's Volt: The Ugly Math of Low Sales, High Costs." *Reuters,* September 10, 2012. http://www.reuters .com/assets/print?aid=USBRE88904J20120910.

Wouk, Victor. "Hybrids: Then and Now." *IEEE Spectrum* 32, no. 7 (1995): 16–21.

Wouk, Victor. "Hybrid Electric Vehicles." *Scientific American* 277, no. 4 (1997): 70–74.

Wouk, Victor. Interview by Judith R. Goodstein. New York, New York, May 24, 2004. Oral History Project, California Institute of Technology Archives. https:// oralhistories.library.caltech.edu/92/.

Wyczalek, Floyd A. "Market Mature 1998 Hybrid Electric Vehicles." *IEEE AES Systems Magazine* 14, no. 3 (March 1999): 41–44.

Wylam, William B. Untitled letter to Subhash Dhar, July 12, 1990. Stanford R. Ovshinsky Papers. Bentley Historical Library, University of Michigan at Ann Arbor.

Yacobucci, Brent D. "The Partnership for a New Generation of Vehicles: Status and Issues." *Congressional Research Service, Report RS20852.* January 22, 2003. https://wikileaks.org/wiki/CRS:_The_Partnership_for_a_New_Generation_of_Vehicles:_Status_and_Issues,_January_22,_2003.

Yoshio, Masaki, Akiya Kozawa, and Ralph J. Brodd. "Introduction: Development of Lithium-Ion Batteries." In *Lithium-Ion Batteries: Science and Technologies*, edited by Masaki Yoshio, Ralph J. Brodd, and Akiya Kozawa, xvii–xxvi. New York: Springer, 2009.

Yost, Charles F. "Memorandum," July 13, 1962. Project Lorraine, Energy Conversion, 1958–1966 Official Correspondence Files, Materials Sciences Office, ARPA. National Archives and Records Administration, College Park, MD.

Yost, Charles F. "Memorandum for Dr. Sproull: Subject: Project Lorraine," September 12, 1963. Project Lorraine, Energy Conversion, 1958–1966 Official Correspondence Files, Materials Sciences Office, ARPA. National Archives and Records Administration, College Park, MD.

Young, Thomas N. "Civil Action 96–70919: Memorandum of Law in Support of Ovonic Battery Company Inc's Motion for Preliminary Injunction." March 3, 1996. Robert C. Stempel Papers. Bentley Historical Library, University of Michigan at Ann Arbor.

Young, Thomas N., and Carl H. von Ende. "Civil Action No. 96–70919: Toyota Motor Sales USA, Inc.'s Memorandum in Support of Motion to Stay This Proceeding Pending Decision of Delaware Court on Motion to Stay, Dismiss, or Transfer." March 12, 1996. Robert C. Stempel Papers. Bentley Historical Library, University of Michigan at Ann Arbor.

Zu, Chen-Xi, and Hong Li. "Thermodynamic Analysis on Energy Densities of Batteries." *Energy and Environmental Science* 4 (2011): 2614–2624.

INDEX

Printed in the United States
by Baker & Taylor Publisher Services